Progress in Reproductive Biology

Vol. 7

Series Editor: P. O. Hubinont, Brussels
Assistant Editor: M. L'Hermite, Brussels

S. Karger · Basel · München · Paris · London · New York · Sydney

7th Seminar on Reproductive Physiology and Sexual Endocrinology, Brussels
May 21–24, 1980

Blastocyst-Endometrium Relationships

Volume Editors
F. Leroy, Brussels, *C. A. Finn,* Liverpool, *A. Psychoyos,* Paris,
P. O. Hubinont, Brussels

94 figures and 30 tables, 1980

S. Karger · Basel · München · Paris · London · New York · Sydney

Progress in Reproductive Biology

Vol. 5: Seasonal Reproduction in Higher Vertebrates. Symposium on the Seasonal Reproduction in Higher Vertebrates, Richmond, Va., 1978. Reiter, R. J., San Antonio, Tex. and Follett, B. K., Bristol (eds.)
VI + 222 p., 84 fig., 7 tab., 1980. ISBN 3–8055–0246–X

Vol. 6: Advances in Prolactin. Satelite Symposium to the 6th International Congress of Endocrinology, Adelaide, 1980. L'Hermite, M., Brussels and Judd, S. L., Bedford Park (eds.)
In Preparation 1980. ISBN 3–8055–0859–X

National Library of Medicine, Cataloging in Publication
Seminar on Reproductive Physiology and Sexual Endocrinology, 7th, Brussels 1980 Blastocyst-endometrium relationships / volume editors, F. Leroy et al. – – Basel ; New York : Karger, 1980.
(Progress in reproductive biology ; vol. 7)
Earlier seminars called International Seminar on Reproductive Physiology and Sexual Endocrinology
1. Blastocyst – physiology – congresses   2. Endometrium – physiology – congresses   3. Nidation – congresses   I. Leroy, F., ed.   II. International Seminar on Reproductive Physiology and Sexual Endocrinology, 6th, Brussels, 1976
III. Title   IV. Series.
W1 PR681B v. 7   QS 645 S471 1980b
ISBN 3–8055–0988–X

© Copyright 1980 by S. Karger AG, P. O. Box, CH-4009 Basel (Switzerland)
Printed in Switzerland by Merkur AG, 4900 Langenthal
ISBN 3–8055–0988–X

Contents

Triggering Stimuli

Endometrial Epithelial Cells

Stroma and the Decidual Cell Reaction

Uterine Sensitization and the Refractory State

Subcellular Regulation

Species Diversity and Implantation in Primates

Control over Human Implantation

Acknowledgments

The Belgian Fund for Medical Research and the Ministry of Education and Culture covered a substantial part of the expenses. Special thanks are due to Cilag-Chemie and Mr. *J. De Backer* for their help. The following firms have also provided financial aid: Organon, Ciba-Geigy, Hoechst, Upjohn and Sandoz.

Foreword

This Seventh Seminar on Reproductive Physiology and Sexual Endocrinology reminds us of the memorable time of May 1968, when in the
midst of academic turmoil, the second meeting of this series was held, on the
same subject. Comparison between the two volumes of Proceedings shows
that remarkable and sometimes unexpected progress has been achieved in
some parts of the field, whilst in others we have to admit that stagnation
has prevailed.

In the present volume we have tried to integrate current lines of research on implantation and to present a comprehensive assessment of
knowledge on this subject, with the ultimate goal of gaining insight into
problems related to human implantation. As shown in the table of contents, the symposium was divided into seven parts each devoted to discussion of a question which is considered important in the field. Each starts
with a keynote review followed by three specialised presentations on the
topic.

During the last decade, notable advances have been obtained from
biochemical investigations of nidatory changes. The molecular biology of
implantation is, however, still very much in its infancy. Although important
information has been gained on the roles of steroid receptors, nucleic acids
and protein synthesis, cyclic nucleotides, prostaglandins, etc., much work
remains to be done before the nidatory network will be unravelled. We
still do not understand the molecular basis for uterine sensitivity and refractoriness. In this case, the investigation of hormone action is complicated, not only because of the well-known heterogeneity of the uterus and
the interplay of its tissular components, but also because of the participation of at least two classes of steroid hormones of which the interactions

await to be deciphered. The intervention of the blastocyst as the other main partner of the reaction, involves, of course, further complexity.

Still another stumbling block is that, for evolutionary reasons, implantation displays a wide interspecies diversity which brings the human closer to some rodents than even to some of the primates studied so far. Whilst human implantation does not lend itself to direct study, it is on the other hand particularly hazardous to extrapolate from animal data in this field.

As happens more often than not, investigation has raised more questions than it has solved. It is hoped, however, that as more information becomes available, human implantation will become more controllable for the benefit of the seemingly extreme aspects of gynaecological practice – the treatment of infertility and the prevention of unwanted pregnancies.

The Editors

Prog. reprod. Biol., vol. 7, pp. 1–13 (Karger, Basel 1980)

Signals Exchanged between the Blastocyst and Endometrium in the Period Leading to Implantation

Marilyn B. Renfree[1]

School of Environmental and Life Sciences, Murdoch University, Perth, W. A.

The mammalian ovum is relatively autonomous during the first days after fertilization, and proceeds through the early cleavage stages without growing much in size. The blastocyst resulting from these cleavage divisions is not usually much larger than the unfertilized egg. Thereafter, the blastocyst appears to require a physiological signal to trigger its expansion and subsequent implantation. The blastocyst lies free in the uterine lumen, but in close contact with the surface of the uterine epithelium, so the uterine secretions must play a crucial role in the mediation of signals from endometrium to blastocyst. In addition, the uterine environment has important influences in the removal of the zona pellucida and the adhesion and attachment phases of implantation. Conversely, the blastocyst must influence the endometrium because there is maternal recognition of pregnancy even before attachment. In this review I propose to discuss some of the signals at the interface between maternal and embryonic tissue in the period leading to implantation.

Early Independent Growth

In most species, the transition from free to implanted blastocyst occurs without interruption. The growth which occurs in the preimplantation period does not appear to depend upon the provision of maternal components of either oviductal or uterine origin, and development up to

[1] I sincerely thank Dr. *D. W. Lincoln* for helpful discussions and constructive criticism of the manuscript.

the early blastocyst stage seems to be regulated by the embryo's own endo-genous programme, rather than by exogenous signals of maternal origin [*Van Blerkom et al.,* 1973]. During the early pre-implantation stage, the embryo requires pyruvate and lactate; and protein synthesis together with other metabolic processes increases many times between the eight-cell and blastocyst stage [*Brinster,* 1971; *Biggers,* 1971]. This increase in meta-bolic activity as the blastocyst stage approaches will, however, decline to a state of dormancy if implantation is delayed, and also in certain con-ditions *in vivo* [*McLaren* 1973a, b]. In non-delayed mouse blastocysts cell division continues at a doubling time of 10 h, but if the mouse is lactating development proceeds to the blastocyst stage and metabolism and cell doubling time declines to 48 h. *In vivo,* diapausing trophoblast cells arrest at the G1 phase, but *in vitro,* if some essential component such as glucose, certain amino acids, or proteins are not present in the medium, blastocysts will stop growing in either the G1 or G2 phase of the cell cycle [*Sherman and Barlow,* 1972]. This biochemical shutdown and arrest of division in the G1 phase has to be viewed as a physiological and not a pathological development, for in many mammalian species diapause is a normal part of the reproductive cycle. In the tammar wallaby, for example, the delay normally persists for 11 months, and may be extended even further by ovariectomy [*Tyndale-Biscoe,* 1979]. This suggests that there must be something particularly stable about the diapausing blastocyst. Alterna-tively, the critical influence may be the uterine environment as controlled by ovarian hormones [*Aitken,* 1979]. Thus, species which exhibit embry-onic diapause assume special relevance with regard to questions concerning the triggering stimuli that pass between maternal and embryonic tissues.

Zona Lysis

In order to adhere to the uterine epithelium, the blastocyst must shed its zona pellucida. However, it should be pointed out that loss of the zona is not necessarily followed immediately by implantation, as in some species which show embryonic diapause such as the laboratory rat and mouse, the roe deer, and the armadillo, the blastocyst remains in delay even though it has lost the zona pellucida [*Aitken,* 1977; *Renfree,* 1978].

In the mouse, lysis of the zona pellucida occurs in the estrogen-sensitized uterus probably through the action of a uterine protease, or possibly through a localized pH effect [*McLaren,* 1973a]. There is a difference in pH values of the uterine fluid between the mesometrial and antimesometrial surfaces of the uterus [*Petzoldt*, 1971], and *Denker* [1978] has suggested that in normal implantation, factors provided by the anti-mesometrial endometrium, like a more alkaline pH, might increase the activity of proteinase, or other factors in the abembryonic pole. The main proteinase activity, as well as the onset of dissolution of blastocyst coverings, including the zona, in the rabbit is always at the abembryonic pole, so that as the blastocyst randomly rotates in the uterus, if the zona enclosed abembryonic pole and uterine anti-mesometrial surfaces come together, dissolution of the blastocyst coverings would begin, and the blastocyst would be trapped in the 'correct' orientation [*Denker*, 1978]. There has been considerable discussion as to the relative importance of uterine versus trophoblast proteinase, some authors suggesting a solely uterine origin [*Pinsker et al.*, 1974], but evidence from the rabbit suggests that a trophoblast proteinase is an important component [*Denker and Hafez*, 1975; *Denker*, 1980]. Clearly both uterine and blastocyst protein-ases are involved, but it is not yet clear whether the blastocyst triggers the uterine reaction or vice versa, or whether they might occur simultaneously. In the mouse, the zona-lytic enzyme of the uterine secretion has been identified as a caseinolytic proteinase [*Pinsker et al.*, 1974], whilst trypsin and chymotrypsin like enzymes have been found in blastocysts [*Dabich and Andary*, 1976].

Relatively little is known about the regulation of implantation as-sociated proteinase activity. Proteinases may be involved in the formation of a cellular contact between blastocyst and endometrium, since *in vitro* studies have shown that the attachment of mouse blastocysts to cell monolayers of mouse uterine, embryo or L cells (NCTC clone 929) may be inhibited by soybean trypsin inhibitor, and stimulated by trypsin [*Kubo et al.*, 1979]. Surprisingly, the zona itself apparently has no role in this process, since zona-free blastocysts elicit the decidual response as well as zona-enclosed blastocysts [*McLaren*, 1969]. Intrauterine administration of proteinase inhibitors interferes with implantation in the mouse [*Dabich and Andary*, 1974] and the rabbit [*Denker*, 1978].

The development of proteinase activity is dependent on progesterone in the rabbit, and could be regulated by an alteration of the synthesis or activation of the enzymes, or by a change in the activity in protease in-hibitors [*Denker*, 1978]. In this context, it is of interest that the rabbit uterine-specific protein, uteroglobin, has been found to have trypsin-inhibiting activity [*Beier*, 1978].

Adhesion

Once hatching from the zona has been achieved, communication between blastocyst and endometrium may be by the secretion of substances, and via direct cell-to-cell interaction. While enclosed in the zona, the blastocyst is an uncharged body, whereas after dissolution of the zona its surface is negatively charged [*Clementson et al.*, 1971] though it is unlikely that adhesion of the blastocyst to the uterus is due to a charge association. *Enders and Schlafke* [1974] suggest that adhesion seems to be accompanied by some localized reduction in thickness of the uterine surface coat. In ovariectomized mice with delayed embryos, on the other hand, the failure of adhesion *(in utero)* appears to be due to the inhospitable surface of the uterine epithelium [*Nilsson*, 1967].

Trophoblast Outgrowth

In vitro, only certain amino acids and serum are essential for hatching and adhesion, but for maximal trophoblast outgrowth almost all essential amino acids are required [*Spindle and Pedersen*, 1974]. In a serum-free medium containing amino acids, vitamins, bovine serum albumin and fetuin, blastocysts hatch, attach and outgrow, though adhesion and trophoblast outgrowth rates are retarded compared to serum containing medium [*Sherman et al.*, 1979]. However, outgrowth of trophoblast cells is influenced by the presence of serum, but the rate of incorporation of [^{3}H]-uridine into RNA *in vitro* is not influenced by the presence of serum [*Weitlauf and Kiessling*, 1980]. This observation suggests that outgrowth, and increased metabolic activity, are regulated differentially and should be regarded as a two-step mechanism in which the increase in metabolic activity may be a necessary preliminary for development of the capability for outgrowth, but in order for this capacity to be expressed, a stimulatory factor, such as is present in serum, must be provided [*Weitlauf and Kiessling*, 1980].

In scanning electron microscope studies of mouse blastocysts induced to implant after delay by injections of estrogens, the cells on the blastocyst surface bulge noticeably, and then smooth out to become closely apposed to the uterine epithelium [*Bergstrom and Nilsson*, 1975] though in normal (non-delayed) implantation alterations in surface morphology are not obvious when blastocysts transform from the non-adherent to adherent stage [*Sherman et al.,* 1979]. Trophoblast cells apparently do not contain any major membrane glycoproteins that are also not represented on the inner cell mass surface [*Sherman et al.,* 1979]. Since collagenase retarded attachment of blastocysts *in vitro,* these workers propose that blastocyst collagen might play a role in trophoblast adhesiveness, although they emphasize that collagen is likely to be only one of several possible factors. Such cell-to-cell contacts might provide one of the early signals for the maternal recognition of pregnancy.

In many mammals, and especially in the ungulates, the blastocyst has a prolonged free-living stage in the uterus and it expands considerably by extensive proliferation of the trophoblast shortly after hatching from the zona pellucida. This proliferation is analogous to trophoblast outgrowth of rodents. In the ungulates, it is accompanied by a morphological transition from a small, round shape to a long (up to 1 m in the pig) threadlike conceptus consisting primarily of trophoblast tissue. In other mammals, the expansion is less dramatic, and the spherical shape of the blastocyst is retained, while the embryo swells considerably by proliferation of the trophoblast and endoderm, as in many marsupials where it expands up to 200,000 times in volume in 17 days [*Renfree,* 1973]. In general, greatest expansion occurs in those species which have less erosive types of implantation, and in most of these species the decidual cell reaction is less well developed or non-existent. Since there is no intimate contact between uterus and embryo at these stages, the constituents of the uterine fluid must be responsible in part for the morphological and biochemical changes which occur. However, no ungulate blastocyst has been cultured intact through the expansion phase, though some success has been achieved in the culture of already expanded opossum vesicles [*New et al.,* 1977]. The specific triggering stimuli for this rapid and complex elongation or expansion are not known, but it may be relevant that it is during this stage of embryonic development that maternal recognition of pregnancy occurs.

Endometrial Response: the Decidual Cell Reaction

After activation and adhesion, the uterine capillaries and wall in the immediate vicinity of the embryo become more permeable to injected dyes, and a local stromal oedema and increase in oxygen tension develops within the uterus. This change can be readily visualized by injection of a high molecular weight dye such as pontamine blue [*Psychoyos,* 1961].

The epithelium becomes disrupted and stromal cells differentiate into decidual cells – the decidual cell reaction. Although the endometrial response produced may vary considerably between species, the reaction is especially marked in rodents and a form of decidual response is also found in members of the orders Anthropoidea and Insectivora, but true decidual cells do not occur in most carnivores or artiodactyls [*Finn and Porter*, 1975]. In most species the transformation occurs only in areas close to the blastocyst, and is dependent on a stimulus from the blastocyst to trigger the reaction [*Finn and Porter*, 1975].

A variety of stimuli can elicit the decidual cell response, but the signal(s) given by the blastocyst to the uterus remain unclear. Intraluminal injection of oil [*Finn and Keen*, 1963] or air [*Orsini*, 1963] causes the decidual reaction, and if the CO_2 levels are increased or decreased beforehand, the air enhances or decreases the reaction. It is suggested that the stimulus given by the blastocyst or oil produces some physicochemical change on the surface of the endometrium [*Finn and Porter*, 1975] and that the blastocyst release CO_2 at implantation which triggers the reaction [*McLaren*, 1970]. However, no direct evidence is yet available to explain the nature of the blastocyst stimulus assumed to initiate permeability changes, and subsequent decidual cell formation and implantation.

Histamine was long ago proposed as the responsible factor involved in the permeability changes [*Shelysnyak*, 1957] but has not gained general acceptance. Recently, however, using histamine H_1 and H_2 antagonists, increased permeability and implantation have been inhibited [*Brandon and Wallis*, 1977], but as *Finn* points out, even if histamine is released from the mast cells on contact with the blastocyst, the problem of the nature of the blastocyst signal remains [*Finn and Porter*, 1975].

The decidual response and permeability increase can also be inhibited by indomethacin, a prostaglandin inhibitor [e. g. *Kennedy*, 1977] but the role of prostaglandins is unclear [*Sauer*, 1979]. Cyclic nucleotides are under the control of prostaglandins and act as controlling factors in cell differentiation and proliferation [*McMahon*, 1974] but it is unlikely that they provide the blastocyst-derived stimulus since they are thought to act only in a 'second messenger' capacity [*Sauer*, 1979]. It is generally accepted that cyclic nucleotides themselves are not deciduogenic but cyclic AMP and dibutryl cyclic AMP can induce implantation in mouse diapause blastocysts, and apparently mimic estradiol in inducing receptivity to implantation [reviewed by *Sauer*, 1979]. Both estradiol and dibutryl cyclic AMP can induce synthesis of specific uterine luminal proteins [*Surani and Webb*, 1977]. Protein synthesis inhibitors such as actinomycin-D do not block the initial reaction between the endometrium and blastocyst in the mouse although later stages of the reaction are blocked, so the signal does not apparently involve transfer of newly formed RNA from the nucleus [*Finn and Martin*, 1972; *Finn and Porter*, 1975].

As pointed out above, the decidual cell reaction is by no means universal among mammals, and perhaps may be found to be representative of only a few mammalian orders – for example it has not been described in any marsupial species so far. However, the decidua has apparently evolved to constrain and regulate growth of the blastocyst in those species where the degeneration of the epithelium is a necessary component of implantation. At the same time, however, the decidual degeneration which occurs could provide a source of nutrition for the embryo [*Finn and Porter*, 1975].

Uterine Secretions

Of the high molecular weight substances, the proteins have been the best studied group, with descriptions of uterine proteins in 20 or more mammals, including man [reviewed by *Daniel*, 1976; *Surani*, 1977, 1980; *Van Blerkom*, 1979; *Beier and Mootz*, 1979]. However, there has been as yet no definitive evidence that any single one of the uterine proteins has a crucial role in promoting blastocyst growth or implantation, but rather it appears that it is the entire uterine environment which is important. Nevertheless, a number of the uterine 'specific proteins' appear to be synthesized *de novo* and they are under the influence of ovarian steroids. In the rabbit, uteroglobin synthesis may be stimulated by progesterone and other progestagens, but not estrogens [*Beier*, 1978]. In the pig, a maternally derived protein, purple protein, accumulates in the allantoic fluids, and if uterine secretory activity is stimulated by estrogen or progesterone, uteri become heavier, and placentae become longer and allantoic fluid more voluminous, suggesting that uterine secretion may enhance placental development in pigs [*Bazer*, 1975]. Pregnancy-specific antigens have been found in sheep uterine flushings [*Findlay et al.*, 1979] but again, no specific function can be assigned to these molecules.

One role for uterine proteins which has been suggested recently is that of provision of a means of maintenance of structural integrity of the trophoblast during the phase of rapid expansion of the embryonic membranes, which is so dramatic in the ungulates [*Cook and Hunter*, 1978]. These workers point out that in the absence of a sufficiently rapid accumulation of fluid in the blastocoel, there would be a collapse of the trophoblast that would be detrimental to the blastocyst, and the uterine proteins might contribute to making available substances which could modify surface tension effects, much in the way surfactant does in the fetal lung.

Apart from proteins, uterine secretions contain a number of other components, all of which may be important as signals from endometrium to embryo. Potassium ions may be tenfold higher in rat luminal fluids than in serum [*Ringler*, 1961] and in the roe deer and wallaby, calcium levels are low during delayed implantation, but rise after activation [*Aitken*, 1974; *Wallace*, 1980]. Zinc concentrations in the rabbit uterus change during early pregnancy [*Lutwak-Mann and McIntosh*, 1969; *McIntosh and Lutwak-Mann*, 1972], and zinc turnover is altered by injury and wound healing, which have analogies with the decidualization process [*Cook and Hunter*, 1978]. Zinc concentrations are greatest in the inter-caruncular regions of the ovine uterus suggesting an association with the sites of secretion rather than the sites of attachment [*Lane and Cook*, cited in *Cook and Hunter*, 1978]. Apart from these relatively few studies on the ionic composition, other components which may influence growth, such as insulin, somatostatin, co-factors, etc., have not been investigated.

Trophoblastic 'Hormones'

Signals of embryonic origin may influence the endometrium either directly, or indirectly via the corpus luteum. We have already seen that progesterone stimulates uterine secretory activity in a number of species, so that during pregnancy the conceptus could effect control of this hormone by secretion of a luteotrophin or anti-luteolysin. Embryonic luteotrophins can be detected as early as days 6–19 in primates [e. g. *Saxena et al.*, 1974], and more recently chorionic gonadotrophins have been described in preimplantation blastocysts of the rat [*Haour et al.*, 1976], mouse [*Wide and Wide*, 1979], rabbit [*Haour and Saxena*, 1974] and pig [*Saunders et al.*, 1980; *Flint*, 1980]. However, it is not clear whether any of these gonadotrophins have luteotrophic activity, and there is no evidence that the primary role of these proteins is to lengthen the life span of the corpus luteum [reviewed by *Heap et al.,* 1979]. The sheep blastocyst may inhibit the production of PGF₂α [*Goding,* 1974] while in the pig, estrogen has a luteotrophic action which suppresses the release of PGF₂α from the uterus [*Heap et al.,* 1979]. Embryos also have the capacity to synthesize steroid hormones. *Huff and Eik-Nes* [1966] showed that rabbit blastocysts contain enzymes capable of synthesizing cholesterol and pregnenolene from acetate. In the pig, *Perry et al.* [1973] demonstrated that the blastocyst actively synthesizes steroids, and particularly estrogen, and their presence is a result of their synthesis *in situ* and not as a result of diffusion from the maternal circulation.

The concept that steroids secreted by the pre-implantation blastocyst provide a local signal facilitating implantation is attractive. In the rat and mouse, hamster and rabbit there is no direct evidence that the blastocyst is capable of steroid synthesis before implantation, and in the rat and mouse there is evidence to the contrary [*Bullock,* 1977]. Direct experimental evidence to establish the essential role of steroids is difficult to obtain. So far, definitive proof of steroid synthesis has only been shown in the pig [*Perry et al.,* 1973], cow and sheep [*Flint et al.,* 1978], although it seems likely that this may be a fundamental property of trophoblast tissue [*Heap et al.,* 1979]. Even so, it remains to be shown whether steroids of trophoblastic origin are involved in attachment of the blastocyst and implantation.

Conclusions

Synchrony between the uterus, the blastocyst and their secretions is essential for embryonic growth. The luminal fluids which surround the free blastocyst convey not one, but many signals between the endometrium and the blastocyst and vice versa, whilst at the same time providing a milieu compatible with growth. Knowledge of the uterine environment, and of the signals exchanged, is not sufficient to allow us to induce embryonic diapause in non-diapausing species, or to culture blastocysts through certain key steps *in vitro*. One characteristic which clearly emerges is that the free blastocyst is far from physiologically inert, and expresses its genome in a dynamic and changing manner so as to not only control its own differentiation, but to convey its presence to the mother and to cue its development to differentiation and changes in the uterine endometrium.

References

Aitken, R. J.: Calcium and zinc in the endometrium and uterine flushings of the roe deer *(Capreolus capreolus)* during delayed implantation. J. Reprod. Fertil. *40:* 333–340 (1974).

Aitken, R. J.: Embryonic diapause; in Johnson, Development in mammals, vol. 1, pp. 307–360 (Elsevier/North-Holland, Amsterdam 1977).

Aitken, R. J.: The hormonal control of implantation; in Whelan, Maternal recognition of pregnancy. Ciba Fdn Symp. 64, pp. 53–74 (Elsevier/North-Holland, Amsterdam 1979).

Bazer, F. W.: Uterine protein secretions: relationship to development of the conceptus. J. Anim. Sci. *41:* 1376–1382 (1975).

Bergstrom, S. and Nilsson, O.: Embryo-endometrial relationship in the mouse during activation of the blastocyst by oestradiol. J. Reprod. Fertil. *44:* 117–120 (1975).

Biggers, J. D.: New observations on the nutrition of the mammalian oocyte and the preimplantation embryo; in Blandau, The biology of the blastocyst, pp. 319–328 (University of Chicago Press, Chicago 1971).

Beier, H. M.: Physiology of uteroglobin; in Spilman and Wilks, Novel aspects of reproductive physiology, pp. 219–248 (Spectrum Publications, New York 1978).

Beier, H. M. and Mootz, U.: Significance of maternal uterine proteins in the establishment of pregnancy; in Whelan, Maternal recognition of pregnancy. Ciba Fdn Symp. 64, pp. 111–132 (Elsevier/North-Holland, Amsterdam 1979).

Brinster, R. L.: Mammalian embryo metabolism; in Blandau, The biology of the blastocyst, pp. 303–318 (University of Chicago Press, Chicago 1971).

Brandon, J. M. and Wallis, R. M.: Effect of mepyramine, a histamine H_1- and burimamide, a histamine H_2-receptor antagonist, on ovum implantation in the rat. J. Reprod. Fertil. *50:* 251–254 (1977).

Bullock, D. W.: Steroids from the pre-implantation blastocyst; in Johnson, Development in mammals, vol. 2, pp. 199–208 (North-Holland, Amsterdam 1977).

Clementson, C. A. B.; Moshfeghi, M. M., and Mallikarjuneswara, V. R.: The surface charge on the five-day rat blastocyst; in Blandau, The biology of the blastocyst, pp. 193–206 (University of Chicago Press, Chicago 1971).

Cook, B. and Hunter, R. H. F.: Systemic and local hormonal requirements for implantation in domestic animals. J. Reprod. Fertil. *54:* 471–482 (1978).

Dabich, D. and Andary, T. J.: Prevention of blastocyst implantation in mice with proteinase inhibitors. Fert. Steril. *25:* 954–957 (1974).

Dabich, D. and Andary, T. J.: Tryptic- and chymotryptic-like proteinases in early and late preimplantation mouse blastocysts. Biochim. biophys. Acta *444:* 147–153 (1976).

Daniel, J. C.: Blastokinin and analogous proteins. J. Reprod. Fertil. *25:* suppl., pp. 71–83 (1976).

Denker, H.-W. and Hafez, E. S. E.: Proteases and implantation in the rabbit: role of trophoblast vs. uterine secretion. Cytobiologie *11:* 101–109 (1975).

Denker, H.-W.: The role of trophoblastic factors in implantation; in Spilman and Wilks, Novel aspects of reproductive physiology, pp. 181–208 (Spectrum Publications, New York 1978).

Denker, H.-W.: Facts and thoughts about the physiological role of certain proteinases in implantation; in Embryonic diapause in mammals. J. Reprod. Fertil. *29:* suppl. (in press, 1980).

Enders, I. C. and Schlafke, S.: Surface coats of the mouse blastocyst and uterus during the pre-implantation peroid. Anat. Rec. *180:* 31–46 (1974).

Findlay, J. K.; Cerini, M.; Sheers, L. D.; Staples, L. D., and Cumming, I. A.: The nature and role of pregnancy-associated antigens and the endocrinology of early pregnancy in the ewe; in Whelan, Maternal recognition of pregnancy. Ciba Fdn Symp. 64, pp. 239–255 (Elsevier/North-Holland, Amsterdam 1979).

Finn, C. A. and Keen, P. M.: The induction of deciduomata in the rat. J. Embryol. exp. Morph. *111:* 673–682 (1963).

Finn, C. A. and Martin, L.: Temporary interruption of the morphogenesis of deciduomata in the mouse uterus by actinomycin D. J. Reprod. Fertil. *31:* 353–358 (1972).

Finn, C. A. and Porter, D. G.: The uterus. Handbooks in reproductive biology, vol. 1 (Elek Science, London 1975).

Flint, A. P. F.: The control of blastocyst growth and the endocrinology of early pregnancy in the pig; in Embryonic diapause in mammals. J. Reprod. Fertil. *29:* suppl. (in press, 1980).

Flint, A. P. F.; Gadsby, J. E., and Heap, R. B.: Blastocyst steroids: their synthesis and actions; in Du Mesnil du Buisson, Psychoyos and Thomas, L'implantation de l'œuf, pp. 177–180 (Masson, Paris 1978).

Goding, J. R.: The demonstration that $PGF_2\alpha$ is the uterine luteolysin in the ewe. J. Reprod. Fertil. *38:* 261–271 (1974).

Haour, F.; Tell, G. et Sanchez, P.: Mise en évidence et dosage d'une gonadotrophine chorionique chez la rat (rCG). C. r. hebd. Séanc. Acad. Sci., Paris *282:* 1183–1186 (1976).

Heap, R. B.; Flint, A. P. F., and Gadsby, J. E.: Role of embryonic signals in the establishment of pregnancy. Br. med. Bull. *35:* 129–135 (1979).

Huff, R. L. and Eik-Nes, K. B.: Metabolism *in vitro* of acetate and certain steroids by six-day-old rabbit blastocysts. J. Reprod. Fertil. *11:* 57–63 (1966).

Kennedy, T. G.: Evidence for a role for prostaglandins in the initiation of blastocyst implantation in the rat. Biol. Reprod. *16:* 286–291 (1977).

Kubo, H.; Spindle, A. I., and Pedersen, R. A.: Possible involvement of protease in mouse blastocyst implantation *in vitro*. Biol. Reprod. *20:* suppl. 1, 50A (1979).

Lutwak-Mann, C. and McIntosh, J. E. A.: Zinc and carbonic anhydrase in the rabbit uterus. Nature, Lond. *221:* 1111–1114 (1969).

McIntosh, J. E. A. and Lutwak-Mann, C.: Zn transport in rabbit tissues. Some hormonal aspects of the turnover of Zn in female reproductive organs, liver and body fluids. Biochem. J. *126:* 869–876 (1972).

McLaren, A.: Transfer of zona-free mouse eggs to uterine foster mothers. J. Reprod. Fertil. *19:* 341–346 (1969).

McLaren, A.: Early embryo-endometrial relationships; in Hubinont, Leroy, Robyn and Leleux, Ova-implantation, human gonadotrophins and prolactin, pp. 18–37 (Karger, Basel 1970).

McLaren, A.: Blastocyst activation; in Segal, Crozier, Corfman and Condliffe, The regulation of mammalian reproduction, pp. 321–328 (Ch. C. Thomas, Springfield, Ill. 1973a).

McLaren, A.: Endocrinology of implantation. J. Reprod. Fertil. *18:* suppl., pp. 159–166 (1973b).

McMahon, D.: Chemical messengers in development: a hypothesis. Science, N. Y. *185:* 1012–1021 (1974).

New, D. A. T.; Mizell, M., and Cockroft, D. L.: Growth of opossum embryos *in vitro* during organogenesis. J. Embryol. exp. Morph. *41:* 111–123 (1977).

Nilsson, O.: Attachment of rat and mouse blastocysts on to uterine epithelium. Int. J. Fert. *12:* 5–13 (1967).

Orsini, M. W.: Induction of deciduomata in hamsters and rats by injection of air. J. Endocr. *28:* 119–121 (1963).

Perry, J. S.; Heap, R. B., and Amoroso, E. C.: Steroid hormone production by pig blastocysts. Nature, Lond. *245:* 45–47 (1973).

Petzoldt, U.: Untersuchung über das anorganische Milieu in Uterus und Blastozyste des Kaninchens. Zool. Jb., Abt. allg. Zool. Physiol. *75:* 547–593 (1971).

Pinsker, M. C.; Sacco, A. G., and Mintz, B.: Implantation-associated proteinase in mouse uterine fluids. Devl Biol. *38:* 285–290 (1974).

Psychoyos, A.: Perméabilité capillaire et décidualisation utérine. C. r. hebd. Séanc. Acad. Sci., Paris *252:* 1515–1517 (1961).

Renfree, M. B.: Proteins in the uterine secretions of the marsupial *Macropus eugenii*. Devl Biol. *32:* 41–49 (1973).

Renfree, M. B.: Embryonic diapause in mammals – A developmental strategy; in Clutter, Dormancy and developmental arrest, pp. 1–46 (Academic Press, New York 1978).

Ringler, I.: The composition of rat uterine luminal fluid. Endocrinology *68:* 281–291 (1961).

Saunders, P. T. K.; Ziecik, A. J., and Flint, A. P. F.: Gonadotrophin-like substance in pig placenta and embryonic membranes. J. Endocr. (in press, 1980).

Sauer, M. J.: Hormone involvement in the establishment of pregnancy. J. Reprod. Fertil. *56:* 725–743 (1979).

Saxena, B. B.; Hassan, S. H.; Haour, F., and Schmidt-Gollwitzer, M.: Radioreceptor assay of human chorionic gonadotrophin: detection of early pregnancy. Science, N. Y. *184:* 793–795 (1974).

Shelesnyak, M. C.: Some experimental studies on the mechanism of ova-implantation in the rat. Recent Prog. Horm. Res. *13:* 269–322 (1957).

Sherman, M. I. and Barlow, P. W.: Deoxyribonucleic acid content in delayed mouse blastocysts. J. Reprod. Fertil. *29:* 123–126 (1972).

Sherman, M. I.; Shalgi, R.; Rizzino, A.; Sellens, M. H.; Gay, S., and Gay, R.: Changes in the surface of the mouse blastocyst at implantation; in Whelan, Maternal recognition of pregnancy, Ciba Fdn Symp. 64, pp. 33–48 (Excerpta Medica, Amsterdam 1979).

Spindle, A. I. and Pedersen, R. A.: Hatching, attachment and outgrowth of mouse blastocysts *in vitro:* fixed nitrogen requirements. J. exp. Zool. *186:* 305–318 (1974).

Surani, M. A. H.: Cellular and molecular approaches to blastocyst uterine interactions at implantation; in Johnson, Development in mammals, vol. 1, pp. 245–306 (Elsevier/North-Holland, Amsterdam 1977).

Surani, M. A. H.: Influence of uterine and embryonic factors in delayed implantation; in Embryonic diapause in mammals. J. Reprod. Fertil. *29:* suppl. (in press, 1980).

Surani, M. A. H. and Webb, F. T. G.: Effect of dibutyryl cyclic AMP and oestradiol on incorporation of [³H]-leucine into the proteins of luminal fluid of the rat uterus. J. Endocr. *74:* 431–439 (1977).

Tyndale-Biscoe, C. H.: Hormonal control of embryonic diapause and reactivation in the tammar wallaby; in Whelan, Maternal recognition of pregnancy. Ciba Fdn Symp. 64, pp. 173–185 (Elsevier/North-Holland, Amsterdam 1979).

Van Blerkom, J.; Chavez, D. J., and Bell, H.: Molecular and cellular aspects of facultative delayed implantation in the mouse; in Whelan, Maternal recognition of pregnancy. Ciba Fdn Symp. 64, pp. 141–163 (Elsevier/North-Holland, Amsterdam 1979).

Van Blerkom, J.; Manes, D., and Daniel, J. C.: Development of preimplantation rabbit embryos *in vivo* and *in vitro*. Devl Biol. *35:* 262–282 (1973).

Wallace, G. I.: in Embryonic diapause in mammals. J. Reprod. Fertil. *29:* suppl. (in press, 1980).

Weitlauf, H. M. and Kiessling, A. A.: Activation of delayed implanting mouse embryos *in vitro;* in Embryonic diapause in mammals. J. Reprod. Fertil. *29:* suppl. (in press, 1980).

Wide, L. and Wide, M.: Chorionic gonadotrophin in the mouse from implantation to term. J. Reprod. Fertil. *57:* 5–9 (1979).

M. B. Renfree, Ph. D., School of Environmental and Life Sciences,
Murdoch University, Perth WA 6150 (Australia)

Prog. reprod. Biol., vol. 7, pp. 14–27 (Karger, Basel 1980)

Blastocyst-Uterine Interactions at Implantation[1]

M. A. H. Surani and S. B. Fishel[2]

ARC Institute of Animal Physiology, Cambridge

Blastocyst implantation requires co-ordination of changes in the uterus and embryos alike since disruption of synchrony leads to failure of the implantation process. The importance and nature of the primary factors involved in the process of blastocyst implantation can be assessed by determining the properties and potential of the two systems. The main difficulty inherent in the study of blastocyst implantation lies, on the whole, in the inaccessibility of studying the process *in situ* and therefore it requires a combination of experimental approaches *in vivo* and *in vitro*. A number of embryonic and uterine factors are probably involved in blastocyst-uterine interactions. Uterine and embryonic secretions can reciprocally influence events at implantation. The synthesis and secretion of uterine proteins can be readily assessed as they undergo modifications with the changes in the endocrine status of the female. Although the influence of uterine secretions on blastocysts has long been suspected, no conclusive evidence has so far emerged to demonstrate how this influence is exerted. However, any influence they may have must relate to the modulation of embryonic metabolic activity and not in directing the pathway of cell differentiation which is dictated by the inherent programme of development lodged in the embryos; there is no need to suspect otherwise since differentiation of embryos can occur in extrauterine sites. The primary necessity is therefore to deduce how the environment affects the

[1] This work was supported by an MRC Project Grant and a grant from the Ford Foundation. *S. B. F.* is a Beit Memorial Research Fellow.

[2] We thank Prof. *C. R. Austin,* Dr. *R. G. Edwards,* Dr. *S. J. Kimber* and Mrs. *S. C. Barton* for their help.

responsiveness of embryos. Changes in the responsiveness of embryos to environmental factors must occur during the course of early development with modifications in the properties of plasma membrane and as the cellular organelles mature to resemble those in somatic cells. Apart from the use of delay of implantation as a model to study the responsiveness of embryos, this phenomenon can also be best studied *in vitro* by precisely manipulating the environmental factors in the medium. Considerable emphasis has been placed on this aspect of regulation of the implantation process. However, the involvement of cell surface interactions between trophectoderm and uterine epithelium should not be minimised since this event is bound to be of increasing importance at the later stage of implantation (as well as during embryonic quiescence) when the two membranes are closely apposed; this leads to specific adhesion and exceptionally, in the rabbit, to fusion between the two cell types. For this event, considerable modifications in the two cell surfaces must occur in terms of rearrangement and synthesis of cell surface components which permit recognition and specific adhesion events to occur. The modifications in uterine epithelium must primarily result from the action of steroid hormones while the trophectoderm cell surface presumably alters as differentiation proceeds. This type of cell-cell interaction is once again difficult to study without disrupting the normal physiological process. However, morphological, ultrastructural and cell surface properties can be studied to a limited extent with the blastocysts *in situ*. In this article, we have mainly emphasized our own studies, limited as they are for this complex process, to assess the nature of blastocyst-uterine interactions during implantation. Supporting evidence and alternative views on the mechanism of blastocyst implantation can be obtained from several of the reviews [*McLaren, 1973; Psychoyos, 1973; Nilsson, 1974; Weitlauf, 1974; Sherman and Wudl, 1976; O'Grady and Bell, 1977; Surani, 1977, 1980; Edwards and Surani, 1978; Renfree, 1978; Aitken, 1979; Enders and Schlafke, 1979; Van Blerkom et al., 1979*].

The uterine luminal environment consists of macromolecules, some of which are serum components which enter the uterus through the process of transudation whilst others are synthesized and secreted by the uterus in response to ovarian steroids. The amount of other metabolites could also fluctuate, and of special significance are the divalent cations and energy substrates which may influence the blastocyst. Major attention has so far been directed towards studying uterine derived macromolecules and most notably, uteroglobin in the rabbit has attracted considerable attention [*Beier and Mootz,* 1979]. Despite concerted efforts, no un-equivocal role can be assigned to this macromolecule and hence it remains a major component in search of a function. Studies in mice [*Pratt,* 1977; *Aitken,* 1979; *Fishel,* 1979], rats [*Surani,* 1977] and hamsters [*Surani and Burling,* 1979] demonstrate that qualitative variations in the *de novo* synthesis and secretion of uterine proteins occur in response to changes in endocrine conditions. Females experiencing delay of implantation show 80–90 % reduction in *de novo* synthesis and secretion of these com-ponents, even in the hamster where embryos do not enter into diapause, the main difference in this species relating to the relatively weak (only 4–5 %) uterotrophic effect of estradiol. The significance of this and other findings for delayed implantation are discussed elsewhere [*Surani,* 1980].

Our contention, as proposed earlier [*Surani,* 1977] is that uterine luminal proteins may possess some components which act in a manner similar to growth factor and they may bind to putative cell surface receptors and induce a metabolic response in blastocysts. The con-sequence of such binding may lead to an influx of essential ions and energy substrates into the cells which then cause metabolic enhancement of blastocysts. Towards this end, the primary necessity is to demonstrate how luminal proteins interact with blastocysts.

In a recent study, we analysed binding of iodinated luminal proteins and other macromolecules to blastocysts in the rat [*Tzartos and Surani,* 1979]. This study demonstrated that the luminal proteins from day 5 of pregnancy, the day of blastocyst implantation, bind to the blastocysts to a greater extent than macromolecules present in pro-estrous uterine fluid or when compared with the binding of rat serum albumin (table I). This observation reflects the higher affinity of interaction that probably occurs between the cell surface and uterine luminal proteins from day 5 of pregnancy. Furthermore, the binding is substantially abolished when heat-denatured samples are used in the assay. Competition assays also show that the unlabelled macromolecules from the pregnant female are most effective in reducing the binding of labelled proteins suggesting a specific interaction. Moreover, the bound day 5 uterine luminal components are readily displaced only by similar unlabelled components and considerably inefficiently by other macromolecules even though displacement is ac-complished in all cases after about 2 h. This association between embryos and proteins cannot be due to their entry into the blastocoelic cavity since collapsed blastocysts in the presence of cytochalasin B show similar results, nor do they enter the cells by phagocytosis since es-

Table I. Binding of ^{125}I-labelled proteins to rat blastocysts

Labelled proteins	Treatments	Counts, cpm/blastocyst		
		total	% of maximum total	% of maximum specific
Day 5	none	236	100	100
Day 5	heat (80 °C)	78	33	0
Day 5	sodium azide (1 mM)	241	102	103
Day 5	cytochalasin B (10 μg/ml)	234	99	**99**
Day 5	EGTA (2.0 μM)	76	32	0
Pro-estrus		68	29	
BSA		57	24	
RSA		47	20	

The blastocysts were incubated for 3 h in ^{125}I-labelled proteins at 37 °C.

sentially similar results are obtained in the presence of sodium azide. The binding also appears to be Ca^{2+}-dependent since the displacement is again very rapid in the presence of EGTA. These results then provide the essential information towards the primary role of uterine luminal proteins in terms of their binding to the cell surface. Because of several technical difficulties it is not yet possible to provide information concerning the response of embryos after binding of these components, nor is it possible at present to determine the binding of individual components. It is likely that some macromolecules may have greater affinity than others. Any metabolic response achieved as a result of binding of macromolecules is reversible as described previously [*Surani,* 1977]. Secondly, the ultimate release of the bound macromolecules in the competition tests suggests that the response may only be achieved as long as the macromolecules are present. Such reversible metabolic enhancement or pleiotypic response as proposed previously can be gauged from many instances of studies on blastocysts *in vivo* and *in vitro.* It should be noted that other interpretations of the significance of the binding of uterine proteins to blastocysts are also possible. For example, such binding may simply bring about changes in the adhesive properties of blastocysts for uterine epithelial cell surface. However, the metabolic response of embryos may not strictly occur only in response to one particular group of macromolecules in the uterine lumen acting as trophic factors since such

Table II. Uptake of radioactive uridine into preimplantation mouse embryos incubated in medium with fetal calf serum (optimal) or bovine serum albumin (suboptimal)

Embryo stages	Number of experiments (number of embryos)	Incorporation of ^{3}H-uridine pmol/embryo/h $\times 10^4$		Difference response %
		optimal medium	suboptimal medium	
2-cell	4 (72)	10.3	8.8	15
4-cell	4 (102)	18.3	17.0	8
8-cell	6 (170)	44.2	33.2	25*
Compacting	7 (154)	49.9	36.0	25**
Early blastocyst	5 (118)	661.4	341.0	49***
Late blastocyst	9 (192)	671.8	336.0	50***

* $p < 0.05$; ** $p < 0.01$; *** $p < 0.0005$ (Student's t test).

a response can be evoked by many other mitogenic substances, for example by the growth factors in fetal calf serum [McLaren, 1973] and, from our preliminary studies, also by the plant lectin, concanavalin A. This phenomenon illustrates that although the response of embryos in situ may occur to one set of physiological growth factors, others can be substituted to evoke a similar response under experimental conditions in vitro. Since this is the case, it is feasible to study the responsiveness of embryos in vitro at different stages of embryonic development and under a variety of environmental conditions.

In a recent experiment, we have tested the responsiveness of embryos to dialysed fetal calf serum from the 2-cell stage onwards [Fishel and Surani, 1978]. The measure of response was deduced from the incorporation of ^{3}H-uridine into RNA (table II). These studies clearly demonstrated that blastocysts show almost 50 % increase in their metabolic response in the presence of dialysed fetal calf serum which contains numerous growth factors when compared with the response to bovine serum albumin. The other interesting feature of responsiveness of embryos to fetal calf serum is related to the absence of response at the 2-cell and the 4-cell stages of development and increases to 25 % at the 8-cell stage. Although the early increases observed in the response appear primarily due to an increased uptake of radioactive uridine, at the blastocyst stage, the increase is due both to enhanced uptake as well as to incorporation of the precursor. Several factors may change during development to give increased responsiveness of embryos to this environmental factor. For example, there are clear indications for maturation of intracellular organelles such as mitochondria and an increase in rough endoplasmic reticulum [*Enders* 1971]. Cell surface changes also occur so that trophic factors can bind to the surface only after differentiation of blastomeres has progressed to a certain extent. Although the suggestion that the G_1 phase of the cell cycle is only introduced later in development at around the 16-cell stage is not unequivocally accepted, there is some evidence for this to be the case [*Mukherjee*, 1976]. As discussed previously, this phase of the cell sycle is crucial since binding of mitogenic substances to the cell surface only occurs at this stage [*Manino and Burger*, 1975]. Also, cell arrest is accomplished readily during this stage of the cell cycle which may be one of the reasons why embryos at an earlier stage of development do not enter into embryonic quiescence [*Surani*, 1977, 1980]. Other important functional changes in plasma membrane also occur during the course of early development. For example, the Na^+-dependent amino acid transport system is detected later in development at the early blastocyst stage [*Borland and Tasca*, 1974]. This is clearly an important observation since the response of blastocysts to growth factors necessitates controlled influx of essential ions, amino acids and energy metabolites as discussed above. The overall, and an important conclusion to be drawn from these studies is that embryos on reaching the blastocyst stage can be readily influenced by environmental factors and hence they come under a rigorous control of any changes in the luminal environment. This probably also implies that a lack of macromolecules as found at the time of delay of implantation may impose an adaptation on the part of the embryo to enter into a state of quiescence under conditions which clearly are not conducive for further development. This hypothesis runs contrary to the one previously proposed suggesting that delay of implantation occurs as a result of the production of uterine inhibitors [*Psychoyos*, 1973; *Weitlauf*, 1974].

Since in our scheme, the metabolic response of embryos to stimulation by macromolecules is dependent upon influx of essential cations and energy substrates into the cell, omission of these metabolites from the medium should prevent the metabolic response of embryos. *Fishel* [1980] has clearly demonstrated that this is indeed the case since omission of either Ca^{2+} or Mg^{2+} from the medium prevents metabolic response of embryos (table III). In Ca^{2+}-free medium, increasing Ca^{2+} concentration

Table III. Effect of divalent cations on responsiveness of mouse blastocysts

Medium	Average acid precipitable dpm/embryo/h ± SEM	% of control	Number of embryos (number of experiments)
Control (1 mM Ca²⁺ 1 mM Mg²⁺)	3,135 ± 70	100	76 (7)
Ca²⁺-deficient			
+1 mM Mg²⁺	1,097 ± 29	35	76 (7)
+20 mM Mg²⁺	3,194 ± 99	102	29 (4)
+1 mM Sr²⁺	3,164 ± 122	101	106 (5)
+10 mM Sr²⁺	3,693 ± 370	118	89 (5)
Mg²⁺-deficient			
+1 mM Ca²⁺	1,693 ± 123	54	35 (3)
+15 mM Ca²⁺	3,075 ± 84	98	57 (4)
Complete medium			
+0.05 mM D600	1,738 ± 94	55	36 (3)
+0.1 mM D600	944 ± 40	30	29 (3)
+0.1 mM papaverine	1,493 ± 83	48	31 (3)
+0.05 mM D600 15 mM Mg²⁺	3,291 ± 110	105	40 (4)

restores the response to the normal value at a concentration of 0.25 mM Ca²⁺. This increase in response appears to be due to both an enhanced uptake and incorporation of ³H-uridine into RNA. Similarly, in medium lacking Mg²⁺, the metabolic response was inhibited but this was restored when Mg²⁺ levels in the medium were increased to 0.2 mM. Moreover, in the medium lacking Ca²⁺, the response could be restored to normal by supranormal levels of Mg²⁺ at 15–20 mM. Similarly, in the medium lacking Mg²⁺, the response could be restored to normal with supranormal Ca²⁺ concentration of 15–20 mM Mg²⁺. Interestingly, Sr²⁺ was also able to overcome the deficiency of Ca²⁺ in the medium. Indeed, as little as 1 mM Sr²⁺ could overcome the deficiency of Ca²⁺ and with 10 mM Sr²⁺ the response increased to values about 20 % higher than the controls. The divalent cations are clearly essential for the enhanced response of embryos although their precise mode of action is not established. Since Ca²⁺ is also essential for binding of macromolecules to cell surface, it was imperative to test whether the lack of response arose out of this effect or primarily because of the necessary entry of Ca²⁺ into the cells. For this purpose

D600 and papaverine were employed, both of which are known to block Ca^{2+} uptake into cells [*Ash et al.,* 1973; *Bayer et al.,* 1975]. With 100 μM D600 or papaverine, the response of blastocysts was substantially reduced. As is the case with Ca^{2+}-deficient medium, the inhibitory effect of the drugs could be overcome by the presence of supranormal levels of Mg^{2+} (15–20 mM) in the medium which suggests a common mechanism for this phenomenon. This study using the drugs shows that it is the influx of Ca^{2+} into the cells which is responsible for inducing the metabolic response of blastocysts.

The overall conclusion to be derived from these studies appears to relate to the fact that the entry of essential metabolites into cells as a consequence of binding of macromolecules to the cell surface may be the mechanism of responsiveness. In the presence of serum in the medium, omission of any other metabolites such as glucose from the medium will inhibit embryonic response as demonstrated recently [*Van Blerkom et al.,* 1979]. As far as physiological mechanisms involved in the control of metabolic activity at implantation or embryonic diapause are concerned, it is true to say that both the restriction of ions and/or macromolecules in the uterine lumen can prevent metabolic response of embryos as seen during delay of implantation. There is some evidence that Ca^{2+} levels are reduced in the lumen of the roe deer during quiescence [*Aitken,* 1979]. However, in view of our findings in rodents showing an almost 90 % reduction in synthesis and secretion of luminal proteins during delay of implantation, these components could well be the primary factor controlling metabolic activity of embryos. However, it has been suggested that perhaps the fluctuations in the metabolites such as Ca^{2+} and glucose in the lumen may be of greater significance in the primary control of metabolic activity [*Heap et al.,* 1979; *Van Blerkom et al.,* 1979]. This conclusion is based on the study [*Van Blerkom et al.,* 1979] where serum was present in the medium and which from our experiments appears to cause an enhanced influx of such metabolites into the cell. Therefore, it appears that the metabolic response of blastocysts is abolished in the face of enhanced influx of these metabolites into the cell, simply by omitting them from the medium and hence depriving the embryos of essential ions and energy substrates. It is likely that the macromolecules act as the primary trigger for entry of these metabolites into cells, the consequence of which gives enhanced response. Hence, macromolecules and ions and other components such as energy substrates act in concert in the regulation of metabolic activity.

Table IV. Synthesis and release of glycoproteins from mouse blastocysts labelled with (^{3}H)-glucosamine

	Exp. 1	Exp. 2	Exp. 3	Exp. 4
Number of blastocysts	99	80	71	76
Incorporation:				
into blastocysts	5,550	4,085	5,280	5,985
into medium	561	389	663	401
Labelled glycoproteins released from blastocysts, %	10.11	9.52	12.56	8.04

The incorporation values are calculated in terms of cpm/blastocyst after incubation for 9 h.

So far, consideration has been given to the role of the uterine effect on blastocysts. However, blastocysts may be actively involved in the process of blastocyst-uterine interactions in implantation. There is now considerable evidence for the synthesis and release of protein macromolecules, especially chorionic gonadotrophin like substances and steroids which may primarily influence maternal recognition of pregnancy [*Heap et al.,* 1979]. Other aspects of embryonic influence on the uterus relate to stimulation of uterine proteins when embryos are present in the lumen; this is most convincingly demonstrated in marsupials [*Renfree,* 1972]. Some secretions are suggested to protect embryos from maternal immunological reaction [*Amoroso and Perry,* 1975]. However, the types of local effects of embryos and their secretions on the uterus are not well understood. In the mouse, blastocysts cultured in microdrops *in vivo* have been shown to synthesise and release a group of *D*-(6-^{3}H)-glucosamine-labelled glycoproteins (table IV) of approximate Mw of 87,000 [*Fishel and Surani,* 1980]. The function of this material released from mouse blastocysts is as yet unclear. Apart from their probable function outlined above, they could conceivably act on the epithelial cell surface to provide a surface conducive for adhesion to proceed between trophectoderm and uterine epithelium. These and other local effects of embryos on the uterus are poorly defined.

Cell surface interactions between trophectoderm and uterine epithelium must assume increasing significance during the later stages of implantation when the zona pellucida is lost. Elegant studies by Enders and colleagues [*Enders and Schlafke,* 1979] clearly show close apposition of membranes at this stage of implantation when several types of cell-cell interactions probably come into play. These cell-cell interactions between the receptive uterus and stimulated blastocysts require specific molecular interactions to ensure synchrony and success of the implantation process and in the rabbit it leads to fusion of trophoblast and epithelial cells. Once again this event is difficult to study because of the inaccessibility of the embryo undergoing implantation *in situ*. Attempts to recreate this event *in vitro* are fraught with difficulties and open to misinterpretation of the process. However, such attempts are required and have been made [*Sherman and Wudl*, 1976] although there are at least two major points which have to be considered. Firstly, the cell types that are used as monolayers on which blastocysts are placed *in vitro* have to be of the type which blastocysts normally encounter during implantation. Secondly, even if uterine epithelial monolayers are used, they do not maintain their regional organization as the cells attach and flatten out in culture and it is the apical surface of the epithelium which is important in such interactions. Furthermore, there could be important reorganization of surface components as well as qualitative modifications occurring with endocrine changes in the sensitized epithelium [*Surani*, 1977] which are difficult to induce precisely *in vitro*. It is also the case that *in situ*, the blastocyst is completely surrounded by the epithelium as it rests in the uterine crypt during the implantation process. All these difficulties necessitate analysis of uterine and embryonic components involved in the process, separately. An assessment of their role in cell-cell interactions can then be made but caution is required in interpreting the results.

Interspecific transfers of embryos can be used in a limited way to deduce if recognition events occur and whether they are species specific. Interspecific transfers of blastocysts between rats and mice show that some embryos can implant although the development of embryos is not entirely normal [*Tarkowski*, 1962; *Surani*, 1977]. However, in our experiments almost 50 % of the embryos fail to implant although they are viable at least up to the 11th day of pregnancy and they evoke a normal decidual response [*Surani and Barton*, unpublished]. This failure of implantation of interspecific embryos is attributed partly to the recognition errors between the foreign trophectoderm and uterine epithelium.

The attachment and outgrowth of trophectoderm cells *in vitro* have been used extensively as a measure of potential of blastocysts to implant [*Jenkinson*, 1977]. This process also reflects the necessary conditions and cell surface components essential for this display which may play a

Table V. Incorporation of leucine and sugars into trichloroacetic acid-insoluble fraction of blastocysts

Precursor	Tunicamycin μg/ml	Number of experiments	Number of blastocysts	Incorporation cpm/blastocyst		% inhibition
				control	experimental	
L-4,5-³H-leucine	0.25	4	87	25,956	21,880	15.70
	0.50	4	101	24,867	19,645	21.01
	1.00	9	252	27,389	22,545	17.69
	2.00	4	76	25,013	17,669	29.36
D-6-³H-glucosamine	0.25	3	99	4,649	4,519	2.80
	0.50	4	78	4,777	3,709	22.35
	1.00	7	236	5,067	3,645	28.06
	2.00	4	112	4,067	2,523	37.96
D-2-³H-mannose	0.25	4	396	345	191	44.63
	0.50	4	415	385	190	50.65
	1.00	10	979	367	75	79.56
	2.00	3	327	341	73	78.59

Blastocysts were preincubated for 6 h in the presence of different concentrations of tunicamycin in the optimal medium with 10 % fetal calf serum. They were then labelled in the glucose-free medium in the presence of radioactive precursors for 6 h. Groups of 10 and 50 embryos, solubilized in 0.1 % SDS and 0.14 M 2-mercaptoethanol in 100 μl samples, were used in each sample. Aliquots of 5 μl were deposited on GFA glass fibre discs and treated with trichloroacetic acid to obtain counts for incorporation of the precursors. All samples were prepared in triplicate.

role in implantation *in situ.* Since cell surface glycoproteins and their oligosaccharide moieties are thought to be involved in recognition events as well as in specific adhesion [*Hughes,* 1976; *Nicolson,* 1976], the importance of these components has been investigated. Several studies have shown important changes in cell surface oligosaccharides and glycoproteins during embryonic development [*Bennett et al.,* 1971; *Pinsker and Mintz,* 1973; *Rowinski et al.,* 1976; *Jenkinson,* 1977]. In our recent study [*Surani,* 1979], we have employed tunicamycin which is known to inhibit dolichol-mediated glycosylation of N-glycosidically linked glycopeptides [*Takatsuki et al.,* 1971; *Hemming,* 1977]. In the presence of 1.0 μg tunicamycin/ml, the incorporation of ³H-leucine is inhibited by only about 18 % and the incorporation of mannose is inhibited by as much as 80 % (table V). Qualitative analysis by disc gel and 2-dimensional gel electrophoresis [*Surani and Braude,* unpublished] reveals that polypeptides are synthesized normally but their glycosylation is impaired. Hence, carbohydrate-deficient polypeptides are synthesized which appear to be translocated normally to the cell surface as judged by the analysis of ^{125}I iodinated cell surface components [*Surani,* unpublished]. However, modification of the cell surface in terms of oligosaccharide moieties is reflected in the fact that concanavalin A binding to the cell surface is considerably reduced. Embryos treated in this way reversibly fail to undergo adhesion and outgrowth *in vitro* [*Surani,* 1979; *Surani and Kimber,* 1980]. Although similar results have not been obtained by others using tunicamycin [*Atienza-Samlos et al.,* 1980], this difference may arise because of the differences in the source of the drug. However, it primarily appears to be due to the use of suboptimal levels of the compound which have no or only a partial effect on inhibiting protein glycosylation.

In conclusion, these studies reveal that blastocyst metabolic activity can indeed be modulated reversibly by environmental factors. This may be achieved by binding of various kinds of trophic factors both *in vivo* and *in vitro* which cause an influx of essential metabolites into the cell. This process is then followed by metabolic response. In this way the uterine environment can exert a considerable influence on the activity of embryos. The embryos also appear to synthesize and release macromolecules which could have a variety of effects including localized effects on the uterus during implantation. The cell-cell interactions at the second stage of the implantation process are difficult to study but with a variety of approaches, some aspects of this process are being analysed. For example, the cell surface glycoproteins and their oligosaccharide moieties appear to be important during implantation. However, care is necessary in devising models of study, and caution should be applied in interpreting the results for their physiological significance.

References

Aitken, R. J.: The hormonal control of implantation; in Maternal recognition of pregnancy; Ciba Fdn. Symp. 65, pp. 53–83 (Excerpta Medica, Amsterdam 1979).
Amoroso, E. C. and Perry, J. S.: The existence during gestation of an immunological buffer zone at the interface between maternal and foetal tissues. Phil. Trans. R. Soc. *271:* 343–361 (1975).

Ash, J. F.; Spooner, B. S., and Wessels, N. K.: Effects of papaverine and Ca $^{++}$-free medium on salivary gland morphogenesis, Devl Biol. *33:* 463–469 (1973).

Atienza-Samlos, S. B.; Pine, P. R., and Sherman, M. I.: Effects of tunicamycin upon glycoprotein synthesis and development of early mouse embryos. Devl Biol. (in press, 1980).

Bayer, R.; Kaufman, R., and Manhold, R.: Inotropic and electrophysiological actions of verapamil and D 600 in mammalian myocardium. Arch. Pharmacol. *290:* 69–80 (1975).

Beier, H. M. and Mootz, V.: Significance of maternal uterine proteins in the establishment of pregnancy; Ciba Fdn. Symp. 65, pp. 111–140 (Excerpta Medica, Amsterdam 1979).

Borland, R. M. and Tasca, R. J.: Activation of a Na $^+$-dependent amino acid transport system in preimplantation mouse embryos. Devl Biol *36:* 169–182 (1974).

Edwards, R. G. and Surani, M. A. H.: The primate blastocyst and its environment. Uppsala J. med. Sci., suppl. 22, pp. 39–50 (1978).

Enders, A. C.: The fine structure of the blastocyst; in Blandau, Biology of the blastocyst (University of Chicago Press, Chicago 1971).

Enders, A. C. and Schlafke, S.: Comparative aspects of blastocyst-endometrial interactions at implantation; in Maternal recognition of pregnancy; Ciba Fdn. Symp. 64, pp. 3–32 (Excerpta Medica, Amsterdam 1979).

Fishel, S. B.: Analysis of mouse uterine proteins at pro-oestrous, during early pregnancy and after administration of exogenous steroids. J. Reprod. Fertil. *55:* 91–100 (1979).

Fishel, S. B.: The role of divalent cations in the metabolic response of mouse blastocyst to serum. J. Embryol. exp. Morph., in the press (1980).

Fishel, S. B., and Surani, M. A. H. Changes in responsiveness of preimplantation mouse embryos to serum. J. Embryol. exp. Morph. *45:* 295–301 (1978).

Fishel, S. B. and Surani, M. A. H.: Evidence for the synthesis and release of a glycoprotein by mouse blastocysts. J. Reprod. Fertil. *59* (in press, 1980).

Heap, R. B.; Flint, A. P. F., and Gadsby, J. E.: Role of embryonic signals in the establishment of pregnancy. Br. med. Bull. *35:* 129–135 (1979).

Hemming, F. W.: Dolichol phosphate, a co-enzyme in the glycosylation of animal membrane bound glycoproteins. Biochem. Soc. Trans. *5:* 1221–1231 (1977).

Hughes, R. C.: Membrane glycoproteins: a review of structure and function (Butterworths, London 1976).

Jenkinson, E. J.: The *in vitro* blastocyst outgrowth system as a model for the analysis of peri-implantation development; in Johnson, Development in mammals, vol. 2, pp. 151–173 (Elsevier/North-Holland, Amsterdam 1977).

Manino, R. J. and Burger, M. M.: Growth inhibition of animal cells by succinylated concanavalin A. Nature, Lond. *256:* 19–22 (1975).

McLaren, A.: Blastocyst activation; in Segal, Crozier, Corfman and Condliffe, The regulation of mammalian reproduction, pp. 321–328 (Thomas, Springfield 1973).

Mukherjee, A. B.: Cell cycle analysis and X-chromosome inactivation in the developing mouse. Proc. natn. Acad. Sci. USA *73:* 1608–1611 (1976).

Nicolson, G. L.: Trans-membrane control of the receptors on normal and malignant cells. Biochim. biophys. Acta *458:* 1–57 (1976).

Nilsson, O.: The morphology of blastocyst implantation. J. Reprod. Fertil. *39:* 187–194 (1974).

O'Grady, J. E. and Bell, S. C.: The role of the endometrium in blastocyst implantation; in Johnson, Development in mammals, vol. 1, pp. 165–244 (Elsevier/North-Holland, Amsterdam 1977).

Pratt, H. P. M.: Uterine proteins and the activation of embryos from mice during delayed implantation. J. Reprod. Fertil. *50:* 1–8 (1977).

Psychoyos, A.: Hormonal control of ovo-implantation. Vitams Horm. *31:* 201–256 (1973).

Renfree, M. B.: Influence of the embryo on the marsupial uterus. Nature, Lond. *240:* 475–477 (1972).

Renfree, M. B.: Embryonic diapause in mammals – a developmental strategy; in Clutter, Dormancy and developmental arrest: experimental analysis in plants and animals, pp. 1–46 (Academic Press, New York 1978).

Sherman, M. I. and Wudl, L. W.: The implanting mouse blastocyst; in Poste and Nicolson, The cell surface in animal embryogenesis and development, pp. 81–125 (North-Holland, Amsterdam 1976).

Surani, M. A. H.: Cellular and molecular approaches to blastocyst uterine interactions at implantation; in Johnson, Development in mammals, vol. 1, pp. 245–305 (North-Holland, Amsterdam 1977).

Surani, M. A. H.: Glycoprotein synthesis and inhibition of protein glycosylation by tunicamycin in preimplantation mouse embryos: compaction and trophoblast adhesion. Cell *18:* 217–222 (1979).

Surani, M. A. H.: Embryonic and uterine factors in delayed implantation. J. Reprod. Fertil., suppl. 29 (in press, 1980).

Surani, M. A. H. and Burling, A.: Uterine growth and differentiation in response to ovarian steroids in the hamster. Biol. Reprod. *21:* 657–666 (1979).

Surani, M. A. H. and Kimber, S. J.: The role of glycoproteins in the development of preimplantation mouse embryos; in Glasser and Bullock, Cellular and molecular aspects of implantation (Plenum Press, New York, in press 1980).

Takatsuki, A.; Arima, A., and Tamura, G.: Tunicamycin, a new antibiotic. I. Isolation and characterization of tunicamycin. J. Antibiot. *24:* 215–223 (1971).

Tarkowski, A. K.: Inter-specific transfer of eggs between rat and mouse. J. Embryol. exp. Morph. *10:* 476–495 (1962).

Tzartos, S. J. and Surani, M. A. H.: Affinity of uterine luminal proteins for rat blastocysts. J. Reprod. Fertil. *56:* 579–586 (1979).

Van Blerkom, J.; Chavez, D. J., and Bell, H.: Molecular and cellular aspects of facultative delayed implantation in the mouse; in Elliott and Whelan, Maternal recognition of pregnancy; Ciba Fdn. Symp. 64, pp. 141–172 (Excerpta Medica, Amsterdam 1979).

Weitlauf, H. M.: Metabolic changes in the blastocyst of mice and rats during delayed implantation. J. Reprod. Fertil. *39:* 213–224 (1974).

M. A. H. Surani, Ph. D., ARC Institute of Animal Physiology, Animal Research Station, 307 Huntingdon Road, Cambridge CB3 OJQ (England)

Prog. reprod. Biol., vol. 7, pp. 28–42 (Karger, Basel 1980)

Role of Proteinases in Implantation[1]

H.-W. Denker

Abteilung Anatomie der RWTH, Aachen

Introduction

During the last years, morphological, biochemical and experimental studies of the implantation process have provided increasing evidence that proteinases are involved in the initiation phase when contact is made between trophoblast and uterine tissues. Most of these studies were performed in the rabbit and the mouse [9–11, 13–15, 17–20, 25, 28, 30, 32, 38, 42]. Coincidentally, recent cell biological and oncological studies performed in other systems have provided ample data which indicate that the action of certain proteinases on cell surfaces has profound effects on various cell biological phenomena like cell adhesion, migration, secretion of other enzymes, and they can cause metabolic changes or mitogenic activation [39, 40, 45]. Such phenomena have most frequently been demonstrated for trypsin, but can interestingly also be elicited by highly specific endopeptidases causing *limited proteolysis,* like thrombin [8, 33].

Naturally, observations like these give strong impact to studies of the role of proteinases in implantation. In fact, the hypothesis that proteinase action is not only somehow related to but essential for implantation initiation, as postulated on the basis of studies of a more analytical type [1, 10–14, 18, 23, 26, 29, 32, 38] has received strong support from experiments on the influence of specific proteinase inhibitors on implantation: after administration *in vivo,* certain proteinase inhibitors can very

---

[1] Dedicated to Professor Dr. *Bent G. Böving,* whose work stimulated me to study the physiology of implantation, in honor of his 60th birthday.

effectively prevent the attachment of the blastocyst in the uterus [19] (see also below). Furthermore, there is evidence that the physiological regulation of initiation of implantation as governed by maternal hormones may at least in part be mediated by changes in protease and/or protease inhibitor activity of the uterine tissues and the trophoblast.

Trophoblast-Dependent Blastocyst Proteinase (Blastolemmase): Occurrence and Physiological Function

A peculiar gelatin-dissolving proteinase is found in the trophoblast and at the surface of implanting rabbit blastocysts. In fact, this enzyme seems to be of great importance for initiation of implantation, as was already assumed at the time of its first description [11] and as could be substantiated by a number of investigations during the following years [for review, cf. 19]. In particular it seems to be involved in dissolution of the extracellular blastocyst coverings and it is therefore named '*blastolemmase*'.

When the highly sensitive gelatin substrate film test is applied which was designed and optimized for this enzyme [12, 16], blastolemmase activity is nondetectable before 6 days post coitum (p. c.) (sometimes traces are found at 5–6 days p. c.), but rises abruptly at $6^2/_3$ days p. c., i. e. few hours before the dissolution of the blastocyst coverings begins. At 7 to $7^1/_2$ days p. c. when the lysis of the coverings is under way and the trophoblast attaches to the uterine epithelium [7, 13], the proteinase activity is restricted to the area in which these processes are going on, i. e. the abembryonic hemisphere of the blastocyst (fig. 1). In contrast, the embryonic pole where the blastocyst coverings remain still intact and where no attachment takes place in this phase does not show any activity worth of mentioning. After completion of the abembryonic dissolution of the coverings and attachment of the trophoblast, the enzyme activity disappears abruptly so that only minor remnants of activity can be traced irregularly in some places between trophoblast and uterine epithelium from 8 days p. c. on.

Some controversy has developed on the question whether this proteinase derives from the trophoblast as suggested by the described observation, or from the uterine secretion as proposed by *Kirchner* [28, 29]. There is much evidence, however, that this proteinase at least *depends* in some way *on the abembryonic trophoblast:* the maximum of enzyme activity and the beginning of the dissolution of the blastocyst coverings are

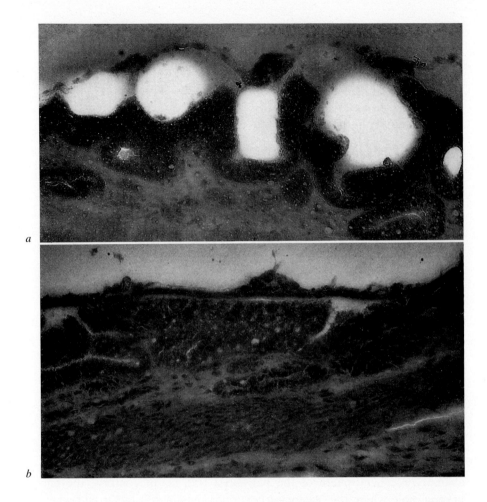

a

b

Fig. 1. Inhibition of blastolemmase activity and of implantation by administration of proteinase inhibitors *in vivo* in the rabbit. Inhibitors were injected into the uterine lumen at $6^1/_2$ days p. c. [for details, cf. 19], and animals were sacrificed at $7^1/_2$ days p. c. Cryostat sections through blastocyst sites, gelatin substrate film test for demonstration of proteinase activity, incubation time 105 min. *a* After administration of 6 mg of ε-aminocaproic acid (which does not inhibit blastolemmase), a completely normal situation is found which corresponds perfectly to the controls. Three trophoblastic knobs are seen in this segment of the abembryonic-antimesometrial region. Blastolemmase activity (bright lysis zones between trophoblast and uterine epithelium) is normal, the blastocyst coverings have been dissolved, and the trophoblastic knobs are establishing contact with the uterine epithelium. $\times 130$. *b* After administration of 6 mg of aprotinin (Trasylol), blastolemmase activity is completely inhibited, the blastocyst coverings (dark band-like structure) are still intact and the trophoblastic knobs (two visible) have not been able to attach to the uterine epithelium. $\times 210$.

always observed at the abembryonic pole of the blastocyst, even if it is abnormally oriented in the uterus, i. e. facing the mesometrial instead of the antimesometrial endometrium (maloriented blastocysts) [17, 19, 20]. Models of blastocyst coverings without trophoblast do not gain any comparable proteinase activity and are not dissolved in the 7- to 8-day uterus [25]. It became very clear from these investigations that the tropho- blast has to provide a factor (or factors) which is (are) essential for the development of this proteinase activity. So far it is not completely clear, however, whether this factor is an activator for a uterus-derived protein- ase or whether it is the gelatinolytic enzyme (or proenzyme) itself. The latter assumption is probably correct because it was possible to demon- strate very high gelatinolytic proteinase activity of the same electro- phoretic mobility in both the trophoblast and the disintegrating blastocyst coverings of the rabbit, while uterine secretion proteinase appears to be a different entity [26]. Recently it became possible to study proteinase activity in trophoblast homogenates using novel chromogenic tri- or tetra- peptide substrates and to define the biochemical properties of the tro- phoblastic enzyme(s) more precisely [23] (see also below).

The main physiological function of blastolemmase seems to be con- nected with the *dissolution of the blastocyst coverings*. One may be tempted, therefore, to compare it with hatching enzymes of lower animals [19]. According to its biochemical properties (particularly its substrate specificity) which will be discussed below, it is probably not respon- sible for complete digestion of the coverings but for softening them due to hydrolysis of just a few peptide bonds. Such a limited hydrolysis might also change cell surface properties and elicit certain cellular responses (see Introduction) forming part of the attachment process.

The role in the dissolution of the blastocyst coverings is most clearly demonstrated by *in vivo inhibition experiments* [19, 20]. Specific proteinase inhibitors which had previously been shown to inhibit blasto- lemmase strongly *in vitro* [18], i. e. aprotinin (Trasylol), antipain, *p*-nitro- phenyl-*p*'-guanidinobenzoate, boar seminal plasma trypsin-acrosin in- hibitor, were administered into the uterus of rabbits half a day before im- plantation (i. e. at $6^1/_2$ days p. c.). Detailed morphological and histo- chemical studies showed that after this treatment, blastolemmase activity is nondetectable or greatly reduced, and the dissolution of the blastocyst coverings is very effectively inhibited (fig. 1–4). Since the coverings re- main as a barrier between the trophoblast and the uterine epithelium, the formation of a cellular contact is not possible. Instead, the blastocyst re-

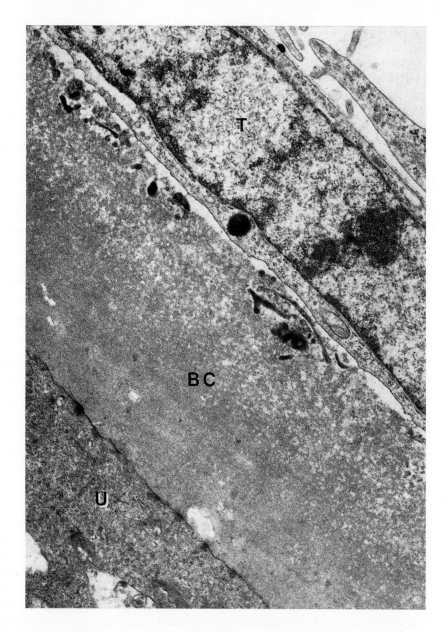

Fig. 2. Inhibition of dissolution of the blastocyst coverings and of implantation in the rabbit by intrauterine administration of 6 mg of aprotinin at $6^{1}/_{2}$ days p. c. Electron micrograph. $\times 16,150$. Stage shown is $8^{1}/_{2}$ days p. c. The blastocyst coverings (BC) are still interposed between trophoblast (T) and uterine epithelium (U) thus preventing formation of a cellular contact. Hemidesmosome-like structures are found at the surface of the uterine epithelial symplasma facing the nondissolved blastocyst coverings.

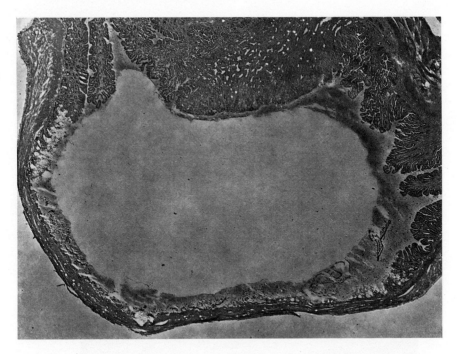

Fig. 3. Inhibition of blastocyst implantation in the rabbit by intrauterine administration of 6 mg of aprotinin at $6^{1}/_{2}$ days p. c. Cryostat section through blastocyst site, $9^{1}/_{2}$ days p. c., gelatin substrate film test for demonstration of proteinase activity, incubation time 105 min. ×9. Even $2^{1}/_{2}$ days after implantation has started in the control the blastocyst lies still free in the inhibitor-treated uterine lumen. The nondissolved blastocyst coverings are mechanically ruptured due to continued expansion of the blastocyst (remnants seen at lower right side). No proteinase activity except for scattered endometrial stroma cells. Protein-rich uterine fluid surrounds the blastocyst, whereas the blastocyst fluid contains very little protein precipitates.

mains free in the uterine lumen. Expansion continues, although at a reduced rate. Due to this continued expansion, the blastocyst coverings are finally ruptured mechanically instead of being dissolved enzymatically. This process bears some resemblance to the hatching from the zona pellucida as seen in the mouse under certain hormonal conditions and *in vitro*. Due to this rupturing, the trophoblast can come into contact with the uterine epithelium in some places. Nevertheless, interestingly, attachment and fusion occurs only focally, i. e. many trophoblastic knobs remain unattached. This may be due to the fact that the trophoblast is now 'out of

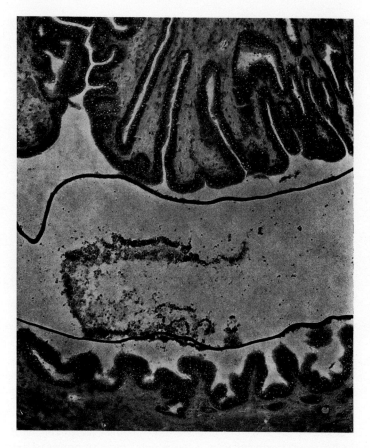

Fig. 4. Inhibition of blastocyst implantation in the rabbit by intrauterine administration of 0.6 mg of *p*-nitrophenyl-*p'*-guanidinobenzoate (NPGB) at 6¹/₂ days p. c. Cryostat section through blastocyst site, 7¹/₂ days p. c., gelatin substrate film test for demonstration of proteinase activity, incubation time 105 min. ×60. The non-dissolved blastocyst coverings are clearly seen as a darkly stained band. No proteinase activity except for two endometrial stroma cells. In contrast to aprotinin (Trasylol), the synthetic inhibitor NPGB shows toxic effects, causing degeneration of blastocyst tissues. Other blastocysts were found to survive NPGB action and to be able to recover after decline of intrauterine inhibitor activity, and to implant normally [19].

phase' with the uterine epithelium which has in the meantime been transformed into a broad symplasma.

In the described experiments, the intrauterine inhibitor concentration had already dropped down considerably at the time at which this delayed focal attachment was observed, due to the fact that only one injection of inhibitor was applied. We are at present investigating whether trophoblast attachment can be completely prevented when the inhibitor concentration is kept high. This would indicate that proteinase action is necessary not only for the dissolution of the blastocyst coverings but also directly involved in the formation of a cellular contact.

Certain differences in the efficacy of the various inhibitors employed were observed and could in part be correlated with differences in elimination rates [19]. The most effective inhibitor in the tested series was aprotinin (Trasylol). On the other hand, ε-aminocaproic acid, an inhibitor of plasminogen activation which does not alter blastolemmase activity *in vitro,* was administered as a control. It was found to be without any effect on the dissolution of the blastocyst coverings and on attachment of the trophoblast even when administered in 100 times higher molar doses than those proven to be effective for aprotinin. This suggests that *plasminogen activators* and *plasmin* probably do not play a major role in the initiation phase of implantation. This does not exclude that this system might be involved in the later process of stromal penetration.

In fact, plasminogen activator activity was described to increase in mouse trophoblast under *in vitro* culture conditions at a phase equivalent to implantation, and the maximum was found at the stage equivalent to the phase shortly after epithelial penetration [42]. However, using fresh, noncultured material from the rat, *Liedholm and Åstedt* [31] observed a decrease of plasminogen activator activity towards implantation in both the embryos and the endometrium. Likewise, no particularly high plasminogen activator activity was detected in rabbit implantation sites, using a comparable technique [*Denker,* unpublished].

Apart from the rabbit, proteinase inhibitors have been administered into the uterus during pregnancy also in the *mouse,* and an antifertility effect was noted [9]. Although no analysis of implantation and early post-implantation stages was carried out in this case, the authors assumed that it was the implantation process which was disturbed also in this species. Recently, it was reported that the attachment of mouse blastocysts to cell monolayers *in vitro* may involve the action of proteinases because addition of soybean trypsin inhibitor inhibited and trypsin stimulated this process [30].

Table I. Biochemical properties of rabbit trophoblast-dependent proteinase (blasto-lemmase)

Substrates	Gelatin; certain obligopeptides with arginyl bonds (no measurable hydrolysis of: low molecular weight trypsin substrates like BANA etc., or comparable chymotrypsin or elastase substrates, or casein or fibrin)
Inhibitors	Aprotinin (Trasylol®), SBTI, antipain, leupeptin, pancreatic secretory trypsin inhibitor (Kazal), α_1-antitrypsin, *p*-nitrophenyl-*p'*-guanidinobenzoate, etc. (not inhibited by EDTA, iodoacetamide or ε-aminocaproic acid)
pH optimum	pH 8.0–8.5 (substrate: gelatin) pH 8.5 (substrate: Tosyl-Gly-Pro-Arg-*p*-nitroanilide)
Active center	related to that of trypsin, but more restricted substrate specificity

Biochemical Properties

Until very recently, investigations of rabbit blastocyst proteinases were generally based on gelatin substrate film tests, particularly on one version which had been specifically designed and optimized for this purpose [12, 16]. Although previously the lack of a kinetic assay was a severe obstacle for detailed biochemical investigations, many interesting data could, nevertheless, already be collected. Experiments on the inhibition of rabbit *trophoblast-dependent blastocyst proteinase (blastolemmase)* by specific proteinase inhibitors *in vitro* [18] revealed that the *active center* of this enzyme is closely related to that of trypsin (table I). However, its *substrate specificity* is much more restricted: the classical chromogenic low molecular weight trypsin substrates like benzoyl-arginine-β-naphthylamide or benzoyl-arginine ethyl ester are not hydrolyzed at any measurable rates. The pH optimum lies in the alkaline range (around pH 8.0–8.5 with the gelatin substrate film test [cf. 19]).

Recently, these findings could be substantiated and more accurate data were obtained using quantitative biochemical assays based on the hydrolysis of novel tri- and tetrapeptide-*p*-nitroanilide substrates [23]. These investigations provided evidence that the proteinase(s) present in rabbit trophoblast homogenates is (are) highly specific insofar as in addition to a specificity for arginyl bonds (P_1), the amino acids in the neighboring positions P_2 and P_3 are also recognized. In fact, substrate specificity is quite comparable to that of highly specific enzymes like thrombin and the kallikreins. This suggests that trophoblast proteinase(s) may *hydrolyse only few peptide bonds* in the physiological substrates, as typical for enzymes which serve a more regulative than a digestive function. Although in addition to blastolemmase, at least one other highly specific endo-peptidase is present in rabbit trophoblast extracts, there is evidence that predominantly blastolemmase was measured in these assays [21]. Thus is appears justified to assume that the physiological role of this enzyme is not to cause complete digestion of the blastocyst coverings but to soften them. It remains to be investigated whether these trophoblast-derived proteinases might also be involved in other, more regulative functions like introducing changes in cell adhesion and in invasive potential, proenzyme activation or liberation of biologically active peptides.

In rabbit implantation, at *uterine secretion* stage, a trypsin-like enzyme is found which appears to resemble trypsin more closely than blasto-lemmase: it is highly basic [19, 28] and shows detectable benzoyl-arginine-β-naphthylamide-splitting activity [26]. It is, like the trophoblast-de-pendent enzyme, being inhibited by the specific high molecular weight trypsin inhibitors. Some first evidence was found that trypsin-inhibiting activity found in the uteroglobin fraction [4] is directed also against this enzyme [26]. Problems of purity of this fraction are discussed elsewhere [21]. Furthermore, blood plasma inhibitors (like α_1-antitrypsin, α_1-anti-chymotrypsin, inter-α-trypsin inhibitor, α_2-macroglobulin) were shown to be present in human endometrium and uterine secretion [41, 43]. Some of them were found to be able to interact with rabbit blastocyst proteinase (blastolemmase) [21]. It appears, therefore, reasonable to assume that, *in vivo,* the activity of blastocyst-derived and of uterine secretion protein-ases is strictly controlled by such inhibitors. Inhibitor activity in the uterine secretion, in turn, was shown to be hormonally regulated [3].

Minor activities of *chymotrypsin-like enzymes* have also been de-tected in rabbit uterine secretion and blastocysts [26] but have not been studied in detail so far.

As far as other species are concerned, some comparable studies have only been performed in the *mouse.* In mouse uterine secretion, a zona pellucida – dissolving activity was reported to increase before implantation ('implantation initiating factor' [32]) and was attributed to a casein-splitting endopeptidase showing parallel changes in activity [38]. Several trypsin-like (BANA splitting) and chymotrypsin-like (GPNA splitting) as well as casein and hemoglobin hydrolyzing enzymes have been detected biochemically in blastocysts and uterine flushings of this species [1, 9]. Since only insignificant endopeptidase activity could be demonstrated in implanting mouse blastocysts using histochemical substrate film tests [5; *Denker,* unpublished] it was assumed that uterine secretion proteinase activity is physiologically more significant in this species [32]. Species differences were also emphasized by *Blandau* [6] who found gelatinolytic activity in late preimplantation blastocysts of the *guinea pig* (which show more invasive properties) but not in those of the *rat* (where the trophoblast seems to be less invasive while the uterine epithelium degenerates more readily). The physiological function of the biochemically detected proteases of mouse blastocysts remains obscure in this context in spite of observations on an increase in activity of some of these enzymes towards implantation [1, 9].

Only histochemical investigations have been performed so far in the guinea pig and the cat. In the *cat,* considerable gelatinolytic activity is found in late preimplantation trophoblast and increases towards implantation. It is particularly high in the girdle-shaped invasion zone at 14 days p. c. where it is located in the trophoblast, between it and the uterine epithelium and in the widened endometrial crypts [22]. Experiments on the pH dependence and on inhibition by various inhibitors revealed that this activity is predominantly due to an enzyme which is totally different from rabbit blastolemmase, namely to an SH-dependent *cathepsin* with an acid pH optimum. It is in many respects similar to cathepsin B, but since no BANA-splitting activity could be demonstrated so far it may also be related to cathepsin L or the 'collagenolytic cathepsin' [2], which is possibly also true for rabbit endodermal cathepsin [19]. In the *guinea pig,* only limited experiments on the pH dependence and on inhibition by various inhibitors have been performed on postimplantation stages [34 to 37]. The obtained results suggest that here, too, the predominant proteolytic activity is due to a cathepsin which might be of the same type. According to the acid pH optimum, an intracellular (intralysosomal) site of action has to be considered for cathepsins, thus differing from the assumed site of action of rabbit blastolemmase (i. e. at the interface between trophoblast and endometrium where an alkaline pH is measured). In both the cat and the guinea pig, however, available data do not exclude the presence of, in addition, a trypsin-like enzyme comparable to blastolemmase particularly in preattachment and attachment stages where it might be involved in the dissolution of the zona pellucida [19]. This point will merit further investigation.

Conclusion

The inhibition of implantation by administration of specific proteinase inhibitors may be considered a new concept possibly applicable for fertility regulation. More extended studies of this aspect including various synthetic low molecular weight inhibitors are in progress in our laboratory. Possible side effects, as increased incidence of dystopic implantation, and ethical aspects should, however, be taken into consideration [19]. It should, furthermore, be kept in mind that whenever proteinase inhibitors are being administered into the uterus as in attempts to prevent increased menstrual blood loss associated with IUD use [44, 46], implantation of blastocysts may be disturbed at the same time.

Acknowledgements

The author wishes to express his gratitude to Mrs. *Gerda Bohr* and Mrs. *Edith Höricht* for excellent technical assistance and Mrs. *Gisela Mathieu* for typing the manuscript. Trasylol® and antipain were kindly provided by Dr. *E. Truscheit* and Dr. *W. Wingender* (Bayer AG, Wuppertal, FRG). These investigations were supported by Deutsche Forschungsgemeinschaft grants No. De 181/1–4 and 7+8 as part of the 'Schwerpunktprogramm Physiologie und Pathologie der Fortpflanzung' and 'Biologie und Klinik der Reproduktion'.

References

1   Andary, T. J.: Trypsin- and chymotrypsin-like enzymes in preimplantation mouse embryos; Diss., Wayne State University Detroit (1974).

2   Barrett, M. C.: Proteinases in mammalian cells and tissues; in Dingle, Research monographs in cell and tissue physiology, vol. 2 (North-Holland, Amsterdam 1977).

3   Beier, H. M.: Hormonal stimulation of protease inhibitor activity in endometrial secretion during early pregnancy. Acta endocr., Copenh. *63:* 141–149 (1970).

4   Beier, H. M.: Immunologische und biochemische Analysen am Uteroglobin und dem Uteroglobin-ähnlichen Antigen der Lunge. Medsche Welt *28:* 788–792 (1977).

5   Bergström, S.: Estimation of proteolytic activity at mouse implantation sites by the gelatin digestion method. J. Reprod. Fertil. *23:* 481–485 (1970).

6   Blandau, R. J.: Embryo-endometrial interrelationship in the rat and guinea pig. Anat. Rec. *104:* 331–359 (1949).

7   Böving, B. G.: Implantation mechanisms; in Hartman, Conference on physiological mechanisms concerned with conception, pp. 321–396 (Pergamon Press, Oxford 1963).

8   Buchanan, J. M.; Chen, L. B., and Zetter, B. R.: Protease-related effects in nor-
    mal and transformed cells; in Schultz and Ahmad, Miami Winter Symp., vol.
    12: Cancer enzymology, pp. 1–23 (Academic Press, New York 1976).

9   Dabich, D. and Andary, T. J.: Prevention of blastocyst implantation in mice
    with proteinase inhibitors. Fert. Steril. 25: 954–957 (1974).

10 Dabich, D. and Andary, T. J.: Tryptic- and chymotryptic-like proteinases in
    early and late preimplantation mouse blastocysts. Biochim. biophys. Acta 444:
    147–153 (1976).

11 Denker, H.-W.: Zur Enzym-Topochemie von Frühentwicklung und Implanta-
    tion des Kaninchens; Diss., Med. Fak. Marburg (FRG) 1969: Enzym-Topoche-
    mie von Frühentwicklung und Implantation des Kaninchens. I. Glykogenstoff-
    wechsel. II. Glykosidasen. III. Proteasen. Histochemie 25: 256–267, 268–285,
    344–360 (1971).

12 Denker, H.-W.: Substratfilmtest für den Proteasennachweis. XIII. Symp. Ges.
    Histochemie, Graz 1969. Acta histochem., suppl. X, pp. 303–305 (1971).

13 Denker, H.-W.: Topochemie hochmolekularer Kohlenhydratsubstanzen in Früh-
    entwicklung und Implantation des Kaninchens. I. Allgemeine Lokalisierung und
    Charakterisierung hochmolekularer Kohlenhydratsubstanzen in frühen Embryo-
    nalstadien. Zool. Jb. Abt. allg. Zool. Physiol. 75: 141–245 (1970).

14 Denker, H.-W.: Topochemie hochmolekularer Kohlenhydratsubstanzen in Früh-
    entwicklung und Implantation des Kaninchens. II. Beiträge zu entwicklungs-
    physiologischen Fragestellungen. Zool. Jb. Abt. allg. Zool. Physiol. 75: 246–308
    (1970).

15 Denker, H.-W.: Blastocyst protease and implantation: effect of ovariectomy and
    progesterone substitution in the rabbit. Acta endocr., Copenh. 70: 591–602 (1972).

16 Denker, H.-W.: Protease substrate film test. Histochemistry 38: 331–338; 39: 193
    (1974).

17 Denker, H.-W.: Trophoblastic factors involved in lysis of the blastocyst cover-
    ings and in implantation in the rabbit: observations on inversely orientated
    blastocysts. J. Embryol. exp. Morph. 32: 739–748 (1974).

18 Denker, H.-W.: Interaction of proteinase inhibitors with blastocyst proteinases
    involved in implantation; in Peeters, Protides of the biological fluids. Proc.
    XXIIIrd Colloquium, Brugge 1975, pp. 63–68 (Pergamon Press, Oxford 1976).

19 Denker, H.-W.: Implantation: the role of proteinases, and blockage of implan-
    tation by proteinase inhibitors (Springer, Berlin 1977): Adv. Anat. Embryol.
    Cell Biol., vol. 53, Fasc. 5.

20 Denker, H.-W.: The role of trophoblastic factors in implantation; in Spilman
    and Wilks, Novel aspects of reproductive biology. 7th Brook Lodge Wokshop on
    Problems of Reproductive Biology, 1977, pp. 181–212 (Spectrum Publications,
    New York 1978).

21 Denker, H.-W.: Inhibitors of trophoblast proteinases; in Beller and Schumacher,
    The biology of the fluids of the female genital tract, pp. 151–161 (Elsevier/
    North-Holland, New York 1979).

22 Denker, H.-W.; Eng, L. A., and Hamner, C. E.: Studies on the early develop-
    ment and implantation in the cat. II. Implantation: proteinases. Anat. Embryol.
    154: 39–54 (1978).

23 Denker, H.-W. and Fritz, H.: Enzymic characterization of rabbit blastocyst pro-

teinase with synthetic substrates of trypsin-like enzymes. Hoppe-Seyler's Z. physiol. Chem. *360:* 107–113 (1979).

24   Denker, H.-W. and Gerdes, H.-J.: The dynamic structure of rabbit blastocyst coverings. I. Transformation during regular preimplantation development. Anat. Embryol. *157:* 15–34 (1979).

25   Denker, H.-W. and Hafez, E. S. E.: Proteases and implantation in the rabbit: role of trophoblast vs. uterine secretion. Cytobiologie *11:* 101–109 (1975).

26   Denker, H.-W. and Petzoldt, U.: Proteinases involved in implantation initiation in the rabbit: microdisc electrophoretic studies. Cytobiologie *15:* 363–371 (1977).

27   Gerdes, H.-J. und Denker, H.-W.: Die strukturelle Entwicklung der Keimhüllen des Kaninchens vom Stadium der Blastozystenbildung bis zur Implantation. Verh. anat. Ges., Jena *73*/Anat. Anz., suppl. 146, pp. 427–434 (1979).

28   Kirchner, C.: Uterine protease activities and lysis of the blastocyst covering in the rabbit. J. Embryol. exp. Morph. *28:* 177–183 (1972).

29   Kirchner, C.; Hirschhäuser, C., and Kionke, M.: Protease activity in rabbit uterine secretion 24 hours before implantation. J. Reprod. Fertil. *27:* 259–260 (1971).

30   Kubo, H.; Spindle, A. I., and Pedersen, R. A.: Possible involvement of protease in mouse blastocyst implantation *in vitro.* Biol. Reprod. *20:* suppl. 1, p. 50A (1979).

31   Liedholm, P. and Åstedt, B.: Fibrinolytic activity of the rat ovum, appearance during tubal passage and disappearance at implantation. Int. J. Fert. *20:* 24–26 (1975).

32   Mintz, B.: Control of embryo implantation and survival; in Raspé, Schering Symp. on Intrinsic and Extrinsic Factors in Early Mammalian Development, Venice 1970; Advances in the Biosciences, vol. 6, pp. 317–342 (Pergamon Press, Oxford 1971).

33   Mosher, D. F. and Vaheri, A.: Thrombin stimulates the production and release of a major surface-associated glycoprotein (fibronectin) in cultures of human fibroblasts. Expl. Cell Res. *112:* 323–334 (1978).

34   Owers, N. O.: Comparison of the proteolytic activity of the implanting rat and guinea-pig blastocyst. Anat. Rec. *166:* 358 (1970).

35   Owers, N. O.: Ingestive properties of guinea pig trophoblast grown in tissue culture: a possible lysosomal mechanism; in Blandau, The biology of the blastocyst, pp. 225–241 (University of Chicago Press, Chicago 1971).

36   Owers, N. O. and Blandau, R. J.: Enzymatic activities of implanting guinea pig embryos. Anat. Rec. *160:* 404 (1968).

37   Owers, N. O. and Blandau, R. J.: Proteolytic activity of the rat and guinea pig blastocyst *in vitro;* in Blandau, The biology of the blastocyst, pp. 207–223 (University of Chicago Press, Chicago 1971).

38   Pinsker, M. C.; Sacco, A. G., and Mintz, B.: Implantation-associated proteinase in mouse uterine fluid. Devl. Biol. *38:* 285–290 (1974).

39   Power, J. A. and Werb, Z.: Protease stimulation of collagenase and plasminogen activator secretion by fibroblasts. J. Cell Biol. *75:* 412a (1977).

40   Ribbons, D. W. and Brew, K.: Proteolysis and physiological regulation. Miami Winter Symp., vol. 11 (Academic Press, New York 1976).

41   Schumacher, G. F. B.; Holt, J. A., and Reale, F.: Approaches to the analysis of

human endometrial secretions; in Beller and Schumacher, The biology of the fluids of the female genital tract, pp. 115–130 (Elsevier/North-Holland, New York 1979).

42 Strickland, S.; Reich, E., and Sherman, M. I.: Plasminogen activator in early embryogenesis: enzyme production by trophoblast and parietal endoderm. Cell 9: 231–240 (1976).

43 Tauber, P. F.: Biochemical components of the human endometrium; in Beller and Schumacher, The biology of the fluids of the female genital tract, pp. 131–150 (Elsevier/North-Holland, New York 1979).

44 Tauber, P. F.; Wolf, A. S.; Herting, W., and Zaneveld, L. J. D.: Hemorrhage induced by intrauterine devices: control by local proteinase inhibition. Fert. Steril. 28: 1375–1377 (1977).

45 Varani, J.; Orr, W., and Ward, P. A.: Cell-associated proteases affect tumour cell migration in vitro. J. Cell Sci. 36: 241–252 (1979).

46 WHO: Special Programme of Research, Development and Research Training in Human Reproduction. 7th Annual Report (1978).

Prof. Dr. med. Dr. rer. nat. H.-W. Denker, Abteilung Anatomie der RWTH,
Melatener Strasse 211, D-5100 Aachen (FRG)

Prog. reprod. Biol., vol. 7, pp. 43–53 (Karger, Basel 1980)

Factors Involved in Implantation-Related Events
in vitro

Michael I. Sherman and Klaus I. Matthaei

Roche Institute of Molecular Biology, Nutley, N. J.

Introduction

Implantation is a complex event that appears to be carefully controlled and timed. We have argued elsewhere that it is likely that both mother and conceptus are individually programmed to undergo implantation-related events [25]. Maternal controls over implantation center about the conditioning of the uterine environment. In particular, the surface epithelium must be primed to accept the implanting blastocyst, whereas the underlying stromal cells must be transformed to form the decidua. There is good reason to believe that these implantation-related events are under hormonal control [see e. g., 4, 10].

For its part, the conceptus appears to participate in implantation by undergoing changes which affect the properties of its surface cells. The outer cells of the preimplantation mouse blastocyst constitute the trophectoderm layer. It is during implantation that these cells begin to acquire properties which we associate with the differentiated phenotype of tropho-blast cells. The presumed implantation-related events which take place in this conversion from trophectoderm to trophoblast are the acquisition of cell surface adhesiveness and, subsequently, of migratory and invasive properties [23, 25]. Unlike the case of implantation-related changes on the part of the uterus, there is no substantive evidence that hormones are involved in the programming of the aforementioned events in murine blastocysts. In fact, trophoblast cells become adhesive and invasive when mouse blastocysts are placed in culture even if steroid hormones have been removed from serum-supplemented medium [12] or if serum-free medium is utilized to culture the embryos [11]. Indeed, the fact that trophoblast cells undergo differentiative steps, including those we believe to be related to implantation, in the absence of contact with maternal cells or even with other cells of the embryo [17, 21] has led us to propose that trophoblast development is intrinsically programmed [11, 17, 25]. Accordingly, we have begun to test this premise. Our initial studies have involved the determination of the timing and the nature of implantation-related events occurring in murine trophoblast cells. These investigations are summarized in the following sections.

Adhesion

Upon hatching from the zona pellucida *in vitro,* the mouse blastocyst initially remains free-floating; the surface cells adhere neither to the culture dish nor to uterine or other cell monolayers. However, 5–15 h later, the blastocysts become adherent to the culture dish surface or to a cellular monolayer [11, 12, 15, 22]. We have attempted to learn more about blastocyst adhesion by interfering with its occurrence through alteration of culture conditions. We have observed that blastocyst adhesion can be blocked by coating the culture dishes with agarose [15, 19]. However, this treatment serves only to deprive blastocysts of a suitable substratum, not to interfere with their developmental program. This is evident from the observation that blastocysts cultured in agarose-coated dishes often stick to each other; furthermore, upon transfer to an uncoated tissue culture vessel, blastocysts previously maintained in agarose-coated dishes become adherent to the surface almost immediately [15].

We have been able to delay, although not to prevent, the onset of adhesion by culturing 4th day blastocysts in suboptimal media. For example, PCM, a simple medium routinely used for the culture of pre-implantation embryos, will permit adhesion, but only after a delay of several hours relative to the time required in cNCTC, a complex, serum-containing medium (table I). Although fetuin (an α-globulin present in fetal serum) promotes outgrowth (see section entitled 'Outgrowth'), it does not improve adhesion rates when added to PCM (PCM+F, table I). In order to determine whether continuous exposure to nutritive medium is necessary for blastocysts to become adherent at optimal rates, we cultured blastocysts in cNCTC medium either for the initial 24 h of culture or from 24 to 48 h and we used PCM+F as the culture medium at other times. The results (fig. 1) indicate that these transient exposures to

Table I. Acquisition of surface adhesiveness by 4th day blastocysts cultured in various media

Culture medium[1]	T_{50}[2]	% of embryos adherent after 90 h culture
PCM	50	93
PCM+F	58	90
PCM+AA+V	—[3]	22
PCM+AA+V+F	72	93
cNCTC	34	100

[1] Details concerning the culture media used are provided by *Sellens and Sherman* [15]. Briefly, PCM is the preimplantation culture medium of *Goldstein et al.* [5], containing salts, glucose, lactate, pyruvate, antibiotics and bovine serum albumin. This medium was supplemented, where indicated, with fetuin (F) and/or amino acids and vitamins (AA+V). It should be noted that vitamins do not influence adhesion times [*Sellens,* unpublished observations]. cNCTC is NCTC-109 medium supplemented with antibiotics and fetal calf serum [18].

[2] T_{50} is the time at which half the embryos became adherent to the culture dish. Data are taken from figure 1 of *Sellens and Sherman* [15].

[3] Less than half of the blastocysts were observed to be adherent in this medium.

cNCTC medium shorten the times required for adhesion compared to those in the continuous presence of PCM+F. However, the 0–24 h treatment is much more effective than the 24–48 h treatment (fig. 1b, c). It should be noted in particular that in the continous presence of cNCTC (fig. 1a), blastocyst adhesion occurs between 24 and 48 h of culture. Thus, exposure to cNCTC is not most beneficial while adhesion actually occurs, but rather *during the time prior to adhesion.* About one-fourth of the blastocysts lag behind the remainder of the population in the time that they become adhesive following a pulse of cNCTC medium from 10 to 24 h (fig. 1b). Since the embryos are not strictly synchronized at the time of collection, it is likely that some of the blastocysts were developmentally too retarded to respond appropriately to the limited exposure to nutritive medium.

We expected that by improving the nutrient value of PCM, we might reduce the time required for the acquisition of adhesive properties by blastocysts. Accordingly, we supplemented PCM with amino acids and vitamins (PCM+AA+V) according to the formulations described by *Spindle and Pedersen* [27]. Contrary to our expectations, only a small

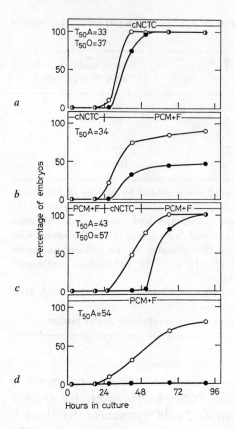

Fig. 1. Kinetics of blastocyst adhesion and trophoblast outgrowth in nutritive and suboptimal media. Blastocysts were obtained on the 4th day of gestation, washed, placed into culture and monitored for adhesion and outgrowth all as described previously [12, 15]. Culture media used were cNCTC (a), PCM+F (d) or both, according to the specified schedules (b, c). Embryos were transferred to new culture dishes when changes were made from one medium to another. Data are averaged from two independent experiments. In both experiments, 30 blastocysts were utilized for each of the culture conditions. $\bigcirc$ = Blastocyst adhesion; $\bullet$ = trophoblast outgrowth. $T_{50}A$ and $T_{50}0$ indicate the times at which half the blastocysts adhered to the substratum and gave rise to trophoblast outgrowths, respectively. $T_{50}0$ values are not given in b and d because fewer than 50% of the embryos gave rise to trophoblast outgrowths during the period of study.

fraction of blastocysts in PCM+AA+V became adherent at all (table I). The addition of fetuin to this medium (PCM+AA+V+F) restored its ability to support blastocyst adhesion; however, the time required for adhesion is substantially longer than it is in unsupplemented PCM (table I).

The above results provide some clues about the acquisition of adhesiveness. Since adhesion occurs in PCM, which contains albumin as the only macromolecular source, we believe that specialized exogenous macromolecules with anchorage-promoting properties are not required for blastocyst adhesion. There is, of course, the caveat that such macromolecules might be present in our albumin preparations, even though we find little evidence from electrophoretic analyses of gross contamination of the albumin samples used [11].

We have observed that blastocysts require exposure to optimal nutritive conditions during at least part of the 24 h period prior to the onset of adhesiveness in order to become adherent at maximal rates. It is, therefore, logical to assume from this observation that synthetic processes essential for adhesion take place during this interval. The nature of the requisite molecules or macromolecules which are presumably synthesized is not clear. From studies with α-amanitin [13, 23], we believe that mRNA transcripts of all genes involved in the acquisition of adhesiveness are already present prior to the midblastocyst stage. We have conflicting data relevant to the question of whether adhesion-related proteins must be synthesized in the 24 h preadhesion period. On the one hand, the values in table I support the view that the synthesis of new protein is not involved because the average adhesion time for blastocysts in PCM (in which there is little increase in total protein content up to the time of adhesion [15]) is less than that in PCM+AA+V or PCM+AA+V+F (in which there is a notable increase in protein content up to the time of adhesion [15]). On the other hand, in preliminary observations we have observed that rates of adhesion can be slowed by cycloheximide at a concentration which inhibits protein synthesis by approximately 60 % [*Matthaei and Schindler, unpublished results*]. We must also consider the possibility that the critical preadhesion period is not actually required for the *synthesis* of new adhesion-related proteins but rather for the *processing* and/or *relocation* at the cell surface of previously synthesized proteins. In this regard, it has been proposed that certain proteins require glycosylation for adhesion to occur [3, 29]; however, studies in our laboratory [1, 20] do not support the view that intact glycoproteins are essential. Finally, one must bear in mind that surface membranes also contain nonprotein components (e. g., lipids); synthesis and/or organization of these other structural entities might have to occur in the period immediately preceding adhesion.

Destabilization

The observation that adhesion in PCM+AA+V or PCM+AA+V+F is inferior to that in PCM or PCM+F (table I) was unexpected. We propose as an explanation for this observation that the surface membranes and/or cytoskeletal components [26] of trophoblast cells become *destabilized* in preparation for outgrowth and that protein synthesis is required for this reorganization. Thus, according to this hypothesis, embryos in PCM or PCM+F, although they are delayed in attaching, remain adherent to the substratum once they attach because they are deprived of exogenous amino acids and cannot undergo destabilization. However, in amino acid-supplemented PCM, changes leading to destabilization occur during the protracted period required for the acquisition of adhesive properties. This creates a conflicting situation between adhesive and destabilizing factors and permanent adhesion is either further delayed or (in the absence of fetuin) prevented. A destabilization period is presumably not obvious in kinetic studies when embryos are cultured in optimal medium (cNCTC) because outgrowth follows relatively quickly upon adhesion.

Support for the existence of a destabilization event, and indeed the reason why we have chosen that word to describe it, derives from the behavior of blastocysts placed in culture after having been maintained for several days in implantation delay in ovariectomized mothers. During delay, protein synthetic rates are low [30, 31] and blastocysts resemble metabolically those cultured in PCM [15]. Upon removal from the uterus and placement into cNCTC medium, previously delayed blastocysts are initially adherent [16], confirming that adhesion factors had been produced under conditions of limiting metabolism. However, within 15 h, presumably after they have resumed normal rates of protein synthesis, the blastocysts have detached from the substratum [16]. We interpret this as being indicative of the onset of destabilization, a phase lasting several hours before the blastocysts reattach, this time permanently. In ultrastructural studies, blastocysts appear to contract when they enter this presumptive destabilization phase. As a result, the cells bulge noticeably and become studded with relatively long microvilli [16]. It is pertinent that although we do not see evidence from kinetic studies of a destabilization period in normal (nondelayed) blastocysts cultured in cNCTC medium, these embryos enter a similar stage of contraction just prior to the onset of trophoblast outgrowth [16, 24, 25].

Our ultrastructural studies do not provide us with information concerning *molecular* changes that occur during this proposed destabilization period. Our working hypothesis is that destabilization reflects a transformation by trophoblast cells from a relatively rigid architecture to a more flexible one which will permit migration. As mentioned above, such a transition could result from alterations in surface structure, underlying cytoskeletal organization, or both.

Outgrowth

We have previously considered at length the phenomenon of trophoblast outgrowth. The following is a summary of the conclusions reached:

(a) Outgrowth is a process occurring 4–8 h after adhesion in which trophoblast cells migrate as a sheet away from the central mass of the embryo [25]. Trophoblast outgrowths are capable of penetrating cell monolayers [2, 3, 12, 22]. Thus, outgrowth is presumably related to the invasive phase of implantation which occurs at equivalent times *in utero*. Trophoblast cells appear to invade predominantly by physical displacement both *in vivo* and *in vitro* [2, 3, 14], although degradative enzymes might also be involved [8, 12, 22]. In this regard, we have demonstrated that trophoblast cells produce the protease plasminogen activator along a temporal schedule paralleling the invasive period [28]. Trophoblast cells lose their invasiveness *in vitro* at about the same time as they do *in utero* [25].

(b) Unlike adhesion, trophoblast outgrowth requires an appropriate facilitating macromolecular component in the medium. This requirement can be met by serum, by the fetuin fraction [6, 11] or, less effectively, by collagen [7, 11].

(c) Also contrary to the process of adhesion, trophoblast outgrowth will not occur in the absence of exogenous amino acids [6, 15, 27] (fig. 1d). Even when PCM is supplemented with fetuin and amino acids, the interval between adhesion and the onset of outgrowth is prolonged [15], suggesting that additional factors are involved. Finally, studies with *α*-amanitin and with embryos from *t*-mutant mice suggest that the expression of some outgrowth-related gene(s) occur(s) prior to the blastocyst stage [23].

On the basis of results from experiments summarized in (c), we have proposed [23] that two synthetic periods are critical for the realization of trophoblast outgrowth. The early period normally occurs between the 8-cell and the blastocyst stages. Nothing is known about the identity of the required molecules or macromolecules synthesized in the early period. The second synthetic period occurs at some time after the blastocyst stage and requires the presence of exogenous amino acids. It seems likely, therefore, that the essential products of the latter period are proteins.

We have attempted to define the time of the second synthetic period for trophoblast outgrowth by restricting the exposure of blastocysts to nutritive medium. First, figure 1d confirms earlier reports that outgrowth will not occur in preimplantation medium [15]. Figure 1b illustrates that when blastocysts are cultured initially for 24 h in cNCTC and then placed in PCM+F, less than half of the embryos give rise to trophoblast outgrowths. However, the time of initiation of outgrowth in the population of competent blastocysts is the same as that of embryos maintained continuously in cNCTC (fig. 1a). Conversely, when embryos are exposed to cNCTC only between 24 and 48 h of culture, all blastocysts give rise to trophoblast outgrowths, although after a substantial delay (fig. 1c). The extent of outgrowth with further culture is, however, inferior to that of embryos exposed to cNCTC medium throughout. Preliminary experiments suggest that the presence of cycloheximide during exposure to cNCTC reduces or, in some cases, eliminates the competence of blastocysts to give rise to trophoblast outgrowths [Matthaei, unpublished observations]. Such observations support the view that the synthesis of new proteins is required for trophoblast outgrowth.

On the basis of the experiments in figure 1, it appears that the second synthetic period required for the initiation of trophoblast outgrowth centers about the period of 24–48 h after the onset of culture (equivalent to the 5th–6th day of gestation); there is, however, some variation, presumably due to the asynchrony of the embryos at the time of collection. It should be noted that the period of exposure to cNCTC medium which is optimal for outgrowth differs from that for adhesion (fig. 1b, c). That is, adhesion is almost complete and prompt when blastocysts are exposed to cNCTC medium in the first 24 h of culture whereas at least half of the blastocysts fail even to initiate outgrowth under this regimen. Finally, it is important to note that neither serum nor amino acids is required during the time of trophoblast outgrowth; presumably, the factors necessary for the initiation of outgrowth, once they are synthesized, permit trophoblast migration in suboptimal medium. Similar conclusions have been reached by *Naeslund* [9] from studies on the activation of ovariectomy-delayed blastocysts. However, since the extent of outgrowth in suboptimal medium is less than that in nutritive medium, it is likely that continuing production of outgrowth factors must occur for maximal trophoblast migration.

Perspectives

We have begun to learn about the scheduling of those events in the early mouse embryo which relate to implantation. As appears to be the case with other developmental processes in these embryos [13], implantation seems to be preprogrammed and under internal control. From the results described here, we believe that blastocyst adhesion and trophoblast outgrowth are separate and independently regulated events. We have also provided indirect evidence for a destabilization phenomenon which presumably involves reorganization of trophoblast surface and/or cytoskeletal structure in preparation for outgrowth. It is probably safe to say that no one has yet identified in an unequivocal manner a single molecule or macromolecule of embryonic origin which plays a direct role in these aforementioned implantation-related processes. Clearly, much work remains to be done in this regard. Indeed, when placed in its proper perspective, the culmination of such studies will represent only the beginning of another, perhaps even more complex, problem. For it will then be necessary to reintegrate the embryo with uterine activities and to learn how the two systems interact to coordinate implantation.

References

1   Atienza-Samols, S. B.; Pine, P. R., and Sherman, M. I.: Effects of tunicamycin upon glycoprotein synthesis and development of early mouse embryos. Devl Biol. (in press 1980).
2   Cole, R. J. and Paul, J.: Properties of cultured preimplantation mouse and rabbit embryos and cell strains derived from them; in Wolstenholme and O'Connor, Preimplantation stages of pregnancy, pp. 82–112 (Academic Press, New York 1965).

3   Glass, R. A.; Spindle, A. I., and Pedersen, R. A.: Mouse embryo attachment to substratum and interaction of trophoblast with cultured cells. J. exp. Zool. *208:* 327–336 (1979).

4   Glasser, S. R.: The uterine environment in implantation and decidualization; in Balin and Glasser, Reproductive biology, pp. 776–833 (Excerpta Medica, Amsterdam 1972).

5   Goldstein, L. S.; Spindle, A. I., and Pedersen, R. A.: X-ray sensitivity of the pre-implantation mouse embryo *in vitro.* Radiat. Res. *62:* 276–287 (1975).

6   Gwatkin, R. B. L.: Defined media and development of mammalian eggs *in vitro.* Ann. N. Y. Acad. Sci. *139:* 79–90 (1966).

7   Jenkinson, E. J. and Wilson, I. B.: *In vitro* studies on the control of trophoblast outgrowth in the mouse. J. Embryol. exp. Morph. *30:* 21–30 (1973).

8   Kirby, D. R. S.: The 'invasiveness' of trophoblast; in Park, The early conceptus, normal and abnormal, pp. 68–74 (University of St. Andrew's Press, Edinburgh 1965).

9   Naeslund, G.: The effect of glucose-, arginine- and leucine-deprivation on mouse blastocyst outgrowth *in vitro.* Upsala J. med. Sci. *84:* 9–20 (1979).

10   Psychoyos, A.: Hormonal control of ovoimplantation. Vitams Horm. *31:* 201–256 (1973).

11   Rizzino, A. and Sherman, M. I.: Development and differentiation of mouse blastocysts in serum-free medium. Expl Cell Res. *121:* 221–233 (1979).

12   Salomon, D. S. and Sherman, M. I.: Implantation and invasiveness of mouse blastocysts on uterine monolayers. Expl Cell Res. *90:* 261–268 (1975).

13   Schindler, J. and Sherman, M. I.: Effects of α-amanitin on programming of mouse blastocyst development (submitted).

14   Schlafke, S. and Enders, A. C.: Cellular basis of interaction between trophoblast and uterus at implantation. Biol. Reprod. *12:* 41–65 (1975).

15   Sellens, M. H. and Sherman, M. I.: Effects of culture conditions on the developmental programme of mouse blastocysts. J. Embryol. exp. Morph. (in press 1980).

16   Shalgi, R. and Sherman, M. I.: Scanning electron microscopy of the surface of normal and implantation-delayed mouse blastocysts during development *in vitro.* J. exp. Zool. *210:* 69–80 (1979).

17   Sherman, M. I.: The role of cell-cell interaction during early mouse embryogenesis; in Balls and Wild, The early development of mammals, pp. 145–165 (Cambridge University Press, London 1975).

18   Sherman, M. I.: Generation of cell lines from preimplantation mouse embryos; in Evans, Perry and Vincent, Tissue culture manual, vol. 1, pp. 199–201 (TCA Inc., Rockville 1976).

19   Sherman, M. I.: Implantation of mouse blastocysts *in vitro;* in Daniel, Methods in mammalian reproduction, pp. 247–257 (Academic Press, New York 1978).

20   Sherman, M. I. and Atienza-Samols, S. B.: *In vitro* studies on the surface adhesiveness of mouse blastocysts; in Ludwig and Tauber, Human fertilization, pp. 179–183 (Thieme, Stuttgart 1978).

21   Sherman, M. I. and Atienza-Samols, S. B.: Differentiation of mouse trophoblast does not require cell-cell interaction. Expl Cell Res. *123:* 73–77 (1979).

22   Sherman, M. I. and Salomon, D. S.: The relationship between the early mouse

embryo and its environment; in Markert and Papaconstantinou, Developmental biology of reproduction, pp. 277–309 (Academic Press, New York 1975).

23 Sherman, M. I.; Sellens, M. H.; Atienza-Samols, S. B.; Pai, A. C., and Schindler, J.: Relationship between the programs for implantation and trophoblast differentiation; in Glasser and Bullock, Cellular and molecular aspects of implantation (Plenum, New York, in press 1980).

24 Sherman, M. I.; Shalgi, R.; Rizzino, A.; Sellens, M. H.; Gay, S., and Gay, R.: Changes in the surface of the mouse blastocyst at implantation. Ciba Fdn Symp. *64:* 33–52 (1979).

25 Sherman, M. I. and Wudl, L. R.: The implanting mouse blastocyst; in Poste and Nicolson, The cell surface in animal embryogenesis and development, pp. 81–125 (North-Holland, Amsterdam 1976).

26 Sobel, J. S.; Glass, R. H.; Cooke, R., and Pedersen, R. A. Cell motility and distribution of actin and myosin during mouse trophoblast development *in vitro;* in Glasser and Bullock, Cellular and molecular aspects of implantation (Plenum, New York, in press 1980).

27 Spindle, A. I. and Pedersen, R. A.: Hatching, attaching and outgrowth of mouse blastocysts *in vitro:* fixed nitrogen requirements. J. exp. Zool. *186:* 305–318 (1973).

28 Strickland, S.; Reich, E., and Sherman, M. I.: Plasminogen activator in early embryogenesis: enzyme production by trophoblast and parietal endoderm. Cell *9:* 231–240 (1976).

29 Surani, M. A. H.: Glycoprotein synthesis and inhibition of glycosylation by tunicamycin in preimplantation mouse embryos: compaction and trophoblast adhesion. Cell. *18:* 217–227 (1979).

30 Van Blerkom, J.; Chavez, D. J., and Bell, H.: Molecular and cellular aspects of facultative delayed implantation in the mouse. Ciba Fdn Symp. *64:* 141–172 (1979).

31 Weitlauf, H. M. and Greenwald, G. S.: Influence of estrogen and progesterone on the incorporation of ^{35}S methionine by blastocysts in ovariectomized mice. J. exp. Zool. *169:* 463–470 (1968).

M. I. Sherman, Ph. D., Roche Institute of Molecular Biology,
Nutley, NJ 07110 (USA)

Endometrial Epithelial Cells

Prog. reprod. Biol., vol. 7, pp. 54–69 (Karger, Basel 1980)

What Roles are Fulfilled by Uterine Epithelial Components in Implantation?

L. Martin[1]

Hormone Physiology Department, Imperial Cancer Research Fund, Lincoln's Inn Fields, London

Introduction

How important is epithelial secretion? How necessary is epithelial absorption? Is the epithelium a barrier; can the trophoblast invade it, come what may or must its cells be programmed to die? Is the epithelium required to transmit messages to the stroma? Is the presence of epithelium necessary for stromal responses? Given the wide interspecies diversity in modes of implantation can one provide answers relevant to the human? I believe so, if one takes 'implantation' in its widest sense and considers not only how uterine epithelia contribute to the final processes whereby embryonic and maternal tissues make intimate contact, but also how they contribute to embryo spacing along the uterus, to embryo placement in the appropriate endometrial niche, and to the timing of all these events.

Implantation is sometimes restricted to specialized endometrial zones. In sheep the chorion of the expanding blastocyst first attaches to endometrial caruncles, areas rich in capillaries but lacking glands [5]. This suggests that, locally, secretion is incompatible with intimate apposition of embryonic and maternal membranes.

In the elephant shrew, *Elephantulus myurus,* implantation is eccentric as in mice, but only one embryo-sized region in each uterine horn is capable of responding to an embryo; so of the many ova shed and fertilized, only those reaching the specialized sites develop [55, 56] and the only mechanism required to ensure the correct number and placing of embryos

---

[1] My thanks to my wife Kay without whose help this would never have been completed.

is sufficient movement to ensure that sooner or later one succeeds. *Elephantulus* is also notable because the 4 cell embryo appears to induce endometrial stromal changes via an intact epithelium, while the *only* embryo to develop *at all* beyond 4 cells is that at the specialized site [54], indicating that local secretions are operative. However, this discussion will concentrate on species without restricted numbers of overtly specialized sites.

Central Implantation

Macropodid marsupials are monovular and their implantation is central and diffuse [57], thus no special mechanisms are needed to space or place the embryo – it simply expands to make contact with the maternal epithelium, which remains intact. For example, in *Macropus eugenii* the ultimate interaction is morphologically equivalent to the initial attachment between blastocyst and uterine epithelium in the mouse, i. e. interdigitation of their microvilli [57]. It is not clear if the maternal epithelium of *M. eugenii* must undergo progestational differentiation before apposition, as seems to be required in mice. On attachment the embryo stimulates endometrial vascular flow and secretion; the message is probably hormonal [96]. Macropodid implantation may be delayed by lactation or season [100]. Delay involves embryonic diapause resulting from inadequate endometrial secretion consequent on a failure of *corpus luteum* development and implantation is initiated by exogenous progesterone or removal of pouch young [107]. During lactational delay in *M. eugenii* uterine secretion is almost undetectable, but after removal of pouch young both amount and protein content rise threefold [97], activating the blastocyst and maintaining its expansion and development to the attachment phase.

Implantation is also central in rabbits and carnivores; blastocysts expand greatly before attachment [5, 114], and the uterine glandular epithelia often undergo spectacular progestational proliferation in early pregnancy/pseudopregnancy [20, 22, 43, 113]. In species with delayed implantation, e. g. mink, badger, the corpus luteum is quiescent and the blastocyst remains small throughout delay [4, 10]. Thus, early blastocyst development and expansion seem to be regulated by progestin-induced secretion. Later stages of implantation go beyond mere attachment, with trophoblastic invasion and destruction of the endometrium. The processes involved are best illustrated by the rabbit in which uterine glandular proliferation and secretion, and ovum implantation can be induced in ovariectomized animals by exogenous progestins [7, 63, 115].

In HCG-induced pseudopregnancy, which mimics pregnancy [20, 21, 104], an initial proliferation of luminal and glandular epithelia produces sufficient secretory cells to maintain blastocyst development and sufficient luminal cells for the folding and relayering of the surface epithelium required for its apposition to the greatly expanded blastocyst. Proliferation then slows (d 4–5) and secretion appears in the gland cells and lumina (d 5–6). Blastocysts expand maximally at this time, and it is generally accepted that the secretion plays a major though undefined role in the process [7]. With expansion each blastocyst distends the myometrium, becoming a source of propagated waves of contraction which, travelling outwards in both directions, space the embryos at regular intervals [11]. Thus, while the final spacing mechanism is myometrial, epithelial secretions play a key role by inducing blastocyst expansion. As this continues the embryos' positions become fixed and the trophoblast fuses with the (now) syncytial maternal surface epithelium [61, 62]. In pseudopregnant animals at this time maternal epithelial cells fuse into multinucleate masses as a result of continuing progestational stimulation; moreover, this development is most rapid and pronounced mesometrially – the site of definitive implantation [20, 21]. This progestationally induced organogenesis probably plays a directive role in trophoblastic invasion. It is pertinent that just before multinucleate cell formation, while gland cells still show signs of secretion, luminal cells develop surface bulges [21] reminiscent of the uterine epithelial pinopods found in rats and mice when embryonic and maternal membranes become apposed [31, 84, 85]. Again, it seems probable that, locally, secretion is incompatible with apposition whereas absorption is necessary.

Trophoblastic fusion induces the multinucleate maternal cells to transform into an acidophilic multinucleate sheet (symplasma); at the same time the connective tissue stroma decidualises [5]. *Hoffman et al.* [51] find that in pseudopregnant rabbits, an intrauterine saline instillation or silastic implant induces symplasma but not deciduomata. Thus, relatively non-specific stimuli, e. g. pressure, induce symplasma formation, which is not *per se* the trigger for decidualization; this requires other perhaps more specific messages. Significantly, implants containing PGE induce decidualization, whereas those with PGF_{2a}, arachidonic acid, histamine, oestradiol, or dcAMP do not. Tricaprylin (a pure triglyceride oil) induces decidualization plus formation and subsequent necrosis of symplasma [19]. Oil also induces the formation and subsequent necrosis of symplasma in the pseudopregnant ferret with hypertrophy though not decidualization of maternal capillaries [6]. It also induces decidualization in rats, mice and hamsters [8, 37, 38] where epithelial breakdown follows rather than causes decidualization. Thus, despite great differences between these species in the gross form of implantation, there may be common pathways at the cellular level – epithelial cells with absorptive capabilities, epithelial cells able to respond to the trophoblast and programmed to die at the appropriate time, epithelial cells able to transmit signals to the stroma.

In the bitch, where the sequence leading up to implantation resembles that in the rabbit, the pre-implantation blastocyst, still in its zona, induces endometrial stromal oedema before any breakdown of the epithelium [53]. Again, the epithelium must transmit signals to the stroma. Might oil also be effective here?

Eccentric and Interstitial Implantation

In species with central implantation there seems little doubt that epithelial secretions, usually induced by progestins[2], play a major role in regulating implantation. Recently it has been suggested that endometrial secretions are important in regulating ovum implantation in species with eccentric or interstitial implantation, where the pre-implantation embryo remains minute, e. g. rat, mouse, hamster and guinea pig.

Here progesterone has a completely different set of actions. It inhibits rather than stimulates uterine epithelial proliferation [13, 72, 75]. It does not stimulate *visible* uterine secretion but suppresses that induced by oestrogen [14, 74, 83] and decreases uterine luminal volume, eventually inducing a closure of the lumen in which the microvilli of the closely apposed surfaces interdigitate [45, 73]. Closure (the 'attachment reaction' of *Nilsson* [78]), occurs at the start of implantation. There are two lines of interest here – How and why? Putative mechanisms include changes in epithelial surface coats, adhesiveness and charge [30, 32, 49, 77], but lu-

---

[2] An exception is the roe deer, *Capreolus capreolus,* which undergoes obligate delay of implantation in which the embryo is small, quiescent and unattached, but the corpora lutea remain active [101]. Delay ends with increased endometrial secretion and blastocyst expansion, but this is not associated with increased levels of progesterone. Oestrogen may be involved [1].

minal epithelial absorption of fluid and electrolytes [14], possibly by endocytosis [31, 66, 84–86], probably plays a major role. During closure in mice the cross-sectional shape of the lumen goes from irregular to a meso/anti-mesometrially aligned slit [67]. This may involve changes in myometrial tone [70] and the patterns of endometrial cell proliferation [41], but what simpler way of producing a slit-like lumen can be postulated than fluid removal – consider a bicycle inner-tube!

In rats and mice implantation sites are regularly spaced [76, 79], probably before implantation. For example, pre-implantation treatment of mice with prostaglandin crowds implantation sites towards the cervix [12], while that with a short-acting non-oestrogenic anti-gestagen, which abolishes luminal closure, crowds them towards the oviduct [71]. In both cases the crowded sites have packed discoid decidua indicating that spacing by post-implantation growth is not significant. *Böving* [11] describes the interactions between myometrium and expanding rabbit blastocysts, which successively spread, space and fix them at definitive implantation sites. How do uteri react to, and space, blastocysts which remain minute?

Spacing in mice occurs from 84 h *post coitum* just as closure develops [60, 92, 98, 102] and when localized areas of intense circular muscle concentration occur [71]. I suggest that initial closure of the lumen enables the uterus to disperse the embryos, then with further closure each blastocyst, by physical presence or chemical message, stimulates the myometrium, becoming a source of propagated waves of circular muscle contraction which space them: final closure fixes each in place[3].

In section the anti-mesometrial cleft of the slit-like lumen usually lies near the centre of the circle formed by the circular muscle[4]. Bodies lying mesometrially would receive asymmetric forces from these muscles and move to the centre of symmetry in the anti-mesometrial cleft [70]. Thus, while the final mechanism of blastocyst spacing and placing is myometrial, epithelial resorption has a key role in closing the lumen and changing luminal shape.

Implantation is delayed in lactating rats and mice but can be precipitated by giving oestrogen. In non-lactating animals it can be delayed by ovariectomy and progesterone treatment. In both types of delay the

---

[3] Throughout this section the morning of finding the vaginal plug (or sperm) = day 1. Hours are timed from 0 h = the midnight or equivalent before finding the plug.

[4] In *Elephantulus,* which implants mesometrially, the opposite applies [56].

endometrium is progestational. The lumen is closed in mice but not in rats [46]. Thus, progesterone alone does not permit implantation, and while closure is probably a prerequisite for implantation it is not the final trigger. Pinocytosis occurs in the uterine epithelia on day 4 of pregnancy in mice and day 5 in rats (i. e. as luminal closure develops), but disappears by days 5 and 6 respectively when implantation is well advanced [84, 85]. In both species pinocytosis occurs during delay but disappears after oestrogenic initiation of implantation [84, 85]. Thus, pinocytosis may be obligatory for implantation but does not precipitate it. Its oestrogen-induced disappearance might be involved in the development of endometrial refractoriness to decidual-inducing stimuli.

By day 6 of pregnancy, implantation in non-lactating mice has culminated in the nidation of actively growing embryos in chambers located anti-mesometrially in the decidualized endometrial stroma [5]. Instillation of oil into the uterine lumen of pseudopregnant or hormonally sensitized females, induces a decidual reaction with formation of implantation chambers indistinguishable from those formed in the presence of an 'invasive' trophoblast [37]. Thus, it seems that at least in the early stages of implantation the blastocyst is largely passive and the hormonally sensitized endometrium, once stimulated, can complete the sequence whether embryos are present or not. Blastocysts implant in extrauterine sites, but not in unsensitized uteri unless the epithelium has been damaged, when a few succeed [16]. Even in normal implantation, trophoblastic invasion does not start before generalized luminal epithelial cell death (see below). Thus, the epithelium seems to be a barrier which can only be invaded after programmed breakdown.

To distinguish between the sequence of events whereby the endometrium becomes sensitized, the sequence of events initiated in the endometrium by the stimulus of the blastocyst/oil droplet, and the sequence of events activating the blastocyst, we need to know when each begins. There is agreement that in mice, luminal closure occurs on the afternoon of day 4. *Potts* and co-workers [60, 92] state that it occurs abruptly at 86 h, while in my own strain, 11/17 animals killed between 87 and 88 h had closed lumina. *Smith and Wilson* [102] found closed lumina at 90 h with the blastocysts spaced out and placed anti-mesometrially, as did others [60, 92], though *Restall and Bindon* [98] found spacing was still in progress at this time. Most workers find blastocysts have intact zonas at 90 h with progressive loss up to 96 h [28, 60, 92, 98, 102]. However, *Orsini and McLaren* [82] found that loss did not start until 95 h. All workers agree that visible dye sites [94] start to appear from 92 to 95 h [40, 82, 98]. Uterine stromal oedema is first evident at 95 h with alkaline phosphatase, an early indicator of decidualization appearing from 96 h on [40]. By 108 h (midday 5) well-differentiated primary decidual zones are present surrounded by numerous mitoses [40, 41, 82, 90] but generalized epithelial breakdown does not occur until 113–119 h [29].

In ovariectomized mice sensitized with a hormone regime which mimics early pregnancy, oedema develops 10 h after intraluminal instillation of oil, followed by primary decidual zones surrounded by mitosis at 18–20 h [17]. Generalized epithelial cell death occurs from 24 h [50]. On this basis, the endometrium is first stimulated by the blastocyst at about 88–89 h, precisely when it is most sensitive to decidual induction by air and oil [48], and towards the end of the final critical period of oestrogen secretion [69], estimated by *Humphrey* [58] to extend from 78 to 90 h. Initial stimulation at 88 h is compatible with the appearance of dye sites 4–8 h later [93]. Thus, the sequence starts within a couple of hours of uterine closure, while the epithelial microvilli simply interdigitate when no secretion is visible in the lumen, 6–7 h before dead epithelial cells or Wilson bodies appear [29, 39, 40, 82, 91, 92] and 12 h before the maternal epithelial microvilli are replaced by blunt processes [90, 91][5].

Stimulation starts just before zona loss but could overlap with its earliest stages. Zona loss may be obligatory for stimulation but alone is not sufficient stimulus – ovariectomy on d 4 prevents implantation but not zona loss [68]. Trophoblastic giant cell transformation occurs after zona loss, from 96 h [28, 60, 92], still during the first phase of luminal closure, though at the same time as endometrial oedema, suggesting that nutrient supply may play a role. However, even at 102–107 h there is no real difference between normal and delayed blastocysts (ovariectomy at 81–84 h), in dry mass or lipid content [47, 112]; increases in both parameters start in control embryos at 112 h – just before generalized maternal epithelial breakdown. Control and delayed blastocysts both increase slowly in size and cell number up to this time; only when decidualization fails to develop do 'delayed' blastocysts lag behind! This suggests that blastocyst growth is not suppressed by secreted inhibitors but by contact or nutritional limitations.

---

[5] 'The second stage of closure' of *Pollard and Finn* [89], who emphasize that it is associated with endometrial insensitivity to decidual-inducing stimuli rather than sensitivity.

Do oestrogen-induced secretions initiate implantation as suggested by *Aitken* [2], who found that the protein content of uterine luminal flushings, taken from pregnant mice, increased greatly between 82–86 and 106–112 h [3]. Firstly, the increases are too late to account for the start of implantation as described above. Secondly, I doubt that they truly reflect luminal contents at a stage of pregnancy when luminal epithelial surfaces are tightly apposed. Flushing mouse uteri with even small volumes of fluid ruptures and removes luminal cells, splits the basement membrane and stroma anti-mesometrially, rupturing and leaching stromal cells and blood vessels. Damage increases with progestation, particularly between 88 and 107 h of pregnancy, apparently because progressive closure of the lumen exacerbates the hydrodynamic stresses caused by flushing [71]. Many 'luminal fluid' proteins are thus derived from luminal and stromal cells, from intercellular fluid and from blood, and changes in their relative abundance in early pregnancy probably reflect only changes in the amount and type of damage caused by 'flushing' as a result of the altered physical state of the lumen induced by hormones and the presence of blastocysts.

Since stromal changes are induced by blastocysts while the epithelium is intact, the epithelium must transmit the stimulus to the stroma. Blastocyst-produced oestrogen has been suggested as the message in some species [27], but this is unlikely in mice, since endometrial responses to blastocysts require as much exogenous oestrogen as those to intraluminal oil or air. Failure to decidualize in the absence of oestrogen is not due to stromal deficiency, because the tissue decidualizes in response to traumatic stimuli, e. g. crushing. Trauma was thought to bypass a deficient epithelium [36], but *Leroy and Lejeune* [65] have evidence that, in the rat, removal of luminal epithelium prevents the stroma responding to trauma. Therefore, the epithelium must be the site of the key events leading to stromal stimulation with a terminal sequence independent of oestrogen and stimulable by trauma. The initial oestrogen-sensitive steps could well involve the cell surface and the adenyl cyclase system. Cyclic AMP induces implantation in ovariectomized progesterone-treated mice and though not deciduogenic itself, supports oil-induced decidualization in such animals [52, 109–111], i. e. it substitutes for oestrogen. However, its role as a second messenger is put in doubt by the fact that non-cyclic AMP is also effective [35]. Prostaglandins have also been proposed as initiating signals; instilled intraluminally they induce decidualization [99] but at heroic doses – 50–100 μg dissolved (how?) in 50 μl aqueous media. Inhibitors of prostaglandin synthesis reduce decidua in size, but not number [33], indicating that prostaglandins are involved only in the later stages of the response.

Biochemical studies on the uterine luminal epithelia are few, because of the difficulties of separating such a small part of the whole organ with any purity [34]. In mice, oestrogen stimulates protein synthesis both in the progestational and non-progestational epithelium [103] although the profiles differ [87, 88]. However, its action in the progestational tissue may involve pathways other than the DNA-dependent transcription of RNA [64]. For example, actinomycin D does not prevent the oestrogenic sensitization of the mouse uterus to a decidual stimulus [42], while drymount autoradiography suggests that oestrogen does not move to the nucleus in the progestational epithelium [105, 106]. However, studies with separated epithelia indicate that oestrogen receptors have the same properties in the progestational and non-progestational tissues including their interaction with the nucleus [95, 103].

In the hamster, implantation follows essentially the same course as in rats and mice [108] but proceeds in ovariectomized adrenalectomized animals given only progesterone [44], i. e. oestrogen is not required. Intraluminal oil and air induce decidua in pseudopregnant animals [8, 80] but I have found no data showing if they are effective in ovariectomized animals given only progesterone. Does the hamster blastocyst secrete oestrogen or is there a real difference in epithelial sensitivity to decidual inducers? Hamster blastocysts are deciduogenic in ovariectomized mice given progesterone and oestrogen [59]; would they be so withut oestrogen? Would mouse blastocysts be deciduogenic in progesterone-treated ovariectomized hamsters?

The guinea pig blastocyst remains small and the uterus exhibits progestational closure [45], but implantation is interstitial, the blastocyst actively penetrating the epithelium to induce a decidual reaction [9, 25]. Perhaps this is why the endocrine requirements for implantation are so loose; it occurs successfully on days 6–7 *post coitum* after ovariectomy on days 3–4 with no progesterone treatment, after ovariectomy and a single dose of 1–4 mg progesterone on day 2, and even after 20 μg oestradiol postovariectomy [23, 24]. Significantly oil and air are ineffective inducers of decidualization in this species [26, 81]. It seems that with an active blastocyst the role of the endometrium may diminish, luminal epithelial differentiation and closure becoming unimportant and programmed epithelial cell death redundant. Guinea pig blastocysts are deciduogenic in ovariectomized mice given oestrogen and progesterone [59]; would they be so without oestrogen?

In the human the blastocyst remains small and one might predict that luminal closure would be important but not secretions. *Potts* [91] states that in paraffin-embedded material the luminal contours of the luteal uterus suggest the walls were in contact during life. *Clemetson et al.* [15] found evidence of increased luminal absorption of water and sodium in the luteal phase, luminal volume decreasing to 39.5 μl. The protein content of uterine flushings varies through the cycle being minimal in the luteal phase. *Datnow* [18] argues that luteal endometrium is wrongly termed 'secretory'. Clear luminal fluids secreted in the follicular phase simply thicken and become visible in the luteal phase as water and electrolytes are resorbed, while the intracellular 'secretory' vacuoles are artefactual spaces due to loss of glycogen stores during fixation. He proposes that luteal gland distension results from occlusion, and functions to increase endometrial height, obliterating the lumen to achieve 'blastocyst grasp'. Human implantation is interstitial; if the blastocyst were as active as that of the guinea pig one might predict that the endocrine requirements for implantation would be loose, and luminal epithelial differentiated function relatively unimportant. Spontaneous decidualization in the human uterus is compatible with this view. However, *Larsen*'s [62] evidence of fusion of human fetal and maternal epithelial cells early in nidation suggests that the situation is not quite that simple.

References

1   Aitken, R. J.: Delayed implantation in roe deer *(Capreolus capreolus)*. J. Reprod. Fertil. *39:* 225–233 (1974).
2   Aitken, R. J.: Embryonic diapause; in Johnson, Development in mammals, vol. 1, pp. 307–359 (North-Holland, Amsterdam 1977).
3   Aitken, R. J.: Changes in the protein content of mouse uterine flushings during normal pregnancy and delayed implantation, and after ovariectomy and oestradiol administration. J. Reprod. Fertil. *50:* 29–36 (1977).
4   Allais, C. and Martinet, L.: Relation between daylight ratio, plasma progesterone levels and timing of nidation in mink *(Muskla vison)*. J. Reprod. Fertil. *54:* 133–136 (1978).
5   Amoroso, E. C.: Placentation; in Parkes, Marshall's physiology of reproduction; 3rd ed., vol. 2, pp. 127–311 (Longmans Green, London 1952).
6   Beck, F.: The development of a maternal pregnancy reaction in the ferret. J. Reprod. Fertil. *40:* 61–69 (1974).
7   Beier, H. M.; Kühnel, W., and Petry, G.: Uterine secretion proteins as extrinsic factors in preimplantation development. Adv. Biosci. *6:* 165–189 (1971).

8 Blaha, G. C.: Effects of age, treatment, and method of induction on deciduo-mata in the golden hamster. Fert. Steril. *18:* 477–485 (1967).

9 Blandau, R. J.: Observations on implantation of the guinea pig ovum. Anat. Rec. *103:* 19–48 (1949).

10 Bonnin, M.; Canivenc, R., and Ribes, O.: Plasma progesterone levels during delayed implantation in the European badger *(Meles meles).* J. Reprod. Fertil. *52:* 55–58 (1978).

11 Böving, B. G.: Biomechanics of implantation; in Blandau, Biology of the blasto-cyst, pp. 423–442 (University of Chicago Press, Chicago 1971).

12 Bronson, R. and Hamada, Y.: The effect of prostaglandins $F_{2\alpha}$ and E_2 on preg-nancy in mice during implantation. Fert. Steril. *30:* 354–361 (1978).

13 Clark, B. F.: The effects of oestrogen and progesterone on uterine cell division and epithelial morphology in spayed, adrenalectomized rats. J. Endocr. *50:* 527–528 (1971).

14 Clemetson, C. A. B.; Verma, U. L., and De Carlo, S. J.: Secretion and reabsorp-tion of uterine luminal fluid in rats. J. Reprod. Fertil. *49:* 183–187 (1977).

15 Clemetson, C. A. B.; Kim, J. K.; De Jesus, T. P. S.; Mallikarjuneswara, V. R., and Wilds, J. H.: Human uterine fluid potassium and the menstrual cycle. J. Obstet. Gynaec. Br. Commonw. *80:* 553–561 (1973).

16 Cowell, T. P.: Implantation and development of mouse eggs transferred to the uteri of non-progestational mice. J. Reprod. Fertil. *19:* 239–245 (1969).

17 Das, D. M. and Martin, L.: Uterine DNA synthesis and cell proliferation dur-ing early decidualization induced by oil in mice. J. Reprod. Fertil. *53:* 125–128 (1978).

18 Datnow, A. D.: A reconsideration of the secretory function of the human en-dometrium. J. Obstet. Gynaec. Br. Commonw. *80:* 865–871 (1973).

19 Davies, J. and Davenport, G. R.: Symplasma formation and decidualization in the pseudopregnant rabbit after intraluminal instillation of tricaprylin. J. Re-prod. Fertil. *55:* 141–145 (1979).

20 Davies, J. and Hoffman, L. H.: Studies on the progestational endometrium of the rabbit. I. Light microscopy, day 0 to day 13 of gonadotrophin-induced pseu-dopregnancy. Am. J. Anat. *137:* 423–446 (1973).

21 Davies, J. and Hoffman, L. H.: Studies on the progestational endometrium of the rabbit. II. Electron microscopy, day 0 to day 13 of gonadotrophin-induced pseudopregnancy. Am. J. Anat. *142:* 335–366 (1975).

22 Dawson, A. B. and Kosters, B. A.: Preimplantation changes in the uterine mu-cosa of the cat. Am. J. Anat. *75:* 1–37 (1944).

23 Deanesly, R.: Implantation and early pregnancy in ovariectomized guinea-pigs. J. Reprod. Fertil. *1:* 242–248 (1960).

24 Deanesly, R.: Further observations on the effects of oestradiol on tubal eggs and implantation in the guinea-pig. J. Reprod. Fertil. *5:* 49–57 (1963).

25 Deanesly, R.: The role of the fertilized egg: reactions in the guinea-pig uterus at ovo-implantation and after thread traumatization. J. Reprod. Fertil. *14:* 243–248 (1967).

26 Deanesly, R.: The differentiation of the decidua at ovo-implantation in the guinea-pig contrasted with that of the traumatic deciduoma. J. Reprod. Fertil. *26:* 91–97 (1971).

27 Dickmann, Z.; Dey, S. K., and Sengupta, J.: A new concept: control of early pregnancy by steroid hormones originating in the preimplantation embryo. Vitams Horm. *34:* 215–242 (1976).

28 Dickson, A. D.: The form of the mouse blastocyst. J. Anat. *100:* 335–348 (1966).

29 El-Shershaby, A. M. and Hinchliffe, J. R.: Epithelial autolysis during implantation of the mouse blastocyst: an ultrastructural study. J. Embryol. exp. Morph. *33:* 1067–1080 (1975).

30 Enders, A. C.: Anatomical aspects of implantation. J. Reprod. Fertil., suppl. 25, pp. 1–15 (1976).

31 Enders, A. C. and Nelson, D. M.: Pinocytotic activity of the uterus of the rat. Am. J. Anat. *138:* 277–300 (1973).

32 Enders, A. C. and Schlafke, S.: Surface coats of the mouse blastocyst and uterus during the preimplantation period. Anat. Rec. *180:* 31–46 (1974).

33 Evans, C. A. and Kennedy, T. G.: The importance of prostaglandin synthesis for the initiation of blastocyst implantation in the hamster. J. Reprod. Fertil. *54:* 255–261 (1978).

34 Fagg, B.; Martin, L.; Rogers, L.; Clark, B., and Quarmby, V. E.: A simple method for removing the luminal epithelium of the mouse uterus for biochemical studies. J. Reprod. Fertil. *57:* 335–339 (1979).

35 Fernandez-Noval, A. and Leroy, F.: Induction of implantation in the mouse by intrauterine injection of adenosine monophosphate. J. Reprod. Fertil. *53:* 7–8 (1978).

36 Finn, C. A.: The reaction of the uterus during implantation in the mouse; in Lamming and Amoroso, Reproduction in the female mammal, pp. 513–530 (Butterworths, London 1967).

37 Finn, C. A. and Hinchliffe, J. R.: Histological and histochemical analysis of the formation of implantation chambers in the mouse uterus. J. Reprod. Fertil. *9:* 301–309 (1965).

38 Finn, C. A. and Keen, P. M.: Studies on deciduomata formation in the rat. J. Reprod. Fertil. *4:* 215–216 (1962).

39 Finn, C. A. and Lawn, A. M.: Transfer of cellular material between the uterine epithelium and trophoblast during the early stages of implantation. J. Reprod. Fertil. *15:* 333–336 (1968).

40 Finn, C. A. and McLaren, A.: A study of the early stages of implantation in mice. J. Reprod. Fertil. *13:* 259–267 (1967).

41 Finn, C. A. and Martin, L.: Patterns of cell division in the mouse uterus during early pregnancy. J. Endocr. *39:* 593–597 (1967).

42 Finn, C. A. and Martin, L.: Temporary interruption of the morphogenesis of deciduomata in the mouse uterus by actinomycin D. J. Reprod. Fertil. *31:* 353 (1972).

43 Hammond, J. and Marshall, F. H. A.: Oestrus and pseudopregnancy in the ferret. Proc. R. Soc. *105:* 607–637 (1929).

44 Harper, M. J. K.; Down, D., and Elliott, A. S. W.: Implantation and embryonic development in ovariectomized-adrenalectomized hamster. Biol. Reprod. *1:* 253–257 (1969).

45 Hedlund, K. and Nilsson, O.: Hormonal requirements for the uterine attachment reaction and blastocyst implantation in the mouse, hamster and guinea-pig. J. Reprod. Fertil. *26:* 267–269 (1971).

46 Hedlund, K.; Nilsson, O.; Reinius, S., and Aman, G.: Attachment reaction of the uterine luminal epithelium at implantation: light and electron microscopy of the hamster, guinea-pig, rabbit and mink. J. Reprod. Fertil. 29: 131–132 (1972).

47 Hensleigh, H. C. and Weitlauf, H. M.: Effect of delayed implantation on dry weight and lipid content of mouse blastocysts. Biol. Reprod. 10: 315–320 (1974).

48 Hetherington, C. M.: The development of deciduomata induced by two non-traumatic methods in the mouse. J. Reprod. Fertil. 17: 391–393 (1968).

49 Hewitt, K.; Beer, A. E., and Grinnell, F.: Disappearance of anionic sites from the surface of the rat endometrial epithelium at the time of blastocyst implantation. Biol. Reprod. 21: 691–707 (1979).

50 Hinchliffe, J. R. and El-Shershaby, A. M.: Epithelial cell death in the oil-induced decidual reaction of the pseudopregnant mouse: an ultrastructural study. J. Reprod. Fertil. 45: 463–468 (1975).

51 Hoffman, L. H.; Strong, G. B.; Davenport, G. R., and Fröhlich, J. C.: Deciduo-genic effect of prostaglandins in the pseudopregnant rabbit. J. Reprod. Fertil. 50: 231–237 (1977).

52 Holmes, P. V. and Bergstrom, S.: Induction of blastocyst implantation in mice by cyclic AMP. J. Reprod. Fertil. 43: 329–332 (1975).

53 Holst, P. A. and Phemister, R. D.: The prenatal development of the dog: pre-implantation events. Biol. Reprod. 5: 194–206 (1971).

54 Horst, C. J. van der: Early stages in the embryonic development of Elephantulus. S. Afr. J. med. Sci., biol. suppl., pp. 55–67 (1942).

55 Horst, C. J. van der and Gillman, J.: The number of eggs and surviving embryos in Elephantulus. Anat. Rec. 80: 443–452 (1941).

56 Horst, C. J. van der and Gillman, J.: Preimplantation phenomena in the uterus of Elephantulus. S. Afr. J. med. Sci. 7: 47–71 (1942).

57 Hughes, R. L.: Morphological studies on implantation in marsupials. J. Reprod. Fertil. 39: 173–186 (1974).

58 Humphrey, K. W.: The induction of implantation in the mouse after ovariectomy. Steroids 10: 591–600 (1967).

59 Kirby, D. R. S.: Immunological aspects of implantation; in Hubinont, Leroy, Robyn and Leleux, Ovo-implantation, human gonadotropins and prolactin, pp. 86–97 (Karger, Basel 1970).

60 Kirby, D. R. S.; Potts, D. M., and Wilson, I. B.: On the orientation of the implanting blastocyst. J. Embryol. exp. Morph. 17: 527–532 (1967).

61 Larsen, J. F.: Electron microscopy of the implantation site in the rabbit. Am. J. Anat. 109: 319–325 (1961).

62 Larsen, J. F.: Electron microscopy of nidation in the rabbit and observations on the human trophoblastic invasion; in Hubinont, Leroy, Robyn and Leleux, Ovo-implantation, human gonadotrophins and prolactin, pp. 38–51 (Karger, Basel 1970).

63 Lee, A. E. and Dukelow, W. R.: Synthesis of DNA and mitosis in rabbit uteri after oestrogen and progesterone injection and during early pregnancy. J. Reprod. Fertil. 31: 473–476 (1972).

64 Leroy, F.: Aspects moléculaires de la nidation; in du Mesnil du Buisson, Psychoyos and Thomas, L'implantation de l'œuf, pp. 81–92 (Masson, Paris 1978).

65 Leroy, F. and Lejeune, B.: Unpublished results (1980).

66 Leroy, F.; Van Hoeck, J. and Bogaert, C.: Hormonal control of pinocytosis in the uterine epithelium of the rat. J. Reprod. Fertil. *47:* 59–62 (1976).

67 McLaren, A.: Early embryo-endometrial relationships; in Hubinont, Leroy, Robyn and Leleux, Ovo-implantation, human gonadotropins and prolactin, pp. 18–37 (Karger, Basel 1970).

68 McLaren, A.: Blastocysts in the mouse uterus: the effect of ovariectomy, progesterone and oestrogen. J. Endocr. *50:* 515–526 (1971).

69 Martin, L.: Decidualization and ovum-implantation in spayed mice treated with oestriol. J. Endocr. *72:* 181–185 (1977).

70 Martin, L.: Early cellular changes and circular muscle contraction associated with the induction of decidualization by intrauterine oil in mice. J. Reprod. Fertil. *55:* 135–139 (1979).

71 Martin, L.: Unpublished results.

72 Martin, L. and Finn, C. A.: Hormonal regulation of cell division in epithelial and connective tissues of the mouse uterus. J. Endocr. *41:* 363–371 (1968).

73 Martin, L.; Finn, C. A., and Carter, J.: Effects of progesterone and oestradiol-17β on the luminal epithelium of the mouse uterus. J. Reprod. Fertil. *21:* 461–469 (1970).

74 Meglioli, G.; Krahenbuhl, C., and Desaulles, P. A.: The action of sex hormones on endometrial secretion in spayed rats sensitized to oestradiol. Experientia *25:* 194–195 (1969).

75 Mehrotra, S. and Finn, C. A.: Cell proliferation in the uterus of the guinea-pig. J. Reprod. Fertil. *37:* 405–409 (1974).

76 Mossman, H. W.: Orientation and site of attachment of the blastocyst: a comparative study; in Blandau, Biology of the blastocyst, pp. 49–57 (University of Chicago Press, Chicago 1971).

77 Nilsson, O.: Estrogen-induced increase of adhesiveness in uterine epithelium of mouse and rat. Expl Cell Res. *43:* 239–241 (1966).

78 Nilsson, O.: Some ultrastructural aspects of ovo-implantation; in Hubinont, Leroy, Robyn and Leleux, Ovo-implantation, human gonadotropins and prolactin, pp. 52–69 (Karger, Basel 1970).

79 O'Grady, J. E. and Heald, P. J.: The position and spacing of implantation sites in the uterus of the rat during early pregnancy. J. Reprod. Fertil. *20:* 407–412 (1969).

80 Orsini, M. W.: Induction of deciduomata in hamster and rat by injected air. J. Endocr. *28:* 119–121 (1963).

81 Orsini, M. W. and Donovan, B. T.: Implantation and induced decidualization of the uterus in the guinea pig, as indicated by pontamine blue. Biol. Reprod. *5:* 270–281 (1971).

82 Orsini, M. W. and McLaren, A.: Loss of the zona pellucida in mice, and the effect of tubal ligation and ovariectomy. J. Reprod. Fertil. *13:* 485–499 (1967).

83 O'Shea, J. D.: Uterine fluid and the duration of pseudopregnancy following transection of the uterus in the rat. J. Reprod. Fertil. *29:* 57–64 (1972).

84 Parr, M. B. and Parr, E. L.: Uterine luminal epithelium: protrusions mediate endocytosis, not apocrine secretion, in the rat. Biol. Reprod. *11:* 220–233 (1974).

85 Parr, M. B. and Parr, E. L.: Endocytosis in the uterine epithelium of the mouse. J. Reprod. Fertil. *50:* 151–153 (1977).

86 Parr, M. B. and Parr, E. L.: Uptake and fate of ferritin in the uterine epithelium of the rat during early pregnancy. J. Reprod. Fertil. *52:* 183–188 (1978).

87 Pollard, J. W. and Martin, L.: Cytoplasmic and nuclear non-histone proteins and mouse uterine cell proliferation. Mol. cell. Endocrinol. *2:* 183–191 (1975).

88 Pollard, J. W. and Martin, L.: The effects of sex steroids on the synthesis of urine non-histone proteins. Mol. cell. Endocrinol. *6:* 223–229 (1977).

89 Pollard, R. M. and Finn, C. A.: Ultrastructure of the uterine epithelium during the hormonal induction of sensitivity and insensitivity to a decidual stimulus in the mouse. J. Endocr. *55:* 293–298 (1972).

90 Potts, D. M.: The ultrastructure of implantation in the mouse. J. Anat. *103:* 77–90 (1968).

91 Potts, D. M.: The ultrastructure of egg implantation; in McLaren, Advances in reproductive physiology, vol. 4, pp. 241–267 (Logos Press, London 1969).

92 Potts, D. M. and Wilson, I. B.: The preimplantation conceptus of the mouse at 90 hours *post coitum.* J. Anat. *102:* 1–11 (1967).

93 Psychoyos, A.: La réaction déciduale est précédée de modifications précoces de la perméabilité capillaire de l'utérus. C. r. Séanc. Soc. Biol. *154:* 1384–1387 (1960).

94 Psychoyos, A.: Perméabilité capillaire et décidualisation utérine. C. r. hebd. Séanc. Acad. Sci., Paris *252:* 1515 (1961).

95 Quarmby, V. E.: The influence of progesterone on oestradiol-17β uptake and distribution in the mouse uterus; PhD thesis, University of London.

96 Renfree, M. B.: Influence of the embryo on the marsupial uterus. Nature, Lond. *240:* 475–477 (1972).

97 Renfree, M. B.: Proteins in the uterine secretions of the marsupial *Macropus eugenii.* Devl Biol. *32:* 41–49 (1973).

98 Restall, B. J. and Bindon, B. M.: The timing and variation of pre-implantation events in the mouse. J. Reprod. Fertil. *24:* 423–426 (1971).

99 Sananes, N.; Baulieu, E.-E., and Le Goascogne, C.: Prostaglandin(s) as inductive factor of decidualization in the rat uterus. Mol. cell. Endocrinol. *6:* 153–158 (1976).

100 Sharman, G. B. and Berger, P. J.: Embryonic diapause in marsupials; in McLaren, Advances in reproductive physiology, vol. 4, pp. 212–240 (Logos Press, London 1969).

101 Short, R. V. and Hay, M. F.: Delayed implantation in the roe deer *(Capreolus capreolus)*; in Rowlands, Comparative biology of reproduction in mammals, pp. 173–194 (Academic Press, London 1966).

102 Smith, A. F. and Wilson, I. B.: Cell interaction at the maternal embryonic interface during implantation in the mouse. Cell Tiss. Res. *152:* 525–542 (1974).

103 Smith, J. A.; Martin, L.; King, R. J. B., and Vertes, M.: Effects of oestradiol-17β and progesterone on total and nuclear-protein synthesis in epithelial and stromal tissues of the mouse uterus and of progesterone on the ability of these tissues to bind oestradiol-17β. Biochem. J. *119:* 773–784 (1970).

104 Staples, R. E.: Blastocyst transplantation in the rabbit; in Daniels, Methods in mammalian reproduction, pp. 290–304 (Freeman, San Francisco 1971).

105 Stumpf, W. E.: Localisation of hormone binding sites by autoradiography. Acta histochem. cytochem. *5:* 209–211 (1972).

106 Tachi, S.; Tachi, C., and Lindner, H. R.: Modification by progesterone of oestradiol-induced cell proliferation, RNA synthesis and oestradiol distribution in the rat uterus. J. Reprod. Fertil. *31:* 59–76 (1972).
107 Tyndale-Biscoe, C. H.; Hearn, J. P., and Renfree, M. B.: Review: control of reproduction in macropodid marsupials. J. Endocr. *63:* 589–614 (1974).
108 Ward, M. C.: The early development and implantation of the golden hamster, *Cricetus auratus,* and the associated endometrial changes. Am. J. Anat. *82:* 231–276 (1948).
109 Webb, F. T. G.: The inability of dibutyryl adenosine 3′,5′-monophosphate to induce the decidual reaction in intact pseudopregnant mice. J. Reprod. Fertil. *42:* 187–188 (1975).
110 Webb, F. T. G.: Implantation in ovariectomized mice treated with dibutyryl adenosine 3′,5′-monophosphate (dibutyryl cyclic AMP). J. Reprod. Fertil. *42:* 511–577 (1975).
111 Webb, F. T. G.: Cyclic AMP and the preparation of the mouse uterus for implantation. J. Reprod. Fertil. *50:* 83–89 (1977).
112 Weitlauf, H. M.: Changes in the protein content of blastocysts from normal and delayed implanting mice. Anat. Rec. *176:* 121–124 (1973).
113 Whitney, J. C.: The pathology of the canine genital tract in false pregnancy. J. small Anim. Pract. *8:* 247–263 (1967).
114 Wimsatt, W. A.: Some comparative aspects of implantation. Biol. Reprod. *12:* 1–40 (1975).
115 Wu, D. H. and Allen, W. M.: Maintenance of pregnancy in castrated rabbits by alpha-hydroxy-progesterone and by progesterone. Fert. Steril. *10:* 439 (1959).

L. Martin, Ph. D., Hormone Physiology Department, Imperial Cancer Research Fund, PO Box 123, Lincoln's Inn Fields, London WC2A 3PX (England)

Prog. reprod. Biol., vol. 7, pp. 70–80 (Karger, Basel 1980)

Electron Microscopic Aspects of Epithelial Changes Related to Implantation[1]

B. Ove Nilsson

Reproduction Research Unit, Biomedical Centre, Uppsala

Implantation is a process which initially comprises a series of interactions between the trophectoderm of the blastocyst and the epithelium of the uterine mucosa. Later, when the uterine epithelium has been broken down, something that does not occur in all species, the interaction instead involves the trophectoderm and the uterine stroma.

Functionally, the early interactions can be regarded arbitrarily as beginning with an epithelial conditioning of the blastocyst environment, i. e. the uterine secretion, to initiate a preparation of the trophectoderm for coming events. This process is called *activation* of the blastocyst. Next, the luminal surface of both the trophectoderm and the epithelium changes to make possible the *attachment* of the blastocyst. Simultaneously, the epithelial cells receive a signal from the blastocyst, which after *transmission* through the epithelium initiates a decidual reaction in the stroma. In those species in which diapause occurs, the uterine epithelium is capable of *delaying* the blastocyst, that is, to keep it inactive but alive for long periods. Further, the uterine epithelium will also assist in removing the zona pellucida [4]. This zona lysis occurs in the mouse already during the delay of implantation [14].

These functional processes are all ultimately governed by the uterine epithelium and they have a structural basis in the epithelial cells [6, 8, 16, 18, 23]. The purpose of the present contribution is to discuss what ultrastructural changes in the cells may be associated with the functional changes. The experiments reported were performed in the rat or the mouse.

[1] Technical assistants: *Barbro Einarsson, Leif Ljung, Marianne Ljungkvist, Sibylle Widéhn.* Financial support: Swedish Medical Research Council (Grant No. B79-12X) and the 'Expressen' Prenatal Research Foundation.

Epithelial Delay of the Blastocyst

Embryonic diapause or delayed implantation is an implantation-preceding state in which the blastocyst is kept inactive in the uterine cavity and at a low metabolism [1]. Neither functionally nor structurally is this state equivalent to the preimplantation stage of a normal pregnancy, where the blastocysts are metabolically more active and rapidly proceed to implantation. Delayed implantation, however, comprises a well-defined state as a point of departure for examination of the changes that follow upon initiation of implantation.

Functionally, the inactive state of the blastocyst during delayed implantation in the mouse is maintained by the epithelium through the production of either a blastocyst-impeding substance or a uterine secretion lacking qualities to support a normal development of the blastocyst [15]. Both these possibilities require secretory activity on the part of the uterine epithelium.

Ultrastructurally, the uterine epithelial cell possesses numerous microvilli and apical protrusions and contains, among other things, a number of apical vesicles, a well-developed endoplasmic reticulum, many coated vesicles and a large Golgi apparatus. In other words, the cell bears signs of being highly productive. However, the apical vesicles, which may reasonably be supposed to be secretory, are separated from the luminal membrane by a dense layer (fig. 1). This subluminal layer, which has a fibrillar, actin-like structure, thus seems to prevent a release of the vesicles into the lumen. Since the coated vesicles are often found connected to the apical vesicles, ultrastructure suggests that some product is delivered to the vesicles from the Golgi apparatus, which reasonably is their place of origin of coated vesides. There are several transitional vesicular structures between the cisternae of the endoplasmic reticulum and the apical vesicles suggesting that the apical vesicles are derived from the endoplasmic reticulum.

The secretory and absorptive activity of a cell can be influenced by vinblastine and colchicine, drugs that disrupt microtubuli. In some types of cell the secretory activity is diminished, while in others it is increased. This dual effect is ascribed to a drug interference with other processes of the cells [22, 24]. On administration of these drugs to animals in a state of delayed implantation, we observed an increase in amount of uterine secretion, resulting in a break of the luminal closure and separation of the trophectoderm from the epithelium. Simultaneously, the Golgi ap-

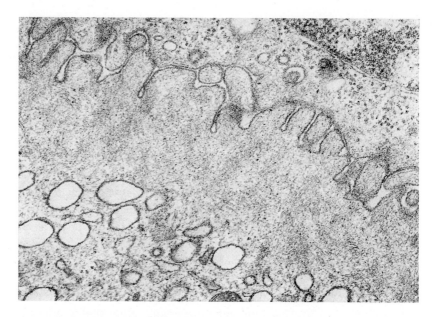

Fig. 1. Uterine luminal epithelium of a mouse in experimental delay of implantation. The borderline between uterine epithelium (below) and trophoblast (above) is seen. A dense and homogeneous subluminal zone separates the apical vesicles of the epithelium from the luminal surface. ×40,000. (From Anat. Embryol, Vol. 155 (1979).

paratus was fragmented and displaced, even appearing basally in the epithelial cells. The uterine fluid produced, however, did not induce any ultrastructural signs of activation of the blastocyst, that is, the fluid neither lost its capacity to impede the blastocyst nor recived new components necessary for blastocyst growth. Analysis of the fluid obtained is in progress.

Epithelial Activation of the Blastocyst

Activation means that the metabolism of the trophoblast cell is increased, caused by a change in the secretory qualities of the epithelium. Various studies have demonstrated alterations in proteins, enzymes, electrolytes and carbohydrates, among other things [2, 3, 5, 21, 25].

Ultrastructurally, the lumen now contains more secretion, resulting in the appearance of a space between the epithelium and the trophectoderm.

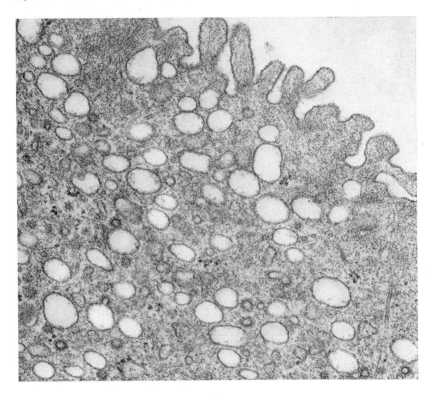

Fig. 2. Uterine luminal epithelium of a mouse in experimental delay of implantation, 6 h after an injection of oestrogen. The epithelial luminal surface is separated from the blastocyst surface by a newly produced secretion. The apical vesicles have approached the cell surface giving the impression of emptying into the uterine lumen. Many coated vesicles are interspaced between the apical vesicles. A microtubule is observed to the right in the picture. ×40,000.

The subluminal layer of the epithelium, which has previously lacked vesicles, now contains several of them, some even in contact with the cell membrane [13] (fig. 2). This is interpreted as a migration of vesicles to the cell surface, followed by emptying of their content into the luminal secretion.

A marked accumulation of the model amino acid, amino-isobuturic acid (AIB), in the uterus occurs at activation. This substance is then taken up by the blastocysts [11].

Also diamino-buturic acid (DAB), like AIB being transported by System A, accumulates in the unterus, specifically in the epithelium

(Appelgren, L.-E., Lindqvist, I., Nilsson, B. O. and Ronquist, G., in preparation). Since various results suggest the presence of an operative System-A carrier in the trophoblast already during delay, the uterine epithelium then has to restrict the availability of the amino acids or of some co-factor, necessary for the transport.

When the production of the uterine fluid was affected by administration of vinblastine or colchicine, the result was a reduction of the amount of glucose of the fluid [*Nilsson and Östensson, in preparation*]. In the uterine epithelium, the agents caused a decrease of vesicles in the subluminal zone and a disorganizing of the apical cell parts with a dislocation of the Golgi apparatus. This indicates an interference with the normal process of release of various cell products. In the blastocyst, no ultrastructural signs of activation were present. It is concluded that the composition of the uterine secretion now is inappropriate for a normal blastocyst function.

Interference with the assumed secretion of glucose by administration of 5-thioglucose, a substance which blocks the synthesis of glucose, left the apical vesicles unchanged. This finding, together with the lack of glucose-6-phosphatase [20] in the vesicle membrane, seems to rule out a function of the apical vesicles in the glucose secretion. Furthermore, they do not contain any appreciable amount of calcium or magnesium ions, as revealed by the potassium pyroantimonate technique (see fig. 5). Thus, the content of the apical vesicles and their role at early implantation is still unknown.

Epithelial Attachment of the Trophectoderm

About 18 h after the injection of oestrogen, the trophoblast cell membrane and the epithelial cell membrane begin to appose, and by 24 h the blastocyst surface is firmly attached to the epithelium. This contact between cell membranes constitutes a marked change in the functional relationship between trophectoderm and epithelium by offering, for instance, means for cellular induction and other short-range influences. Destruction of the epithelium now begins, and an incompatible trophoblast, transplanted into a host uterus, will not until this stage suffer immunological damage [10]. These features indicate that the two tissues now interact more intensely. But a close contact alone is not sufficient for these processes to occur – during the close contact of delay, the

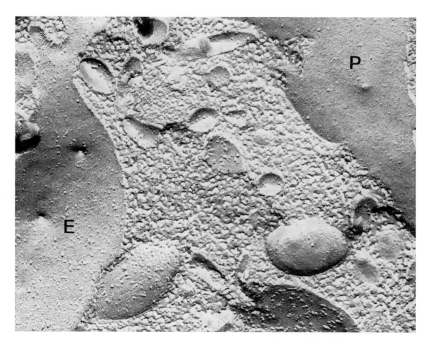

Fig. 3. Freeze-etched uterine luminal surfaces of a mouse in experimental delay of implantation, 24 h after an injection of oestrogen. The protoplasmic (P) face
is seen to the right, the external (E) one to the left. A few apical vesicles are
noticed in the middle of the picture. The specimen was prepared by a perfusion
fixation with glutaraldehyde, rinsed and soaked in a solution of 15 % glycerol in
water for 20 h. The freezing was made in Freon, cooled by liquid nitrogen. After
cleavage at −100 °C, sublimation was allowed during 2 min and then followed the
conventional procedure for replication. ×110,000.

tissues are unaffected. This lack of reaction may, of course, be due to
cellular incompetence to signal or to respond, but it may also be due to
an insulating property of the cell surfaces, specific for the membrane
contact during delay.

Several differences do exist in the surfaces of epithelial cells during
delay and attachment, for instance in ultrastructure [17], in the properties
of the extracellular coat [7, 17, 23], and in surface charge [9]. Also the
distribution of intramembraneous particles differs in delay and attachment, respectively. Our freeze-etch studies show that the number of

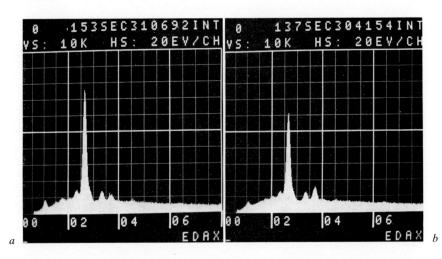

Fig. 4. Energy dispersive X-ray microanalyses of Sephadex beads which have absorbed uterine secretion when lying in the uterine lumen. The analyses were performed with an Edax-Edit equipment attached to a Philips 400 electron microscope using 40 kV and 2×10^{-9} A to reach a total integral of 3×10^6 counts. *a* Spectrum of a bead from a mouse in experimental delay of implantation. *b* Spectrum of a bead from a mouse in experimental delay of implantation, 18 h after an injection of oestrogen. The Ca peak at attachment is found to be higher than that of delay. The Na and K peaks, however, are rather unchanged.

particles in the P face remain at about 1,500 particles/μm² while those of the E face decreases from about 3,000 particles/μm² at delay to about 2,000 particles/μm² at attachment (fig. 3). Further, the proportion of large particles to small particles seems to increase at attachment. The meaning of this change for the attachment of the blastocyst is unknown, however.

The first patches of cellular contact between trophoblast cells and epithelial cells appear about 18 h after administration of oestrogen. Since calcium is known to affect cellular adhesion, we examined the changes in the content of calcium in the uterine secretion [*Nilsson and Ljung*, in preparation]. Using Sephadex beads to absorb uterine fluid and a microprobe to analyse their content of electrolytes [19], we found a marked increase in the calcium concentration of the secretion at the time of blastocyst attachment (fig. 4). When applying the sodium pyroantimonate technique, which reveals the presence of calcium and magnesium in tissues, we observed ultrastructural signs of these ions between the

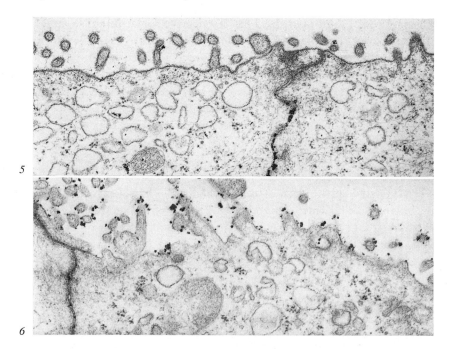

Fig. 5. Apical parts of uterine luminal epithelium of a mouse in experimental delay of implantation. The tissue was prepared according to a potassium pyroantimonate technique. Precipitates indicating presence of calcium and/or magnesium ions are observed intercellularly but neither at the luminal surface nor in the apical vesicles. ×32,000.

Fig. 6. Apical parts of uterine luminal epithelium of a mouse in experimental delay of implantation, 18 h after an injection of oestrogen. The tissue was prepared according to a potassium pyroantimonate technique. Precipitates are present intercellularly and now, also at the luminal cell surface. ×32,000.

epithelial cells and at their luminal cell surfaces at implantation (fig. 5, 6). Thus in the mouse, as in humans, the calcium content of the uterine secretion is high when the blastocyst is about to implant.

Epithelial Transmission of a Deciduogenic Signal

About 8–12 h after the injection of oestrogen, there are signs of decidual transformation of the stromal cells in the area surrounding the blastocyst [12]. Still later, by 24 h, a further stromal response appears in the form of a Pontamine blue reaction. But evidently, well in advance of early attachment, a signal from the blastocyst reaches the epithelium to be transmitted into the stroma.

From a structural point of view, few clues to the mechanisms are yielded by examination of the uterine epithelium. It is true that apically, occasional trophoblast-epithelial contacts and some endocytotic vesicles can be observed and that basally, a few membrane-surrounded granules are present. None of these structures, however, can as yet be associated with an epithelial uptake or release of a signal substance.

The decidual response of the stromal cells, however, can be blocked by indomethacin, which interferes with the prostaglandin synthesis. If it is assumed that this is the action of indomethacin also during decidualization, then prostaglandins may be involved as a trigger substance in one or several processes occurring in the trophoblast, the uterine epithelium, the stromal cells or the endothelium. In an attempt to shed some light on how indomethacin blocks implantation, we have examined its ultrastructural effects [13]. We found that indomethacin did not affect the early implantatory changes of the blastocyst, i. e. activation and attachment of the trophoblast, but later on, blastocyst degeneration occurred. The normal decidual changes of the stroma and the Potamine blue reaction never appeared. Only in the uterine epithelium were indomethacin-induced ultrastructural changes in the form of large apical spaces containing a secretory substance observed. This change suggests a disturbance of epithelial activity and might indicate that indomethacin interferes with transmission of the signal from the blastocyst to the stroma.

References

1   Aitken, R. J.: Embryonic diapause; in Johnson, Development in mammals; vol. 1, pp. 307–359 (North-Holland, Amsterdam 1977).
2   Aitken, R. J.: The hormonal control of implantation; in Maternal recognition of pregnancy. Ciba Fdn Symp. 64, pp. 53–83 (Excerpta Medica, Amsterdam 1979).
3   Bazer, F. W.; Roberts, R. M., and Sharp, D. C., III: Collection and analysis of female genital tract secretions; in Daniel, Methods in mammalian reproduction, pp. 503–528 (Academic Press, New York 1978).
4   Dickmann, Z.: Shedding of the zona pellucida; in McLaren, Advances in reproductive physiology, vol. 4, pp. 187–206 (Logos Press, London 1969).

5   Edwards, R. G. and Surani, M. A. H.: The primate blastocyst and its environment. Upsala J. med. Sci. Suppl. *22:* pp. 39–50 (1978).

6   Enders, A. C.: Anatomical aspects of implantation. J. Reprod. Fertil. *25:* 1–15 (1976).

7   Enders, A. C. and Schlafke, S.: Surface coats of the mouse blastocyst and uterus during the preimplantation period. Anat. Rec. *180:* 31–46 (1974).

8   Enders, A. C. and Schlafke, S.: Comparative aspects of blastocyst-endometrial interactions at implantation; in Maternal recognition of pregnancy. Ciba Fdn Symp. 64, pp. 3–32 (Excerpta Medica, Amsterdam 1979).

9   Hewiit, K.; Beer, A. E., and Grinnell, F.: Disappearance of anionic sites from the surface of the rat endometrial epithelium at the time of blastocyst implantation. Biol. Reprod. *21:* 691–707 (1979).

10   Håkansson, S.; Lundkvist, Ö., and Nilsson, B. O.: Survival of rat blastocyst transplanted into the uterus of hyperimmunized mice during delay of implantation. Int. J. Fertil. *23:* 148–151 (1978).

11   Lindqvist, I.; Einarsson, B.; Nilsson, O., and Ronquist, G.: The *in vivo* transport of ^{14}C-α-aminoisobuturic acid into mouse blastocysts during activation for implantation. Acta physiol. scand. *102:* 477–483 (1978).

12   Lundkvist, Ö.: Ultrastructural studies of the endometrial stromal cells in rats during estradiol-induced implantation after an experimental delay. Biol. Reprod. *18:* 306–316 (1978).

13   Lundkvist, Ö. and Nilsson, B. O.: Ultrastructural changes of the trophoblast-epithelial complex in mice subjected to implantation-blocking treatment with indomethacin. Biol. Reprod. (in press).

14   McLaren, A.: The fate of the zona pellucida in mice. J. Embryol. exp. Morph. *23:* 1–19 (1970).

15   McLaren, A.: Blastocyst activation; in Segal, Crozier, Corfman and Condliffe, The regulation of mammalian reproduction, pp. 321–328 (Thomas, Springfield 1973).

16   Nilsson, O.: The morphology of blastocyst implantation. J. Reprod. Fertil. *39:* 187–194 (1974).

17   Nilsson, O.: Changes of the luminal surface of the rat uterus at blastocyst implantation. Z. Anat, EntwGesch. *144:* 337–342 (1974).

18   Nilsson, B. O.; Bergström, S.; Håkansson, S.; Lindqvist, I.; Ljungkvist, I.; Lundkvist, Ö., and Naeslund, G.: Ultrastructure of implantation. Uppsala J. med. Sci. Suppl. *22:* 27–38 (1978).

19   Nilsson, O. and Ljung, L.: Electron probe micro-X-ray analyses of electrolyte composition of fluid microsamples by use of a Sephadex bead. Upsala J. med. Sci. *84:* 1–2 (1979).

20   Nilsson, O. and Lundkvist, Ö.: Ultrastructural and histochemical changes of the mouse uterine epithelium on blastocyst activation for implantation. Anat. Embryol. *155:* 311–321 (1979).

21   Nilsson, O.; Östensson, C.-G.; Eide, S., and Hellerström, C.: Utilization of glucose by the implanting mouse blastocyst. Endocrinology (in press).

22   Oliver, J. M. and Berlin, R. D.: Microtubules, microfilaments and the regulation of membrane functions; in Secretory mechanisms. Symposia of the Society for Experimental Biology; No. 23, pp. 277–298 (University Press, Cambridge 1979).

23 Schlafke, S. and Enders, A. C.: Cellular basis of interaction between trophoblast and uterus at implantation. Biol. Reprod. *12:* 41–65 (1975).

24 Stephens, R. E. and Edds, K. T.: Microtubules: structure, chemistry, and function. Physiol. Rev. *56:* 709–767 (1976).

25 Van Blerkom, J.; Chavez, D. J., and Bell, H.: Molecular and cellular aspects of facultative delayed implantation in the mouse; in Maternal recognition of pregnancy. Ciba Fdn Symp. 64, pp. 141–172 (Excerpta Medica, Amsterdam 1979).

B. O. Nilsson, MD, Reproduction Research Unit, Box 571, Biomedical Centre, S-751 23 Uppsala (Sweden)

Prog. reprod. Biol., vol. 7, pp. 81–91 (Karger, Basel 1980)

Endocytosis in the Uterine Epithelium during Early Pregnancy

Margaret B. Parr

Unit of Human Morphology, School of Medicine,
The Flinders University of South Australia, Adelaide, S. A.

Introduction

Endocytosis is a widespread cellular function that regulates the uptake of exogenous molecules from the cell's environment via plasma membrane derived vesicles and vacuoles. Both soluble (pinocytosis) and particulate (phagocytosis) substances may be interiorized, destined either for the lysosome system and intracellular digestion or transport through the cytoplasm and subsequent exocytosis. Endocytosis is particularly prominent in neutrophils, macrophages, capillary endothelium, thyroid follicle cells, yolk sac endoderm, and oocytes, where it is involved in host defence, macromolecular transport, hormone transformations and cellular nutrition. The purpose of the present report is to describe the endocytotic activity in yet another cell type, the uterine luminal epithelial cell, and to explore its role during early pregnancy.

Endocytosis at the Apical Surface

Endocytosis in the rat uterus during progestational stages was first reported by *Vokaer* [21], who observed that dyes, such as trypan blue, injected into the uterine lumen were localized in the luminal epithelium after 15–20 h. The endocytotic properties of the luminal epithelium were supported by *Ljungkvist* [11], who noted the presence of numerous structures usually associated with the uptake for foreign materials, such as coated vesicles, multivesicular bodies and pleomorphic bodies in progesterone-treated ovariectomized rats. In 1971, *Psychoyos and Mandon*

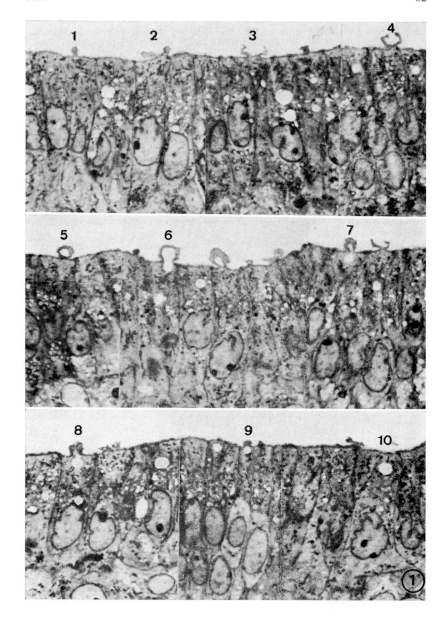

[18] suggested that the uptake of luminal fluids during the luteal phase could be mediated by fungal-like projections observed on the surface of the uterine epithelium. Such structures have been observed both with scanning and transmission electron microscopy in human, rat, mouse ad rabbit uteri, but most investigators indicated that the protrusions were secretory and pinched off from the cells to release cellular materials into the lumen [for references see 15]. However, experimental evidence in the rat (fig. 1) and mouse showed that these structures are involved in endocytosis, not apocrine secretion [3, 15, 16]. Tracers such as ferritin and horseradish peroxidase introduced into the uterine lumen were rapidly incorporated into vacuoles up to 3 μm in diameter formed by the apical protrusions, or pinopods and into small coated pinocytotic vesicles. 10 min after their administration, the tracers were located in two kinds of endocytotic vesicles, in apical vacuoles up to 3 μm in diameter that appeared to be derived from the pinopod vacuoles, and in multivesicular bodies. 60 min later the tracers were present in a heterogeneous group of multivesicular and dense bodies, most of which contained acid phosphatase. Thus, the material taken into the epithelial cells by endocytosis was channelled into the lysosome system [17]. Lysosomes containing ferritin were essentially all located in the apical half of the cells 1 h after administration of the tracer, but at 2 h they were distributed throughout the cell. The movement of lysosomes was dependent on microtubules, since colchicine treatment blocked the migration of ferritin-containing lysosomes to the basal halves of the cells [13, 14]. Lysosomes containing ferritin were often observed near the base of the cells, but no exocytosis of tracer into the stroma was observed by light or electron microscopy either in rats on day 5 of pregnancy [17] or in ovariectomized rats treated with progesterone [9]. This contrasts with reports that trypan blue was transported

---

Fig. 1. This illustration is a collection of light micrographs to show the sequence of steps in the formation of endocytotic vacuoles by protrusions or pinopods from the apical surface of uterine luminal epithelial cells. At first, a small, irregular protrusion appears (1); it is larger than the microvilli but does not involve the entire surface of the cell. The protrusion then spreads out (2) and invaginates (3) to form an irregular cup-shaped structure. When the cup closes (4) it sequesters (5) from the uterine lumen a vacuole, 0.5–3.0 μm in diameter. The vacuole then passes downward (6), out of the projection (7) and into the apical portion of the cell (8). The projection then regresses (9), and the vacuole migrates deeper into the cell (10). $\times$1,200. (Reprinted with permission of Biol. Reprod.).

to the stroma after uterine trauma [21, 22] and in animals given daily injections of progesterone and estrogen [19]. Endocytosis mediated by pinopod formation essentially identical to that seen in the rat has also been reported in mouse uterine epithelium on day 4 of pregnancy and on day 10 of delayed implantation [16]. It remains to be demonstrated whether fungal-like protrusions on the apical surfaces of uterine epithelial cells in other species, including humans [12], are also vehicles of endocytosis.

In rats, the distribution of uterine cells that take up tracers or display pinopods is irregular. With scanning electron microscopy the lining of the uterine lumen showed areas free of projections, isolated individual pinopods, or pinopods arranged in clusters and rows [3]. More cells at the antimesometrial side of the uterus, the site of implantation, showed fungal-like protrusions [18] and took up trypan blue [21] and ferritin [17] than those at the mesometrial side on day 5 of pregnancy and during delayed implantation. However, no differences in the distribution of endocytotic cells were noted along the length of uterine horns, and pinopods were observed both adjacent to unimplanted blastocysts (fig. 2) and in inter-embryonic areas.

The endocytotic activity of the rat uterus was observed only during progestational periods. The uptake of trypan blue occurred on days 5 and 6 of pregnancy [21], during pseudopregnancy [21], and at sites of local progesterone injection [22], but not during the estrus cycle or after ovariectomy [21]. Other tracers such as ferritin were shown to be endocytosed by uterine epithelial cells during early pregnancy, with a distinct peak of activity on day 5 in the rat [3, 15, 17] and day 4 in the mouse [16]. A day later in each case there was a marked decrease in the number of cells containing tracer molecules. Treatment of ovariectomized rats with progesterone alone, or with estradiol followed by progesterone, caused uptake of intraluminally administered trypan blue and ferritin, while estradiol alone did not [9]. Endocytosis of ferritin also occurred during lactational delayed implantation [3] and during delayed implantation obtained by ovariectomy during early pregnancy followed by daily injections of progesterone in the rat [unpubl. observations] and the mouse [16]. The administration of nidatory estradiol to animals in delayed implantation had no effect on endocytosis 24 h later hut after 48 h the endocytosis had ceased [unpubl. observations; 16].

The function of endocytosis in the preimplantation and implantation period remains unknown. Although selective uptake of materials cannot be excluded, it seems likely that pinopods mediate bulk uptake of macro-

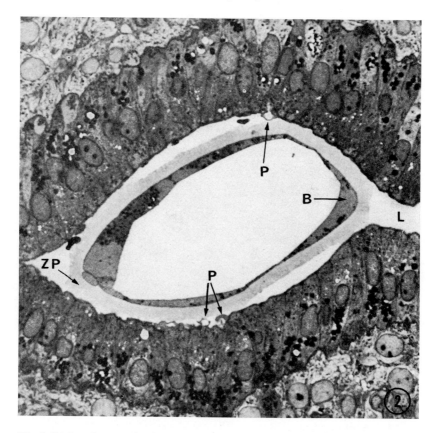

Fig. 2. Light micrograph of a rat uterus on day 5 of pregnancy showing a blasto-cyst (B) surrounded by its zona pellucida (ZP) lying in the uterine lumen (L). Note the pinopods (P) protruding from the apical surface of several epithelial cells ad-jacent to the unimplanted blastocyst. × 850.

molecules from the lumen which are hydrolyzed in secondary lysosomes. Digestion products could then diffuse back into the uterine lumen to alter the molecular environment of the blastocysts, or they could pass to the underlying stroma to provide an indirect mechanism for transferring in-formation from the lumen to the stroma. Direct transepithelial passage of macromolecules appears not to occur, since tracer injected into the uterine lumen did not appear in the stroma [17]. In addition to possible functions already suggested [3, 9, 15], endocytosis may be part of a process that regulates implantation by controlling the properties of the epithelial cell

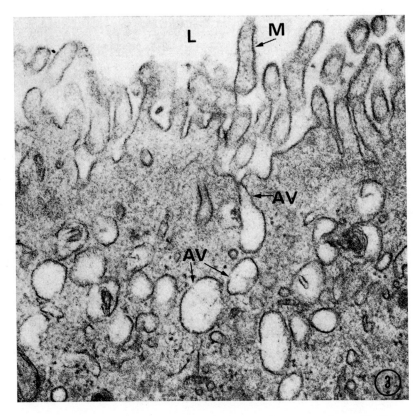

Fig. 3. Electron micrograph of the apical part of a rat uterine epithelial cell on day 5 of pregnancy showing electron-transparent apical vesicles (AV). The membrane of one apical vesicle appears to be continuous with the surface membrane at the base of the microvilli (M) bordering the uterine lumen (L). × 58,000.

apical membrane. The fate of membrane that is internalized by endocytosis is not yet fully understood, but numerous studies have demonstrated that endocytosis is coupled to exocytosis or secretion in a variety of cells [for references see 4, 17]. Thus, as surface membrane is taken in by endocytosis it is replaced by secretory vesicle membrane. Electron microscopic images of rat uterine epithelial cells on day 5 of pregnancy suggest the secretion of a class of electron-transparent vesicles. The vesicles appear to originate from the Golgi complex, accumulate in the apical part of the cell, and may fuse with the apical surface membrane at the base of the microvilli (fig. 3). Several other observations are con-

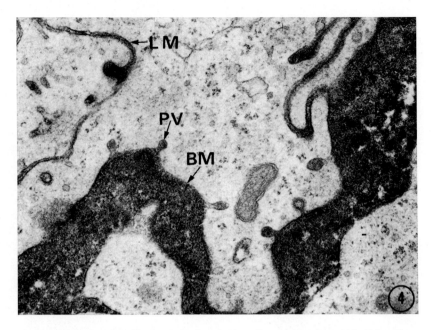

Fig. 4. Electron micrograph of the basal part of a rat uterine epithelial cell on day 5 of pregnancy showing peroxidase reaction product below the basal membrane (BM), between the lateral membranes (LM) and in pinocytotic vesicles (PV). Unstained section. ×35,000.

sistent with this view. The vesicles develop and disappear in conjunction with endocytosis but do not take up ferritin. A similarity of the vesicle and apical membrane is suggested by periodic acid-silver proteinate staining. In addition, the apical vesicles and Golgi cisternae are normally located in the apical halves of the epithelial cells and both are shifted together to the basal portions of the cells by colchicine treatment, providing further evidence that the apical vesicles are derived from the Golgi complex [13]. Endocytosis and its coupled exocytosis may maintain the apical membrane of epithelial cells in a preparatory condition for implantation. This is consistent with the occurrence of endocytosis on days 4 and 5 of pregnancy and throughout delayed implantation. After the estrogen secretion on day 4 of pregnancy or after estrogen administration to animals in delayed implantation endocytosis gradually decreases, the properties of the apical membrane may change, and implantation can proceed.

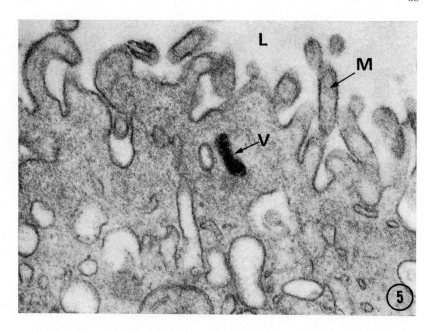

Fig. 5. Electron micrograph of the apical part of a rat uterine epithelial cell on day 5 of pregnancy showing peroxidase reaction product in a vesicle (V) close to the apical surface membrane. Unstained section. ×67,500. (Reprinted with permission of J. Reprod. Fertil.).

Endocytosis at the Basal and Lateral Surfaces

In addition to endocytosis at the apical surface of rat uterine epithelial cells, uptake of blood-borne substances occurs at the lateral and basal membranes by means of coated and noncoated pinocytotic vesicles [13a]. 20 min after intravenous injection of horseradish peroxidase on day 5 of pregnancy the tracer was present between epithelial cells up to the junctional complex, in invaginations in their basal and lateral membranes (fig. 4), and in pinocytotic vesicles in the periphery of the cells. 60 min after injection the tracer was found in the above-mentioned sites and in vesicles distributed throughout the epithelial cell cytoplasm (fig. 5). 2 h after injection the tracer was found only in a few vesicles in the apical region of the cells, and some of these vesicles appeared to be fusing with the apical cell membrane. Pinocytotic invagina-

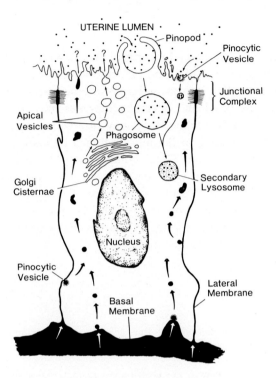

Fig. 6. A diagram of a rat uterine epithelial cell on day 5 of pregnancy show-ing two endocytotic pathways and the possible movement of apical vesicles from the Golgi complex to the cell surface.

tions in the basal membranes were 4- to 5-fold more numerous on day 5 of pregnancy than on days 1–4 and 7, suggesting a role for this activity during early implantation. These observations, as well as others [1, 10], suggest that there is intracellular transepithelial movement of materials from the blood and/or stroma into the uterine lumen. Such an intra-cellular pathway could account for the presence of macromolecules, such as plasma proteins, in the uterine fluid [7, 20]. The transported macro-molecules may be required for endometrial sterility [8], protection of the blastocyst against immunological rejection [2], or blastocyst implantation and development [5, 6].

Conclusions

Uterine epithelial cells exhibit two endocytotic pathways on days 5 and 6 of pregnancy (fig. 6). Pinopod vacuoles and pinocytotic vesicles formed at the apical surface are channelled into secondary lysosomes, while pinocytotic vesicles formed at the basal and lateral membranes move across the cell towards the luminal surface where they may fuse with the apical membrane to release their contents into the uterine lumen by exocytosis. In addition, vesicles derived from the Golgi complex appear to move toward the apex of the cell and fuse with the surface membrane. Such cellular activities may regulate the structure of the apical plasma membrane and modify the molecular environment of the blastocyst during the preimplantation and implantation stages. The precise nature and significance of these processes await further study.

References

1 Anderson, W.; Kang, Y., and de Sombre, E.: Endogenous peroxidase: specific marker enzyme for tissue displaying growth dependency on estrogen. J. Cell Biol. *64:* 668–681 (1975).

2 Bernard, O.; Ripoche, M., and Bennett, D.: Distribution of maternal immunoglobulins in the mouse uterus and embryo in the days after implantation. J. exp. Med. *145:* 58–75 (1977).

3 Enders, A. C. and Nelson, D. M.: Pinocytotic activity of the uterus of the rat. Am. J. Anat. *138:* 277–300 (1973).

4 Farquhar, M. G.: Recovery of surface membrane in anterior pituitary cells. Variations in traffic detected with anionic and cationic ferritin. J. Cell Biol. *77:* R 35 (1978).

5 Fishel, S. B. and Surani, M. A. H.: Changes in responsiveness of preimplantation mouse embryo to serum. J. Embryol. exp. Morph. *45:* 295–301 (1978).

6 Gwatkin, R. B. L.: Nutritional requirements for post-blastocyst development in the mouse. Int. J. Fert. *14:* 101–110 (1969).

7 Hawegawa, Y.; Sugawara, S., and Takeuchi, S.: Studies on the permeabilities of serum protein into uterine lumen and specific protein in uterine fluid in rat. Tohoku J. agric. Res. *25:* 67–76 (1974).

8 Kelly, J. K. and Fox, H.: The local immunological defence system of the human endometrium. J. Reprod. Immunol. *1:* 39–45 (1979).

9 Leroy, F.; Hoeck, J. van, and Bogaert, C.: Hormonal control of pinocytosis in the uterine epithelium of the rat. J. Reprod. Fertil. *47:* 59–62 (1976).

10 Ljungkvist, I.: Attachment reaction of rat uterine luminal epithelium. II. The effect of progesterone on the morphology of the uterine glands and the luminal epithelium of the spayed virgin rat. Acta Soc. Med. upsal. *76:* 110–126 (1971).

11 Ljungkvist, I.: Attachment reaction of rat uterine luminal epithelium. IV. The cellular changes in the attachment reaction and its hormonal regulation. Fert. Steril. *23:* 847–865 (1972).

12 Nilsson, O. and Nygren, K. G.: Ultrastructure of human uterine epithelium at the time of implantation after postovulatory administration of norethindrone Uppsala J. med. Sci. *79:* 65–71 (1974).

13 Parr, M.: A morphometric analysis of microtubules in relation to the inhibition of lysosome movement caused by colchicine. Eur. J. Cell Biol. *20:* 189–194 (1979).

13a Parr, M. B.: Endocytosis at the basal and lateral membranes of rat uterine epithelial cells during early pregnancy. J. Reprod. Fertil. *60:* (in press 1980).

14 Parr, M. B.; Kay, M. G., and Parr, E. L.: Colchicine inhibition of lysosome movement in the rat uterine epithelium. Cytobiologie *18:* 374–378 (1978).

15 Parr, M. B. and Parr, E. L.: Uterine luminal epithelium. Protrusions mediate endocytosis not apocrine secretion in the rat. Biol. Reprod. *11:* 220–233 (1974).

16 Parr, M. B. and Parr, E. L.: Endocytosis in the uterine epithelium of the mouse. J. Reprod. Fertil. *50:* 151–153 (1977).

17 Parr, M. B. and Parr, E. L.: Uptake and fate of ferritin in the uterine epithelium of the rat during early pregnancy. J. Reprod. Fertil. *52:* 183–188 (1978).

18 Psychoyos, A. et Mandon, P.: Etude de la surface de l'epithélium utérin au microscope électronique à balayage. Observations chez la ratte au 4e et au 5e jour de la gestation. C. r. hebd. Séanc. Acad. Sci., Paris *272:* 2723–2725 (1971).

19 Sartor, P.: Athrocytose du bleu trypan par l'endomètre de la ratte. Manifestation du processus en fonction du contexte hormonal. C. r. Séanc. Soc. Biol. *163:* 2564–2567 (1969).

20 Surani, M. A. H.: Qualitative and quantitative examination of the proteins of rat uterine luminal fluid during pro-estrus and pregnancy and comparison with those of serum. J. Reprod. Fertil. *50:* 281–287 (1977).

21 Vokaer, R.: Recherches histophysiologiques sur l'endomètre du rat, en particulier sur le conditionnement hormonal de ses propriétés athrocytaires. Archs Biol., Liège *63:* 1–84 (1952).

22 Vokaer, R. and Leroy, F.: Experimental study on local factors in the process of ovimplantation in the rat. Am. J. Obstet. Gynec. *83:* 141–148 (1962).

M. B. Parr, PhD, Unit of Human Morphology, The Flinders University of South Australia, Bedford Park, S. A. 5042 (Australia)

Prog. reprod. Biol., vol. 7, pp. 92–101 (Karger, Basel 1980)

Role of the Uterine Epithelium in Inducing the Decidual Cell Reaction[1]

B. Lejeune and F. Leroy

Laboratory of Gynecology and Research on Human Reproduction,
St. Pierre Hospital, Free University, Brussels

The mechanism by which the decidual cell reaction (DCR) is induced, represents a central problem in the physiology of implantation. So far, only partial and controverted explanations have been provided [3, 6, 14].

Many artificial stimuli were shown capable of inducing decidualization during pseudopregnancy as well as in ovariectomized animals prepared by an adequate sequence of estrogen and progesterone injections. Scratching of the uterine horn [13], electric stimulation of the myometrium [11], intraluminal instillation of histamine [14], systemic injection of pyrathiazine [15], intraluminal instillation of oil [8], or carbon dioxide [9], or prostaglandins [10], etc . . . , have all been successfully used.

In rats and mice, the physiological DCR induced by the blastocyst, as well as that elicited by intraluminal oil instillation, require the previous intervention of luteal estrogen in addition to progesterone [7]. This observation suggests that luteal estrogen acts on the epithelium by sensitizing it to contact with the blastocyst or with oil, hence allowing the transmission of deciduogenic information towards the stroma. In contrast, trauma, which is capable of inducing the DCR in animals treated with progesterone alone, would bypass the epithelium and act on the stroma directly. However, the methods usually employed for traumatizing the uterus must of necessity injure some epithelial cells at least. The validity of the short-circuit theory may therefore not be asserted on the sole basis of such experiments. Some published data indicate, to the contrary, that

[1] This work was supported by a grant from the Belgian FRSM. *F. L.* is 'Chercheur Qualifié' at the Belgian FNRS.

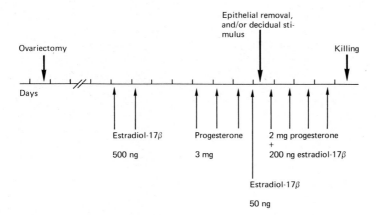

Fig. 1. General experimental schedule.

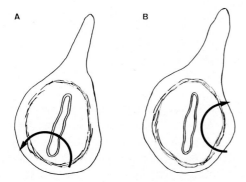

Fig. 2. A Right horn: thread piercing the epithelium, perpendicularly to the meso-antimesometrial plane. *B* Left horn: thread passing in a parallel direction to the meso-antimesometrial plane and avoiding injury to epithelial cells.

trauma would only be effective insofar as epithelial cells *are* indeed injured. *Fainstat* [4] stated that passing a thread into the uterine lumen, thus in contact with the epithelium, would induce a DCR whereas no decidualization would occur by piercing only the stroma. Similar results were obtained by *Ferrando and Nalbandov* [5] by selectively freezing the uterine epithelium without damaging the stroma. In this latter case, scratching of the horn did not induce a DCR. The uterine epithelium would thus appear as an obligatory intermediate in any type of 'decidual' stimulation whatsoever.

Table I. Deciduomata induced by sutures at 3 levels of each horn (a, b, c)

Rats	Right horn (fig. 2A)			Left horn (fig. 2B)		
	a	b	c	a	b	c
1	+ +	+ +	+ +	+	+	+ + +
2	+ +	+ +	+ +	+	+ +	0
3	+ +	+ +	+ + +	0	+ +	0
4	+	+ +	+ +	0	0	+ +
5	+ +	+ + +	+ + +	+ + +	+ + +	+ +
6	+ + +	+ + +	+ +	0	+ +	0
7	+ +	+	+ +	0	+	0
8	+ +	+ +	+ + +	0	0	0
9	+ + +	+ + +	+ +	0	0	+ +
10	+ +	+ +	+ +	+ +	+ +	+

0 = No deciduoma; + = deciduoma, diameter $\leq$ 4 mm; + + = deciduoma, diameter $\geq$ 5 mm; + + + = fused deciduomata.

We have tried to verify this hypothesis by different methods aiming at separating the contribution of the epithelium from that of the stroma in the mechanism of decidualization.

In the paper by *Fainstat* [4], it is not indicated how the results were evaluated, nor are detailed data provided. Therefore, we have tried to repeat similar experiments and assess the results semiquantitatively. In ovariectomized rats, submitted to an adequate hormonal treatment (fig. 1), at the time of maximal uterine sensitivity, the uterine lumen is closed and oriented as a meso-antimesometrial slit. In one horn, threads were passed perpendicularly to this plane in order to pierce the epithelium and make sure that some of its cells were injured (fig. 2A). On the heterolateral side, threads were inserted in a direction parallel to the meso-antimesometrial plane, hoping to cross only the stroma while avoiding epithelial damage as much as possible (fig. 2B). The results show that the amount of deciduomata induced by sutures is twice as large when threads cross the uterine lumen (table I). Although not clear-cut, this observation is consistent with *Fainstat*'s statement and is compatible with the hypothesis of an obligatory role of the uterine epithelium in the triggering of decidualization.

To further analyze the role of the epithelium, we have used a method which was derived from that used by *Bitton-Casimiri et al.* [1] to remove the uterine epithelium from dissected horns. When rats are maintained

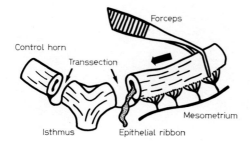

Fig. 3. Removal of uterine epithelium *in vivo.*

under progesterone treatment, it is possible to selectively remove the uterine epithelium without damaging the stroma by gently squeezing the horn from one end to the other. It is possible to apply this method *in vivo* after transverse section of uterine horns near the isthmus and to collect the detached epithelium (fig. 3). Histological controls showed that the removed tissue is pure epithelium while the stroma remaining *in situ* looks morphologically intact (fig. 4A, B). If animals treated that way are left alone, the epithelium regenerates from residual glands in approximately 1 week (fig. 5) [12]. Despite trauma which is necessarily linked to the method of epithelial ablation, we did not observe a significant ensuing DCR, although the animals were maintained under adequate hormonal treatment (fig. 8, group 5). However, if uterine horns are extensively crushed with a strong forceps but without detaching the epithelium from the stroma (fig. 8, group 3), the DCR is as massive as that observed after longitudinal intraluminal scratching or localized pinching (fig. 8, groups 1 and 4). Squeezing out the epithelium can thus not be held responsible for preventing decidualization through simultaneous destruction of stromal cells which is bound to be worse after extensive crushing. The mere disruption of topographical relationships between epithelium and stroma is sufficient to prevent the DCR since squeezing the uterine horn without pulling out the epithelial sheet was enough to leave the uterus unresponsive to 'decidual' stimuli (fig. 8, group 6). Ablation of the epithelium performed immediately before or after longitudinal scratching, as well as its removal 4 h after intraluminal oil instillation, also prevented the DCR from occurring (fig. 8, group 8, 7 and 9). Contrarily, in animals in which the epithelium had regenerated at 8 days after ablation, hormonal resensitization allowed to induce, by scratching,

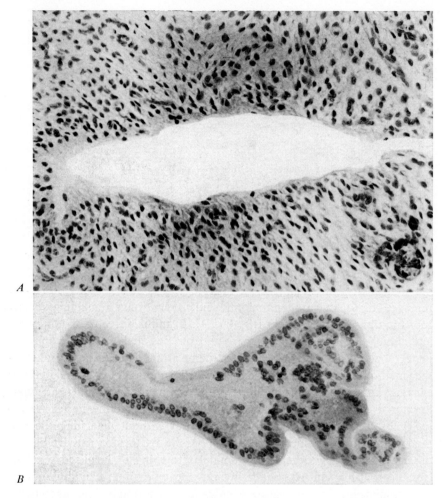

Fig. 4. A Progestational uterus after mechanical removal of surface epithelium.
B Removed epithelium.

a DCR amounting to 65 % in weight of the maximal response observed in intact horns (fig. 8, group 14).

From this first series of experiments, it may be concluded that even widespread and global damage to uterine horns does not alter the mechanism or the quantitative capacity for decidualization. On the other hand, removal of the epithelium or even the mere disruption of its topographical relationships with the stroma is enough to prevent the decidual

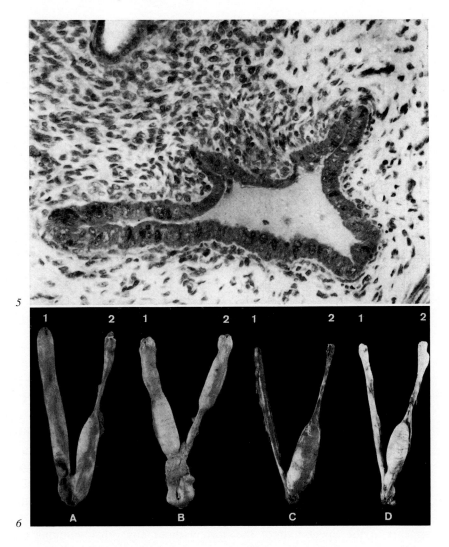

Fig. 5. Regenerated epithelium, 4 days after epithelial ablation.

Fig. 6. Typical effects on DCR of partial *in vivo* deinsertion or removal of uterine epithelium: A_1 and B_1, scratched control horn with massive DCR. Deepithelization of upper (A_2) and lower (B_2) half with DCR in intact adjacent segment. Epithelial ablation (C_2) or deinsertion (D_2) in upper half of left horn with massive DCR in intact half but without extension to the untouched contralateral cornu (C_1 and D_1).

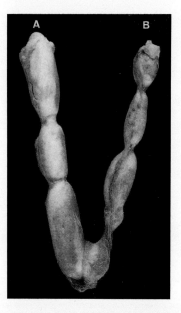

Fig. 7. A Massive DCR after pinching the horn at two points. *B* Massive DCR after extensive crushing of the horn.

reaction. It is therefore clear that trauma cannot act through bypassing the epithelium which appears to be indispensable in the initiation of a DCR. The epithelium would act as a transducer which delivers the same deciduogenic message to the stroma regardless of the nature of its initial stimulation (apical contact either with oil or blastocyst, scratching, crushing, pinching, threads, electric stimulation, etc). Longitudinal propagation appears to be possible since pinching the horn at two points entails a massive longitudinal response (fig. 7A). Consistent with a propagative mechanism linked to epithelial function is the observation that a DCR appears in the lower half of the horn without any other stimulus than ablation of the epithelium in the upper part (fig. 6, C_2 and D_2). The above-mentioned considerations lead to different possible explanations: (1) injury to epithelial cells liberates substances in the uterine cavity, which stimulate the intact epithelial cells to transmit their deciduogenic message to the stroma; (2) trauma entails a physical phenomenon (depolarization wave?) along the epithelium which is necessary to decidualization. *Ferrando and Nalbandov* [5] have shown that, when the epithelium is frozen in half a uterine horn, the DCR induced by scratching the intact part,

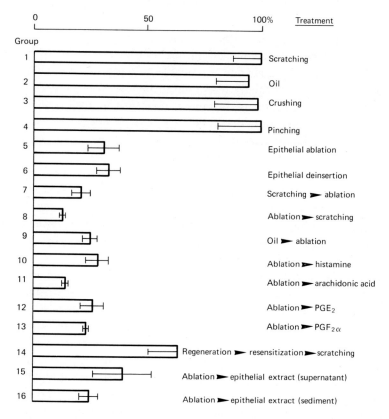

Fig. 8. Diagram of DCR results obtained by different procedures in relation to *in vivo* removal of uterine epithelium.

extends to the treated segment; this observation would be in favour of a physical membrane phenomenon, still capable of occurring along the frozen epithelial layer.

Physiologically, however, the delivery of the epithelial message to the stroma must be locally restricted, since the DCR induced by the blastocyst is limited to this latter's vicinity. Also when the epithelium is removed from the upper half of the horn, and the DCR induced by scratching the lower half, it does not extend to the deepithelized part (fig. 6, A₂).

The possible nature of the signal which is physiologically transmitted to stromal cells to induce them to decidualize has been a longlasting matter of debate.

According to the histamine theory [14], this latter substance is only effective when instilled into the uterine lumen. Moreover, histamine restores the capacity of decidualization previously suppressed by freezing the epithelium [5]. Pyrathiazine is capable of inducing the DCR after systemic injection, by allegedly liberating histamine from mast cells [15]. In our hands, the injection of histamine dissolved in PBS, either under ligature or in implants, in a deepitheliated horn of adequately sensitized animals, does not induce any DCR (fig. 8, group 10).

Prostaglandins, particularly PGE$_2$, have been found effective for the induction of decidualization in the rabbit, but it seems that their presence in the uterine lumen must be of long enough duration. Indeed, PGE$_2$ is active only when administered in implants [10] but not as an aqueous solution [16]. We have tested the action of PGE$_2$ and PGF$_{2\alpha}$ after epithelial ablation. Intraluminal instillation of those compounds did not elicit a significant DCR (fig. 8, groups 11, 12 and 13).

One of the key problems of the mechanism for decidual induction would seem to be the duration of the stimulating effect occurring at the level of stromal cells. Contrary to the mechanism of initiation at the level of the epithelium which can be obtained by brief stimulation (scratching, electric stimulation), it seems that thereafter, the metabolic message inducing stromal commitment has to be sustained for at least several hours. Our results would confirm this interpretation since epithelial ablation 4 h after oil instillation also prevents DCR from occurring (fig. 8, group 9). It thus appears that the epithelial message has not effectively stimulated the stroma at that time.

Even particulate or supernatant epithelial extracts, (fig. 8, groups 15 and 16) obtained from adequately sensitized donor rats, failed to induce a significant DCR in deepitheliated horns. The inefficiency of these extracts might in principle be explained as follows: (1) too low amounts of deciduogenic substances present in donors' epithelium at the time of its removal; (2) too short duration of action. As suggested by *Tobert* [16], experiments performed with substances dissolved in water are questionable, since these latter are not retained any longer than half an hour in the uterus [2].

In conclusion, our results confirm the obligatory role of the uterine luminal epithelium. They also suggest a longitudinal transmission of the induction stimulus along the epithelial layer, which entails sustained and localized liberation of deciduogenic substances by the epithelial cells towards the stroma.

References

1   Bitton-Casimiri, V.; Rath, N. C., and Psychoyos, A.: A simple method for separation and culture of rat uterine epithelial cells. J. Endocr. *73:* 537 (1977).

2   Conner, E. A. and Miller, J. W.: The distribution of selected substances into rat luminal fluid. J. Pharmac. exp. Ther. *184:* 291–298 (1973).

3   De Feo, V. J.: Decidualization. In Wynn, Cellular biology of the uterus, pp. 191–290 (North-Holland, Amsterdam 1967).

4   Fainstat, T.: Extracellular studies of the uterus. Disappearance of the discrete collagen bundles in endometrial stroma during various reproduction states in the rat. Am. J. Anat. *112:* 337–369 (1963).

5   Ferrando, G. and Nalbandov, A. V.: Relative importance of histamine and estrogen on implantation in the rats. Endocrinology *83:* 933–937 (1968).

6   Finn, C. A.: Biology of decidual cells. Adv. reprod. Physiol. *5:* 1–26 (1971).

7   Finn, C. A.: The implantation reaction; in Wynn, Biology of the uterus, pp. 245–308 (Plenum Press, New York 1977).

8   Finn, C. A. and Keen, P. M.: Studies on deciduomata formation in the rat. J. Reprod. Fertil. *4:* 215–216 (1962).

9   Hetherington, C. M.: Induction of deciduomata in the mouse by carbon dioxide. Nature, Lond. *219:* 863–864 (1968).

10 Hoffman, L. H.; Strong, G. B.; Davenport, G. R., and Frolich, J. C.: Deciduogenic effect of prostaglandins in the pseudopregnant rabbit. J. Reprod. Fertil. *50:* 231–237 (1977).

11 Krebiehl, R. H.: Cytological studies of the decidual reaction in the rat during early pregnancy and in the production of deciduomata. Physiol. Zool. *10:* 212–234 (1977).

12 Leroy, F.; Lejeune, B., and Galand, P.: Regeneration of uterine epithelium after experimental ablation in the rat (in press).

13 Loeb, L.: The production of deciduomata and the relation between the ovaries and the formation of the decidua. J. Am. med. Ass. *50:* 1897–1901 (1908).

14 Shelesnyak, M. C.: Some experimental studies on the mechanism of ovo-implantation in the rat. Recent. Prog. Horm. Res. *13:* 269–322 (1957).

15 Shelesnyak, M. C. and Kraicer, P. F.: A physiological method for inducing experimental decidualization of the rat uterus: standardization and evaluation. J. Reprod. Fertil. *2:* 438–446 (1961).

16 Tobert, J. A.: A study of the possible role of prostaglandins in decidualization using a non-surgical method for the instillation of fluids into the rat uterine lumen. J. Reprod. Fertil. *47:* 391–393 (1976).

Dr. B. Lejeune, M. D., Laboratoire de gynécologie, Hôpital Saint-Pierre, rue Haute 322, B-1000 Bruxelles (Belgium)

Prog. reprod. Biol., vol. 7, pp. 102–114 (Karger, Basel 1980)

Role of Stromal Cell Function Diversity in Altering Endometrial Cell Responses[1]

Stanley R. Glasser[2] *and Shirley A. McCormack*

Department of Cell Biology, Baylor College of Medicine, Houston, Tex.

Growth, morphogenesis, differentiation and expression of functional activity of urogenital organs are regulated postnatally by estrogens, androgens and/or progestins [1]. Studies of the biochemical and molecular bases of hormone action have focused on the responses of epithelial tissues because epithelium is responsible for secretory and other functions of organs derived from urogenital anlage. Results from experiments with whole organs are often subjectively considered as epithelial responses because the latter are deemed so important.

By contrast, the uterine stroma, except for the already transformed decidual cell, has been regarded as an inactive matrix involved in maintaining histiotypic organization of the epithelium. Morphological and biochemical effects of hormones upon stromal elements appear subtle and difficult to interpret. Consequently the role(s) of stroma in growth and differentiation has not been explained in depth [1]. Those studies of stroma not concerned with decidualization have emphasized the regressive changes associated with remodeling postpartum and diestrous uteri [2, 3].

What is Stroma?

Two factors contribute to the difficulty of studying the role of specific stromal cell types during cyclic and gestational states of the endometrium. The origin and fate of individual stromal cells is not clear [4]. There are 11 or more cell types in the stroma (table I) and they cannot be precisely

[1] Supported by NIH grants HD-12964, HD-07495, and CA-20853.

[2] This work would not have been possible without the technical skills of Ms. *Joanne Julian.*

Table I. Cell types of the uterine stroma (based on observations in mouse, rat, oppossum and human)

Resident cells	Transient cells

'Fixed cells'
  Fibroblasts
  Stem cells
	Constitutive
	  Decidual cells
	  Endometrial granular cells

Lymphocytes (T and B)
Mast cells
	Ameboid
	  Monocyte-macrophage
	  Lymphocytes (T and B)
	  Eosinophil
	  Heterophils (neutrophils)
	  Plasma cells

The data are based on the studies of *Padykula* and co-workers [2, 3, 5] and *Feyrter* [4].

identified in routine light microscope preparations. Selective light microscopic histochemical analyses have proved necessary, along with electron microscopy, to recognize the overall distribution of cell types.

The composition of the uterine stroma is dynamic. Cell types and the differentiative state of both fixed and transient cell populations vary constantly. This activity occurs within an extracellular matrix (ECM) that is also subject to progressive modification and suggests interaction between stromal cells and ECM. The stromal macrophage has been the chief subject of previous study because of interest in uterine pathology [4] and remodeling of the postpartum uterus [5]. These catabolic functions are shared to some extent by the fibroblast which is also active in subsequent stromal cell renewal [5]. In consideration of the fact that the anabolic processes required to organize the stroma for pregnancy have received little study this essay will focus on the possible roles of the stromal fibroblast in the pre- and peri-implantation rat uterus.

What Can Stroma Do?

Fibroblasts play an active role in the synthesis of precursors of the ECM framework [5], i. e., collagen, elastin, proteoglycans. This synthetic activity is modulated by estrogen (E) and progesterone (P). Treatment with 17β-estradiol (E$_2$) produces an enlargement of the cytoplasmic protein synthesizing and transporting system and increases the permeability of the endometrial microvasculature. The number of intracellular collagen fibrils and the synthesis and secretion of enzymes involved in ECM modeling (procollagenase, proteoglycan splitting enzymes) are also increased. This is compatible with the observation [6] that various cells of mesenchymal origin (fibroblasts, smooth muscle cells, osteoblasts) contain collagen and are involved in the production and maintenance of ECM.

The possible role of the fibroblast in intracellular degradation of collagen is only a recently developed concept. A dual fibroblast enzyme system for the cyclic remodeling of collagen in uterine ECM, that may be receptive to hormonal regulation, could be based on the (a) intracellular lysosomal system (proteases, peptidases) and procollagenase and (b) extracellular latent collagenase. This proposal is consistent with the finding that significant amounts (30%) of newly synthesized collagen are degraded intracellularly by fibroblasts in culture [7]. Fibroblasts may also possess a post translational mechanism for regulating collagen release from the cell.

Inductive Stromal-Epithelial Interactions

The roster of stromal cell capabilities is extended by the research of *Cunha* and co-workers [1, 8] on epithelial-stromal interactions during differentiation of the urogenital system. Normal determination of epithelial differentiation of organs dependent on gonadal steroids (E, P) occurs through the action of the native (homotypic) stroma which induces and specifies epithelial morphogenesis and cytodifferentiation [1]. Thus dissociated embryonic or neonatal epithelium, grown in the absence of stroma (table II, A), fails to undergo morphogenesis and usually exhibits limited cytodifferentiation. Isolated stroma or mesenchyme (table II, B) may survive for an extended period when cultured *in vitro* or *in vivo* but morphogenesis is severely restricted. Recombination of uterine or vaginal epithelium with their respective native stromata, soon after these tissues have been dissociated, will yield the normal epithelial response after *in vivo* culture in the anterior chamber of the eye (table II) [1, 8].

Once determined, urogenital epithelia continue to require a relatively specific interaction with urogenital stroma, either heterotypic or homotypic, for full epithelial expression. Recombination of 2- to 3-day uterine stroma with 2- to 3-day vaginal epithelium yields an uterine-like epithelial

Table II. Developmental response in female hosts of recombinants composed of uterine and vaginal epithelium and stroma from neonatal and adult mice[1]

Group	Recombinants		Developmental response		
	stroma	epithelium	uterus	vagina	mixture (ut/vag)
A	–	uterine	–	–	–
B	uterine	–	–	–	–
C	uterine 2–9 days	uterine 2–9 days	55/55	–	–
D	vaginal 2 days to adult	vaginal 2 days to adult	–	59/59	–
E	uterine	vaginal			
1	2–3 days	2–3 days	24/24	–	–
2	5–7 days	5–7 days	10/54	40/54	4/54
3	1–9 days	9 days to adult	–	68/68	–
F	vaginal	uterine			
1	2 days to adult	2–5 days		94/94	
2	6–9 days	6–9 days	8/52	22/52	22/52

[1] Adapted from Cunha and co-workers [1, 8].

response (table II, E₁). The reciprocal reassociation produces an epithelial response that is vagina-like in terms of morphogenesis, cytodifferentiation and function (table II, F₁).

Competence is the physiological state of epithelium which permits it to react in a morphogenetically specific way to determinative cues from stroma [9]. The competence of urogenital epithelium is related to trophic steroid hormones. However, the acquisition and loss of hormone sensitivity occur at specific developmental periods and depend on relatively specific stromal requirements [1]. The presence of a trophic hormone, per se, is not sufficient to develop an epithelial response. Although the responding epithelium may passively or actively modulate inductive stimuli, the presence of stromal components or factors is essential. These results, while implicating the stroma in epithelial responses, do not directly demonstrate the mechanism(s) by which stroma functions in hormone action.

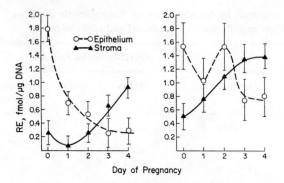

Fig. 1. Estrogen receptor concentrations (fmol/μg DNA $\pm$ SEM) in high speed cytosol (a) and nuclear (b) fractions of epithelial and stromal cells separated from pregnant rat uteri on successive days of gestation. Day 0 designates presence of vaginal sperm. From *Glasser and McCormack* [11].

Separate Endometrial Cell Types

Methods which provide clean separation of epithelium and stroma prior to analysis would be required to identify which of the characteristics assigned to the stroma function to define the role of the fibroblast in blastocyst-endometrial interactions. Efficient techniques which separate clean, viable populations of intact stromal and epithelial cells have been developed [10, 11]. Data derived from the use of these methods elucidates the part each cell type plays in the adult function of organs differentiated from urogenital anlage. It is clear that epithelial and stromal cells, separated from uteri on successive days of pregnancy (fig. 1), respond to a given hormonal milieu in a dissimilar if not independent manner.

During the first days of gestation, as epithelial cells responded to proestrus plasma E by undergoing cell division, the concentration of total estrogen receptor (RE) fell in these cells [11]. The fall was greater in cytosol (RcE, 70%) than in the nuclear compartment (RnE, 40%) suggesting prenidatory E at least influenced the translocation of epithelial RE. Stromal RE did not appear responsive to proestrus E but both RcE and RnE increased at the time (day 2) the stromal cell becomes a responsive E target [12] because of rising plasma P.

Since valid separation methods are new, we are uncertain about the interpretation of this data. These procedures do emphasize the limitations

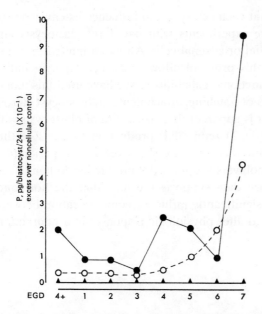

Fig. 2. Daily production of progesterone by rat blastocysts co-cultured with monolayers of uterine epithelial (○) or stromal (▲) cells or on plastic dishes (●). Equivalent gestation days (EGD) are the days in culture added to day 4 of pregnancy when the blastocysts were collected. Steroid measurements are corrected for control values obtained by analysis of incubated cell-free media. From *Glasser and McCormack* [11, 14].

of whole organ analysis in deciphering changes that relate to differential responses of individual cells. To further clarify the regulatory biology of stromal-epithelial interactions requires data on the effect of mitosis on the ontogeny of specific steroid receptors.

Studies of cell surface and secreted proteins of each cell type are also feasible because of cell separation methods. The secretion of molecules which may carry morphogenetic information and may thereby alter either the stromal-epithelial interface, e. g., basal lamina [1] or the composition of the extracellular matrix [5] deserves examination. Following blastocyst attachment, the first projections through the basal lamina appear to be from the decidual cells rather than the trophoblast [13]. This may be related to P-directed changes in the nature of the basal lamina which alters its association with the epithelial cells [5].

Preliminary evidence that each cell type can influence its environment was derived from co-culture experiments with day 4 rat blastocysts and monolayers of either epithelial or stromal cells. Although limited in its use as an implantation analog this protocol allows each cell type, including trophoblast, to display its functional capabilities. We have used this model tu study the rate and success of hatching, attachment and blastocyst steroid hormone production [14]. It is apparent that co-culture of blastocysts with epithelial cell monolayers (fig. 2) reduced P production to 52 % of that produced by blastocysts cultured on plastic dishes over a 7-day period (4 + 0 to 4 + 7). Stromal cell monolayers suppressed P production to nondetectable levels. This rudimentary data supports the idea that the hormone-sensitive stromal cell could significantly influence communication between cells and cell environments to alter physiologic responses in a nonrandom manner.

Decidualization

During decidualization the fibroblast-like stromal cell differentiates to a large polyhedral, polyploid cell rich in glycogen and lipid. In some animals this process occurs naturally in response to stimuli from the developing embryo. It may be mimicked experimentally by exposing the hormonally sensitive endometrium to a variety of physical and chemical agents. Decidualization and the decidual cell have been exhaustively examined in terms of cytodifferentiation [15], biochemistry and physiology [15, 16] but the inductive process which initiates decidualization is not well studied or understood.

The role of decidual tissue remains undefined. The oldest concept of decidual function [17] survives as the most tenable, i. e., decidualization may be viewed as a conservative mechanism by which trophoblast invasion of the endometrium is controlled and limited during establishment of the placenta. This hypothesis is virtually untested [18]. Whatever its function, e. g., blastocyst receptivity, attachment and/or invasion, decidualization reflects an informed change in the endometrial substrate. Which of the capabilities attributed to the stromal fibroblast are those that are critically involved in stromal transformation and function as decidual tissue?

Of all the latent stromal cell functions (see above) those which alter endometrial cell expression, in a manner essential to deciduogenesis and ovum receptivity, are ordered by a precise program of hormone-specific induction of informative proteins [16]. Thus, the proportion of the genome available for transcription increases with each successive preimplantation day. A similar increase can be stimulated in the castrate uterus by daily injections of P [19]. The progressive selection and translation of information by which the stromal cell gains competence for its changing role in the preimplantation uterus forms the basis for the following hypothetical scheme.

Hypothetical Scheme for Stromal Cell Transformation

The neutral, unresponsive uterus contains stromal cells (An) in the resting or 'A' state (fig. 3a) [20]. Exposure of An cells to daily injections of P, in the absence of E, increases the number of DNA sequences available for initiation of RNA synthesis in a uterus where there is essentially no cell proliferation. Three injections of P (P$\times$3) complete the sequence required to convert the neutral uterus to one sensitive to deciduogenic stimuli (fig. 3b, c; As, As (D) cells) but not receptive to the subtler stimulus of the blastocyst. Continued P injections (P$\times$4, P$\times$5, etc.) will maintain sensitized stromal cells with their enhanced population of RNA initiation sites [19].

The role of E in these events is of interest. The sensitivity of stromal cells exposed to P and E$_2$ simultaneously (fig. 3d) is developed in a different cellular environment. Uterine template capacity is significantly enhanced but, like the pregnant uterus, to a lesser extent than with P alone. When the P+E$_2\times$3 sequence is completed the cells are not only sensitive to deciduogenesis (As (D)) but receptive to the blastocyst (fig. 3d; A$_S$ (I)).

It is the nature of the receptivity that eludes definition. The epithelial cells associated with the development of As (I) stroma have undergone division [21]. How do these epithelial cells differ functionally from those associated with As (D) stroma? Is the replicated epithelial cell the obligatory trigger for ovum receptivity proposed by *Leroy and Lejeune* [22]. This is not a simple hypothesis; much significant data cannot be fitted. Physical removal of uterine epithelial cells, as late as 4 h after stimulation, prevents decidualization. While this data confirms an obligative role for the epithelial cell the time frame allows two more elaborate theoretical schemes to be considered: (a) synthesis of a required intermediary [23] of epithelial origin or (b) an active primary role for stromal cell secretions that would minimally modulate if not induce or specify the epithelial response [1] during blastocyst attachment.

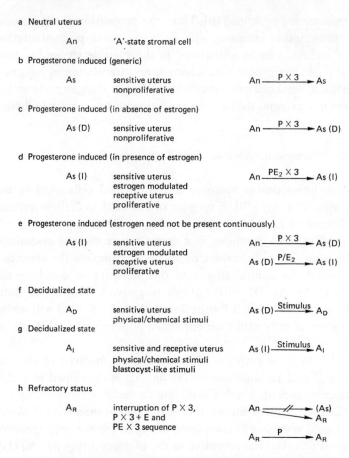

a Neutral uterus

    An   'A'-state stromal cell

b Progesterone induced (generic)

    As   sensitive uterus   An $\xrightarrow{P \times 3}$ As

                nonproliferative

c Progesterone induced (in absence of estrogen)

    As (D)   sensitive uterus   An $\xrightarrow{P \times 3}$ As (D)

                nonproliferative

d Progesterone induced (in presence of estrogen)

    As (I)   sensitive uterus   An $\xrightarrow{PE_2 \times 3}$ As (I)

                estrogen modulated

                receptive uterus

                proliferative

e Progesterone induced (estrogen need not be present continuously)

    As (I)   sensitive uterus   An $\xrightarrow{P \times 3}$ As (D)

                estrogen modulated

                receptive uterus   As (D) $\xrightarrow{P/E_2}$ As (I)

                proliferative

f Decidualized state

    A_D sensitive uterus   As (D) $\xrightarrow{Stimulus}$ A_D

                physical/chemical stimuli

g Decidualized state

    A_I sensitive and receptive uterus As (I) $\xrightarrow{Stimulus}$ A_I

                physical/chemical stimuli

                blastocyst-like stimuli

h Refractory status

    A_R interruption of P × 3,   An $\xrightarrow{\;\;/\!/\;\;}$ (As)

                P × 3 + E and A_R

                PE × 3 sequence

                                    A_R $\xrightarrow{P}$ A_R

Fig. 3. The hormonally directed development of a receptive uterus from a neutral environment involves the structural and functional differentiation of the stromal cell. This provisional scheme designates the functional variants involved in stromal cell transformation. From *Glasser and McCormack* [26].

Both these hypotheses involve the synthesis of new biochemical information. The essentiality of new RNA synthesis is confirmed by the complete blockade of deciduogenesis by nontoxic, intraluminal doses of actinomycin D [15]. Given alone, in place of nidatory E, systemic doses of this RNA inhibitor are reported to initiate implantation [24]. Recent data about the role of nidatory E in uterine gene transcription may resolve this apparent contradiction [19].

While the induction of As (I) status requires the presence of E_2 (fig. 3d) its presence need not be continuous; E_2 can be given as a single injection at the end of the $P \times 3$ sequence (fig. 3e). While the status of cell proliferation is different than $P + E_2 \times 3$ there are no discernable differences in these two groups in terms of their deciduogenic or nidatory responses. The significant difference between As (D) and AS (I) stroma is the E-induced differentiation of a cell competent to interact with the blastocyst.

E_2, by itself, is uterotrophic but not deciduogenic. Single doses will stimulate cell proliferation and induce E_2-specific mRNAs and growth-related proteins. E_2, given as a single dose before the completion of the $P \times 3$ sequence will bar uterine sensitivity and ablate the increase in the number of RNA initiation sites. Any interruption of the $P \times 3$ sequence (fig. 3h) will produce stromal cells refractory to any stimulus (A_R). The refractory status is not a null position but is actively maintained as long as the uterus is being hormonally stimulated. Only complete cessation of hormone stimulation, for at least 48 h, will return the endometrial cells to a state of neutrality ($A_R \rightarrow An$) [23].

If a single dose of E_2 is given after the completion of the $P \times 3$ sequence (fig. 3e) implantation can be initiated in the sensitive uterus during the next 24 h although the number of RNA initiation sites decay from 20,000 to 4,000/pg DNA within 4 h [19]. If implantation is not initiated within that period the uterus loses sensitivity, 24–30 h after E_2, and becomes refractory (equivalent to A_R status of stromal cells; fig. 3h). Thus, while the intervention of E_2 initiates the processes of receptivity and implantation it also signals the program of events whereby the uterus loses sensitivity to any deciduogenic stimulus and becomes refractory (As (D), As (I) $\rightarrow A_R$).

However, if the endometrium is engaged by either a deciduogenic agent or blastocyst before the uterus loses its sensitivity (< 24 h?) the uterus is 'rescued' from the refractory state. There is a striking restoration of template capacity; the proportion of the genome available for initiation of RNA synthesis is increased significantly beyond the $P \times 3$ point. Our experiments support the notion that the action of P in inducing sensitivity is at the level of transcription [19]. Unique to these experiments is the apparent restriction of the P-stimulated genome by E_2. This qualitative alteration in gene expression permits information to be specified that directs activation of the blastocyst and induction of stromal-epithelial interactions that transform the sensitive uterus to one receptive to the activated blastocyst. Attachment (embryonic signal?) reverses the retrograde action of E_2

and allows implantation to occur before the loss of uterine sensitivity. Qualitative transient gene restriction may prove to be a requirement for implantation and its sequelae. This would rationalize the putative action of actinomycin D in facilitating implantation [24].

At the present time a catalog of presumptive stromal fibroblast functions could include: induction, specification and mediation of morphogenetic responses, regulation of steroid hormone receptor ontogeny, maintenance of epithelial structure and function, synthesis of anabolic, metabolic and catabolic enzymes as well as precursors of ECM. The stromal cell grows, proliferates, differentiates, invades and is invaded. Our knowledge of how and when each of these functions becomes operational is so insecure that only a limited amount of this data could be used to provisionally designate functional stromal cell variants (fig. 3).

Only careful study of each of these cataloged responses will aid in extending our concept of the nature and significance of stromal cell functions in implantation and decidualization. Thus, regardless of the outcome, future experiments should challenge the idea that the embryonic role of the stroma in inducing, specifying and mediating epithelial responses may be carried over into adult life. This is the suggested basis for the cyclic renewal of hormonally regulated reproductive organs [1, 8]. It is a provocative, exciting concept that should be examined without prejudice.

Implantation should also be viewed in a more general manner, as the differential expression of stromal cell function which is a part of the continuing process of uterine morphogenesis. Thus, surface changes of endometrial cells should be rigorously studied as basic cellular phenomena involved in cell motility, cell adhesion and attachment [25]. Deciduogenesis then becomes endowed with special significance to the implantation process of certain species because of the role played by the transformed stromal cell in remodeling the ECM as the reactive substrate for the trophoblast [26].

References

1   Cunha, G. R.; Chung, L. W. K.; Shannon, J. M., and Reese, B. A.: Stromal-epi-
     thelial interactions in sex differentiation. Biol. Reprod. *22:* 19–42 (1980).
2   Padykula, H. A. and Taylor, J. M.: Cellular mechanisms involved in cyclic re-
     newal of the uterus. I. The opposum, *Didelphis virginiana.* Anat. Rec. *184:* 5–
     26 (1976).
3   Padykula, H. A. and Campbell, A. G.: Cellular mechanisms involved in cyclic
     renewal of the uterus. II. The albino rat. Anat. Rec. *184:* 27–48 (1976).

4 Feyrter, F.: Origin and particular morphology of stromal cells; in Schmidt-Matthiesen, The normal human endometrium, pp. 65–85 (McGraw-Hill, New York 1963).

5 Padykula, H. A.: Shifts in uterine stromal cell populations during pregnancy and regression; in Glasser and Bullock, Cellular and molecular aspects of implantation (Plenum, New York 1980).

6 Brandes, D. and Anton, E.: Lysosomes in uterine involution: intracytoplasmic degradation of myofilaments and collagen. J. Geront. *24:* 55–61 (1969).

7 Bienkowski, R. S.; Baum, B. J., and Crystal, R. G.: Fibroblasts degrade newly synthesized collagen within the cell before secretion. Nature, Lond. *276:* 413–416 (1978).

8 Cunha, G. R.: Epithelial-stromal interactions in development of the urogenital tract. Int. Rev. Cytol. *47:* 137–194 (1976).

9 Holtfreter, J. and Hamburger, V.: Amphibians; in Willier, Weiss and Hamburger, Analysis of development, pp. 230–296 (Saunders, Philadelphia 1955).

10 McCormack, S. A. and Glasser, S. R.: Differential response of individual uterine cell types from immature rats treated with estradiol. Endocrinology (in press, 1980).

11 Glasser, S. R. and McCormack, S. A.: Separated endometrial cell types as analytical tools in the study of decidualization and implantation; in Glasser and Bullock, Cellular and molecular aspects of implantation (Plenum, New York 1980).

12 Tachi, C.; Tachi, S., and Lindner, H. R.: Modification by progesterone of oestradiol-induced cell proliferation, RNA synthesis and oestradiol distribution in the rat uterus. J. Reprod. Fertil. *31:* 59–76 (1972).

13 Enders, A. C.; Chavez, D. S., and Schlafke, S.: Trophoblast outgrowth during implantation *in utero* and *in vitro*; in Glasser and Bullock, Cellular and molecular aspects of implantation (Plenum, New York 1980).

14 McCormack, S. A. and Glasser, S. R.: Hormone production by blastocysts and mid-gestation trophoblasts *in vitro;* in Kimball, The endometrium. 8th Brook Lodge Symposium on Problems in Reproductive Physiology (Spectrum, New York 1980).

15 Glasser, S. R.: The uterine environment in implantation and decidualization; in Balin and Glasser, Reproductive biology, pp. 776–833 (Excerpta Medica, Amsterdam 1972).

16 Glasser, S. R. and Clark, J. H.: A determinant role for progesterone in the development of uterine sensitivity to decidualization and ovoimplantation; in Markert and Papaconstantinou, Developmental biology of reproduction, pp. 311–345 (Academic Press, New York 1975).

17 Bryce, T. H. and Teacher, J. H.: Contributions to the study of early development and imbedding of the human ovum (MacLehose, Glasgow 1908).

18 Cowell, T. P.: Control of epithelial invasion by connective tissue during embedding of the mouse ovum; in Tarin, Tissue interactions during carcinogenesis, pp. 435–463 (Academic Press, London 1972).

19 Glasser, S. R. and McCormack, S. A.: Estrogen-modulated uterine gene transcription in relation to decidualization. Endocrinology *104:* 1112–1118 (1979).

20 Smith, J. A. and Martin, L.: Regulation of cell proliferation; in Padillo, Came-

ron and Zimmerman, Cell cycle controls, pp. 43–60 (Academic Press, New York 1974).

21 Martin, L. and Finn, C. A.: Hormonal regulation of cell division in epithelial and connective tissues of the mouse uterus. J. Endocr. *41:* 363–371 (1968).

22 Leroy, F. and Lejeune, B.: The uterine epithelium as a transducer for the triggering of decidualization in rats; in Glasser and Bullock, Cellular and molecular aspects of implantation (Plenum, New York 1980).

23 Psychoyos, A.: Hormonal control of ovo-implantation. Vitams Horm. *32:* 201–256 (1974).

24 Finn, C. A. and Bredl, J. C. S.: Studies on the development of the implantation reaction in the mouse uterus: influence of actinomycin D. J. Reprod. Fertil. *34:* 247–253 (1973).

25 Sherman, M. I. and Wudl, L. R.: The implanting blastocyst; in Poste and Nicholson, Cell surface interactions in embryogenesis, pp. 81–125 (North-Holland, Amsterdam 1976).

26 Glasser, S. R. and McCormack, S. A.: Functional development of rat trophoblast and decidual cells during establishment of the hemochorial placenta. Adv. Biosci. *25:* 165–197 (1979).

S. R. Glasser, Ph. D., Department of Cell Biology, Baylor College of Medicine, Houston, TX 77030 (USA)

Prog. reprod. Biol., vol. 7, pp. 115–124 (Karger, Basel 1980)

Cell Proliferation and Decidual Morphogenesis

P. Sartor

Laboratoire des Interactions cellulaires, Université de Bordeaux II, Bordeaux

The deciduomata is a rapidly growing tissue which appears normally during the implantation process and of which the characteristics have been previously described by *Loeb* [1907], *Krehbiel* [1937] and many other workers [for review, se *De Feo, 1967; Glasser, 1972, Glasser and Clark, 1975*]. In 1955, *Sachs and Shelesnyack* tried to determine the ploidy level of the decidual cells by the squash technique and later *Leroy* [1974] and *Dupont et al.* [1974] by cytophotometric methods clearly demonstrated that a high level of polyploidy was reached (32 n). After the initial *in vivo* experiments of *Shelesnyack and Tic* [1963], *Lobel et al.* [1965a, b], *Finn and Hinchliffe* [1965], many workers have approached the biochemical aspects of its development [for review, see *Psychoyos, 1973*].

Recently the study of this tissue *in vivo* and *in vitro* allowed *Sananes et al.* [1976], *Vladimirsky et al.* [1977] and *Acker et al.* [1979] to analyze the factors of cell differentiation, its growth features and some of its cytochemical properties. Whereas, as stated by *Glasser* [1972], 'the decidual reactions is not the perfect analog for the study of implantation, recognition and utilization of these differences, rather than their similarities, may prove to be more instructive in understanding these processes.' During the course of our experiments we have been using the traumatic as well as the physiologic deciduomata, to try to separate the hormonal from the fetal component of growth and organization of this tissue.

Some Aspects of the Morphology of Decidual Tissue in the Rat

In traumatic deciduomata the cells in the mesometrial part of the endometrium cannot get a ploidy level higher than 4n [*Dupont et al.,*

1975a], whereas the antimesometrial part may contain nuclei with a 32 n ploidy level [*Leroy*, 1974; *Dupont et al.*, 1974, 1975b]. The radioauto-graphic study of the [^{3}H]-thymidine incorporation demonstrate that this is achieved by a different pattern of cell division in each part. In mesometrial cells, the occurrence of mitosis as a general feature is assessed by the rapid dilution of radioactivity in a greater number of nuclei [*Dupont et al.*, 1978] than the antimesometrial cells, where in turn, endomitosis seems to be a commonly occurring fact during the growing period. It would be inter-esting to know why the chorial cells of each part of the endometrium have a different evolution. Perhaps the explanation may partly be approached by our experiments on the action of high doses of oestradiol on the de-velopment of the decidual reaction [*Sartor et al.*, 1978].

When 10 μg of oestradiol are injected at 10 a. m. from the 4th day of pseudopregnancy (day 1 is the day where the vaginal plug is found), the decidual reaction is inhibited as judged by the weight study, the [^{3}H]-uridine uptake and the morphological analysis [*Sartor et al.*, 1978]. The cytophotometric study of this tissue shows that the ploidy level, in the antimesometrial part of the endometrium, cannot reach more than 4 n values in most of the cells. Very few 8n nuclei are seen [*Sartor et al.*, 1978].

Comparison of the cell cycle in normal and treated pseudopregnant rats (10 μg of oestradiol on day 4, 10 a. m.) demonstrates an increase in G2 phase nuclei of the treated pseudopregnant rat uteri on day 4 at 4 p. m.

Two hypotheses arise from these results: (1). The inability of meso-metrial-endometrial cells to reach a ploidy level higher than 4 n is due to an increased blood and thus oestradiol supply in this uterine part which is a consequence of the vascular anatomy. (2). The increasing amount of endogenous oestradiol, in normal day 4 pregnant rats, leads to the receptive phase for implantation and to the maximum of uterine sensitivity to the deciduogenic trauma.

Increasing the amount of endogenous oestradiol available, from an injection of 10 μg of oestradiol on day 4 at 10 a. m., we can try to ad-vance the period of maximal sensitivity to develop the deciduomata. The first results obtained seem to show that the period of maximum sensitivity in our system is shortened, but not advanced from D 5 to D 4 [*Acker*, unpublished; the trauma was achieved by a cotton thread, in one uterine horn]. In 1963, with chemical inducers, *De Feo* has obtained the same shortening. In a different system (needle scratch) *Yoshinaga and Greep* [1970] were able to induce a precocious uterine sensitization.

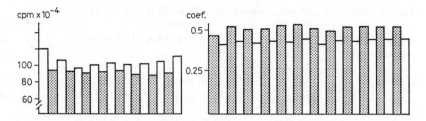

Fig. 1. Uptake and retention of [³H]-oestradiol and [³H]-progesterone along a sectioned uterine horn in the rat. *a* The animal was killed 2 h after [³H]-oestradiol injection on the 6th day (10 a. m.). Implanted (stippled) and unimplanted (open) sites are located as in the living animal. *b* The animal was killed 30 min after the [³H]-progesterone injection on the 6th day (10 a. m.). Implanted (stippled) and unimplanted (open) sites are located as in the living animal.

The role of Hormones in the Organization of the Growing Deciduomata of Pregnant Rats

Our first experiments are designed to analyze the hormonal uptake in implanted and unimplanted sites of day 6 pregnant rat uteri [for details of the technique, see *Sartor,* 1974, 1977]. The uptake of [³H]-progesterone is higher in implanted sites of normal 6th day pregnant rats (fig. 1b) than in unimplanted areas. On the contrary, the [³H]-oestradiol is more retained by the unimplanted sites in the same conditions (fig. 1a). In pregnant females, ovariectomized on day 4 at 10 a. m., receiving daily 5 mg of progesterone until sacrifice (day 6) and 0.3 μg of oestradiol on day 4 at 4 p. m., the injection of [³H]-oestradiol on the 6th day does not give any difference in uptake between implanted and unimplanted sites [*Sartor,* 1977]. These results show that the retention of both hormones is different in the areas where the blastocysts are implanted and decidual tissue is growing. The values obtained in implanted sites of normal pregnancies with [³H]-oestradiol reflect higher retention of the endogenous hormone. The report of *Ward et al.* [1978] supports our data.

The study of receptor levels and their cellular localizations for both hormones in day 6 endometrium has been performed to give us more information [*Logeat et al.,* 1980]. The total of cytosol and nuclear receptors for both hormones does not vary to a great extent between implanted and unimplanted sites, but their nuclear translocation is higher in the areas where the blastocysts have attached the luminal epithelium (table I). Two

Table I. Oestradiol and progesterone receptor levels in day 6 pregnant rat uteri

Receptors in endometrium		IS, pmol/mg DNA ± SE	UIS, pmol/mg DNA ± SE	p
Oestradiol	cytosol	11.33 ± 1.73	12.11 ± 1.05	NS
	nuclei	8.83 ± 0.34	4.62 ± 0.33	< 0.001
Progesterone	cytosol	1.41 ± 0.04	1.74 ± 0.39	NS
	nuclei	0.86 ± 0.07	0.45 ± 0.02	< 0.001

The implanted sites were determined by the trypan blue technique and separated from the unimplanted ones. The endometrium of each part was gently scraped away using a razor blade and kept in liquid nitrogen until receptor measurement. Exchange techniques were used to measure the nuclear receptor of oestradiol and the nuclear and the cytosolic receptors of progesterone. The cytosolic oestradiol receptor was assayed by a 2 h incubation at 0 °C in the presence of 10 nM [^{3}H]-oestradiol or 10 nM [^{3}H]-oestradiol + 10 mM cold diethylstilbestrol.

Table II. Oestradiol and progesterone receptor levels in castrated pregnant rat uteri

Receptors in endometrium		IS, pmol/mg DNA ± SE	UIS, pmol/mg DNA ± SE	p
Oestradiol	cytosol	4.49 ± 0.38	2.99 ± 0.81	NS
	nuclei	5.49 ± 0.45	2.42 ± 0.33	< 0.001
Progesterone	cytosol	4.17 ± 0.09	2.23 ± 0.26	< 0.05
	nuclei	1.11 ± 0.02	0.67 ± 0.05	< 0.001

Castration was performed on the morning of day 5 and the animals were sacrificed on day 6 at 10 a. m. The endometrium was prepared and treated as in the first receptor measurement.

hypotheses may be suggested: (1) the blood supply of hormones is increased in implanted sites; (2) the blastocyst has a local effect on the cellular localization of the hormonal receptors and the cellular events thus initiated.

To test these hypotheses, the same measurements have been done in pregnant female rats castrated on the morning of the 5th day, and left without progesterone until the 6th day. Normal ovoimplantation occurs on the morning of the day 6. The cellular localization of both hormonal receptors is the same as that of normal 6 day pregnancies. The nuclear binding in implanted sites is also twofold higher than in unimplanted areas, whereas the total quantity of receptor is diminished for oestradiol but is enhanced for progesterone (table II). Thus, the blood supply of hormone does not seem to be responsible for the higher nuclear content of receptors.

Effect of the Blastocyst on the Development of the Physiological Deciduomata

On day 5 of pseudopregnancy one uterine horn is traumatized to induce decidualization. On day 6 the receptors are measured in the endometrium of decidualized and control horns. The nuclear content of both oestradiol and progesterone receptors are similar in both horns (table III).

As a consequence of the absence of the blastocyst, the disappearance of the higher nuclear receptor translocation can be seen in the endometrium where the decidual reaction is growing. This confirms the results of the previous experiment. A number of other authors have measured receptors in decidual tissue [*Armstrong et al.*, 1977, 1978; *Talley et al.*, 1977; *Peleg et al.*, 1979]. The wide variations in experimental conditions used precludes any valid comparison with the above data. Morphological and physiological analyses of the first stages of implantation in rats ovariectomized on the morning of the 5th day also supports the evidence of a local effect of the blastocyst [*Sartor and Duluc*, 1979]. Table IV shows that the castration performed on the morning of the 5th day does not greatly influence the number of implantation sites recovered 24 h later.

The rate of [³H]-thymidine incorporation is affected only on day 6 at 6 p. m. in the growing deciduomata of implanted sites. Prior to that time, the lowered endogenous hormonal level (only adrenal progesterone and oestradiol are secreted) results in a lower uptake on the afternoon of day 5 in both implanted and nonimplanted sites [*Sartor and Duluc*, 1979]. From day 5 to day 6 the slopes of the uptake curves are similar in castrated and control rats. Histological study of the implanted sites of castrated females reveal normally growing deciduomata on the morning of day 6. At this time most of the blastocysts are well attached in their implantation chambers [*Sartor and Duluc*, 1979]. Therefore, even in the absence of ovarian steroids the first steps of the implantation reaction occur. These results clearly demonstrate that the blastocyst has a local effect on the physiological deciduomata. *Mester et al.* [1974], *Martel and Psychoyos*

Table III. Oestradiol and progesterone receptor levels in pseudopregnant rat uteri

Receptors in endometrium		Decidualized horns pmol/mg DNA	Control horns pmol/mg DNA
Oestradiol	cytosol	1.23 ± 0.12	1.33 ± 0.17
	nuclei	1.43 ± 0.14	1.38 ± 0.16
Progesterone	cytosol	2.25 ± 0.21	2.27 ± 0.23
	nuclei	0.94 ± 0.09	0.93 ± 0.05

The oviduct was sectioned on day 1 of pregnancy. On the morning of day 5, a cotton thread was introduced in one uterine lumen, the animal being sacrificed on the morning of day 6. The preparation of the endometrium and the receptor assays are as described in table I.

Table IV. Implantation sites in normal and castrated pregnant rats

	Total number of animals	Animals with implantation sites	% of fertility	Total number of ovoimplantations	Mean number of ovoimplantations ± SEM
Control	553	484	87.5	5,490	11.34 ± 0.132
Castrated	365	302	82.7	3,031	10.14 ± 0.218

The implantation sites were recovered by means of an intravenous injection of trypan blue 30 min before sacrifice, on the morning of day 6. Controls were normal pregnant rats, castrated ones were ovariectomized on the morning of the 5th day, and then left without hormonal injection until sacrifice.

[1976] and *Vu-Hai et al.* [1978] have presented results, in pseudopregnant or pregnant uterine receptors, before or during implantation. The technique of separation of implanted and unimplanted sites proved very useful in demonstrating differences related to the implantation process.

Discussion

As in regenerating liver and normal pituitary, the deciduoma has to respond to drastic situations. Polyploidy seems to be the preferred means to attain higher synthesis with the minimum of cellular organelles. *Glasser and McCormack* [1979] have reported evidence that oestradiol reduces the transcriptive activity of chromatin in the sensitive uterus. Oestradiol does not promote the real growth of the deciduomata which is under progesterone control. Our results indicate a high level of translocation of the progesterone receptor in the decidualized and control uterine horns, whereas the oestradiol receptor translocation is low in both horns compared to normal or castrated pregnant endometrium. This agrees well with *Glasser's* hypothesis. In physiological deciduomata, oestradiol and blastocyst action could be the components which enhance the defenses of the mother. As emphasized by *Bernard* elsewhere in these proceedings, the deciduomata may have an important place in the immune system of the couple uterus-blastocyst. In control horns the potential exists for decidual development, owing to the nuclear receptor content. Without trauma this potential remains unfulfilled. The possible role of prostaglandins in decidualization and implantation has been suggested by *Kennedy* [1977], *Kennedy and Zamecnik* [1978] and *Sananes et al.* [1976]. In addition, *Dupont et al.* [1977] found a modification in the histones binding affinity for dyes, immediately after the decidualizing trauma.

In which way does the blastocyst act in the implantation and decidualization process? *Dickman et al.* [1976] have proposed a fundamental role for steroid hormones originating from the blastocyst.

We can say that our results bring the evidence that the blastocyst participates in the local delivery of oestradiol and progesterone to the endometrium [*Logeat et al.,* 1980]. Whereas the regression of the decidual tissue observed on day 6 at 6 p. m. in pregnant female rats castrated on day 5 [*Sartor and Duluc,* 1979] suggests a release of stored maternal hormones, rather than synthesis originating from the blastocyst. Nevertheless, this does not exclude the possibility that the blastocyst secretes a factor which enhances receptor translocation [*Logeat et al.,* 1980].

References

Acker, G. M.; Lieberherr, M.; Bourguignon, J., and Dubuisson, L.: Study of decidual cells in primary cultures: action of sex steroids on growth histological and ultrastructural morphology and acid and alkaline photosphatases activities. Abstr. 61st Annu. Meet. Endocr. Soc. 1979.

Armstrong, E. G.; Tobert, J. A.; Talley, D. J., and Villee, C. A.: Changes in progesterone receptor levels during deciduomata development in the pseudopregnant rat. Endocrinology *101:* 1545–1551 (1977).

Armstrong, E. G., jr., and Villee, C. A.: Estrogen receptor in purified nuclei from deciduomata of the pseudopregnant rat. J. Steroid. Biochem. *9:* 1149–1154 (1978).

De Feo, V. J.: Determination of the sensitive period for the induction of deciduomata in the rat by different inducing procedures. Endocrinology *73:* 488–497 (1963).

De Feo, V. J.: Decidualization; in Wynn, Cellular biology of the uterus, pp. 192–290 (Appleton Century Crofts, New York 1967).

Dickman, Z.; Dey, S. K., and Sen Gupta, J.: A new concept: control of early pregnancy by steroid hormones originating in the preimplantation embryo. Vitams Horm. *34:* 215–242 (1976).

Dupont, H.; Dupont, M. A.; Sartor, P.; Esnault, C. et Mayer, G.: Déterminisme de l'évolution des fibroblastes en G1 et G2 dans la muqueuse utérine chez la ratte au cours de la décidualisation. Biol. Cell. *30:* 8a (1977).

Dupont, H.; Esnault, C.; Duluc, A. J. et Mayer, G.: Evolution de la polyploidie dans le deciduome expérimental chez la ratte en grossesse unilatérale. Etude cytophotométrique. C. r. hebd. Séanc. Acad. Sci., Paris *279:* 501–504 (1974).

Dupont, H.; Esnault, C.; Duluc, A. J. et Mayer, G.: Etude cytophotométrique de l'acide desoxyribonucléique dans les cellules mésométrales du déciduome expérimental chez la ratte en grossesse unilatérale. C. r. Séanc. Soc. Biol. *169:* 935–936 (1975a).

Dupont, H.; Esnault, C.; Duluc, A. J. et Mayer, G.: Evolution du déciduome expérimental et du blastocyste en léthargie chez la ratte. I. Le déciduome expérimental chez la ratte en grossesse unilatérale. Ann. Biol. anim. Biochim. Biophys. *15:* 765–769 (1975b).

Dupont, H.; Sartor, P.; Dupont, M. A.; Duluc, A. J. et Mayer, G.: Evolution du déciduome expérimental chez la ratte: Filiation cellulaire et morphogenèse. Biol. Cell. *32:* 215–222 (1978).

Finn, C. A. and Hinchliffe, J. R.: Histological and histochemical analysis of the formation of implantation chambers in the mouse uterus. J. Reprod. Fertil. *9:* 301–309 (1965).

Glasser, S. R.: The uterine environment in implantation and decidualization; in Balin and Glasser, Reproductive biology, pp. 776–833 (Excerpta Medica, Amsterdam 1972).

Glasser, S. R. and Clark, J. H.: A determinant role for progesterone in the development of uterine sensitivity to decidualization and ovo-implantation; in Markert and Papaconstantinou, The developmental biology of reproduction, pp. 311–345 (Academic Press, New York 1975).

Glasser, S. R. and McCormack, S. A.: Estrogen-modulated uterine gene transcription in relation to decidualization. Endocrinology *104:* 1112–1118 (1979).

Kennedy, T. G.: Evidence for a role for prostaglandins in the initiation of blastocyst implantation in the rat. Biol. Reprod. *16:* 286–291 (1977).

Kennedy, T. G. and Zamecnik, J.: The concentration of 6-keto-prostaglandins FIα is markedly elevated at the site of blastocyst implantation in the rat. Prostaglandins *16:* 599–605 (1978).

Krehbiel, R. H.: Cytological studies on the decidual reaction in the rat during early pregnancy and in the production of deciduomata. Physiol. Zool. *10:* 212–233 (1937).

Leroy, F.: Etude fonctionnelle de l'acide desoxyribonucléique dans l'endomètre du rongeur. Analyse morphologique semi quantitative (Arscia, Brussels 1974).

Lobel, B. L.; Tic, L., and Shelesnyak, M. C.: Studies on the mechanism of nidation. XVII. Histochemical analysis of decidualisation in the rat. 2. Induction. Acta endocr., Copenh. *50:* 469–485 (1965a).

Lobel, B. L.; Tic, L., and Shelesnyak, M. C.: Studies on the mechanism of nidation. XVII. Histochemical analysis of decidualisation in the rat. 3. Formation of deciduomata. Acta endocr., Copenh. *50:* 517–536 (1965b).

Loeb, L.: Über die experimentelle Erzeugung von Knoten von Deciduagewebe in dem Uterus des Meerschweinchens nach stattgefundener Copulation. Zentbl. allg. Path. path. Anat. *18:* 563–565 (1907).

Logeat, F.; Sartor, P.; Vu-Hai, M. T., and Milgrom, E.: Local effect of the blastocyst on estrogen and progesterone receptors in the rat endometrium. Science, N. Y. (in press, 1980).

Martel, D. and Psychoyos, A.: Endometrial content of nuclear estrogen receptor and receptivity for ovoimplantation in the rat. Endocrinology *99:* 470–475 (1976).

Mester, J.; Martel, D.; Psychoyos, A., and Baulieu, E. E.: Hormonal control of oestrogen receptor in uterus and receptivity for ovoimplantation in the rat. Nature, Lond. *250:* 776–778 (1974).

Peleg, S.; Bauminger, S., and Lindner, H. R.: Oestrogen and progestin receptors in deciduoma of the rat. J. Steroid Biochem. *10:* 139–145 (1979).

Psychoyos, A.: Endocrine control of egg implantation; in Greep and Astwood, Handbook of physiology; Endocrinology, vol. II, pp. 187–215 (1973).

Sachs, L. and Shelesnyack, M. C.: The development and suppression of polyploidy in the developing and suppressed deciduoma in the rat. J. Endocr. *12:* 146–151 (1955).

Sananes, N.; Baulieu, E. E., and Le Goascogne, C.: Prostaglandin(s) as inductive factor of decidualization in the rat uterus. Mol. cell. Endocrinol. *6:* 153–158 (1976).

Sartor, P.: Contribution à l'étude de l'ovoimplantation chez la ratte. Etude de la participation de l'utérus à la réalisation de la nidation. Thèses Bordeaux (1974).

Sartor, P.: Exogenous hormone uptake and retention in the rat uterus at the time of ovoimplantation. Acta endocr., Copenh. *84:* 804–812 (1977).

Sartor, P. et Duluc, A. J.: Déroulement des premiers stades de l'ovoimplantation chez des rattes castrées le cinquième jour de la gestation, avant midi. C. r. hebd. Séanc. Acad. Sci., Paris (in press, 1979).

Sartor, P.; Dupont, H.; Dupont, M. A.; Duluc, A. J., and Mayer, G.: The action of high doses of oestradiol on implantation and decidual reaction in the rat. Anim. reprod. Sci. *i:* 93–96 (1978).

Shelesnyak, M. C. and Tic, L.: Studies on the mechanism of decidualization. IV. Synthetic processes in the decidualizing uterus. Acta endocr., Copenh. *42:* 465–472 (1963).

Talley, D. J.; Tobert, J. A.; Armstrong, E. G., jr., and Villee, C. A.: Changes in estrogen receptor levels during deciduomata development in the pseudopregnant rat. Endocrinology *101:* 1538–1544 (1977).

Vladimirsky, F.; Chen, L.; Amsterdam, A.; Zor, U., and Lindner, H. R.: Differentiation of decidual cells in cultures of rat endometrium. J. Reprod. Fertil. *49:* 61–68 (1977).

Vu-Hai, M. T.; Logeat, F., and Milgrom, E.: Progesterone receptors in the rat uterus: variations in cytosol and nuclei during the oestrous cycle and pregnancy. J. Endocr. *76:* 43–48 (1978).

Ward, W. F.; Frost, A. G., and Ward Orsini, M.: Estrogen binding by embryonic and interembryonic segments of the rat uterus prior to implantation. Biol. Reprod. *18:* 598–601 (1978).

Yoshinaga, K. and Greep, R. O.: Precocious sensitization of the uterus in pseudo-pregnant rats. Proc. Soc. exp. Biol. Med. *134:* 725–727 (1970).

P. Sartor, Dr. Es Sciences, Laboratoire des Interactions cellulaires,
Université de Bordeaux II, 146, rue Léo-Saignat, F-33076 Bordeaux Cédex
(France)

Prog. reprod. Biol., vol. 7, pp. 125–134 (Karger, Basel 1980)

Decidualization *in vitro*

Effects of Progesterone and Indomethacin[1]

N. Sananès, S. Weiller, E. E. Baulieu and C. Le Goascogne

Unité de recherches sur le métabolisme moléculaire et la physio-pathologie des stéroïdes de l'INSERM, Hôpital de Bicêtre, Bicêtre

Introduction

The ability to grow *in vitro* rat endometrial cells and to obtain their differentiation into decidual cells in the cell culture conditions [11, 13] appears to be convenient to further deepen insight into the mechanism of decidualization. It has been shown that the progesterone treatment of the ovariectomized rats, which is an absolute need to induce the decidual reaction *in vivo,* is also a prerequisite for the development and differentiation of endometrial cells *in vitro.* On the other hand, although oestradiol is necessary for blastocyst implantation, the decidual response can be induced experimentally by a traumatism of the endometrium of rats submitted to progesterone alone, without oestradiol. Similarly, the decidual cell differentiation has been obtained *in vitro* with endometrial cells from animals treated by progesterone alone. The decidualized cells in culture share several features with the deciduoma in the rat; especially they are binucleated, accumulate filaments and have the same life span.

There are evidences to show that prostaglandins intervene in the decidual response [5, 7, 8, 10, 12] and we have suggested [11] that the mechanical trauma of the endometrium, either *in vivo* or during the isolation of the cells, induces the decidual reaction, at least partially through prostaglandin release, since it is, *in vivo,* inhibited by indomethacin. Then it was of interest to test the effect of indomethacin in culture. We have also examined the effect of serum and progesterone. Electron microscopic observations have been especially focused on the components of the cytoskeleton.

[1] Supported by INSERM (CL No. 78.5.029.4).

Material and Methods

Animals

Adult female rats of the Sprague-Dawley strain were ovariectomized under ether anaesthesia 2 weeks before the experiments. Then, the basic hormonal preparation of the spayed animals consisted in progesterone (5 mg in 0.25 ml sesame oil/day) for 4 successive days and oestradiol (0.15 μg in 0.15 ml sesame oil on the 4th day), both hormones being given subcutaneously. Animals were killed by bleeding under ether anaesthesia 18 h after oestradiol injection. In one experiment the hormonal treatment was reduced to 2 progesterone injections. In another series of experiments animals were injected subcutaneously with indomethacin (gift from Merck, Sharp & Dohme; 500 μg in oil) at 24, 18 and 2 h before sacrifice.

Cell Culture

The preparation of isolated endometrial cells was as described before [11] except that pellets of cells were resuspended in serumfree medium at 0 °C. The cells were grown in 25 cm^2 plastic dishes, in Dulbecco's modified Eagle's medium (DME) supplemented with 10 % heat-inactivated calf serum, at 37 °C, in an atmosphere containing 5 % CO_2. In some experiments, serum (0.1 or 10 %) was charcoal-dextran extracted. When used, progesterone was 0.5 μM and indomethacin 270 μM. The culture medium was renewed every day.

Observations

Living cells were observed under the phase contrast microscope. For light microscopy, the cells washed in phosphate-buffered saline 0.2 M, pH 7.2 (PBS), were fixed *in situ* for 30 min in 1 % osmium tetroxide in PBS (O_sO_4), dehydrated and stained with Giemsa. For electron microscopy, the cells were fixed 15 min either in O_sO_4 or in 3 % glutaraldehyde in cacodylate buffer followed by a 15-min postfixation in O_sO_4. After dehydration, the cells were embedded *in situ* in the epoxy resin Epon. Cells were sectioned parallel to the substratum and contrasted with uranyl acetate and lead citrate.

Fig. 1. Comparative growth of rat endometrial cells under various experimental conditions. Cells are from animals treated with progesterone and oestradiol. The same number of cells ($\sim 10^6$) has been plated in each dish. *a–d* Cells at 24, 48, 72 and 96 h in medium supplemented with 10 % calf serum and progesterone. *e–h* Cells at 24, 48, 72 and 96 h in the same culture medium as (a) but in the presence of indomethacin (270 μM). *i, j* Cells at 24 and 48 h grown in serumfree medium until 24 h, then with 0.1 % extracted calf serum. *k, l* Cells at 72 and 96 h, grown in the same conditions as (i, j) and supplemented with 10 % calf serum and progesterone at 48 h. Phase contrast micrographs of fixed cells stained with Giemsa. Arrows point to binucleated cells. Bar = 100 μm.

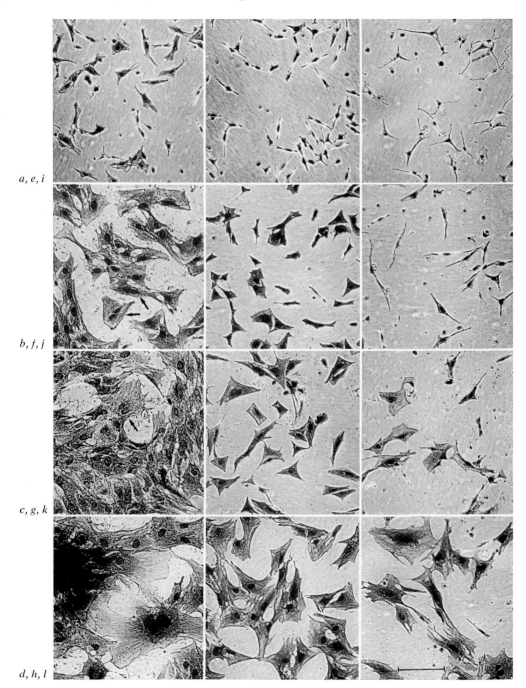

a, e, i

b, f, j

c, g, k

d, h, l

Results

Controls

The endometrial cells of rats hormonally prepared by progesterone and oestradiol in order to maximally sensitize the uterus to a deciduogenic stimulus, grew well *in vitro* when the culture medium was supplemented with 10 % calf serum, as illustrated in figure 1a–d. The cells were well attached at 24 h (fig. 1a) and spread at 48 h (fig. 1b). When plated at a density of 10^6 cells per dish, they reached confluency on the third day (fig. 1c) and began to overlap one another, indicating lack of contact inhibition. On the fourth day, this process was exaggerated and aggregating cells formed dense clusters (fig. 1d) leaving empty zones in between. The percentage of binucleated cells increased exponentially from 48 to 96 h (fig. 2). It was noticed that many cells remained flat throughout their mitotic cycle. The size of the nuclei of either the mononucleated or the binucleated cells varied in the same culture, from 48 h onwards, supposedly in connection with the polyploidization process which has been well documented *in vivo* [1, 4, 9]. The nucleoli were 1–4 in number but occasionally more numerous. In binucleated cells, the number of nucleoli was often the same in both nuclei. At the ultrastructural level, the granular component of nucleoli was more abundant than the fibrillar one, and the chromatin was almost completely dispersed.

In addition to the microtubules (25 nm diameter) (fig. 4a) and the 5 nm diameter microfilaments which were seen running along the cytoplasmic membrane (fig. 4a), radiating from anchorage zones (fig. 4b) or crossing the cytoplasm as stress fibers, the third cytoskeletal component (10 nm intermediate filaments) was also observed (fig. 4c). In zones when intermediate filaments were arranged in loose arrays as in figure 4c, polysomes could be seen associated with them through thinner filaments. In some areas, the abundance of intermediate filaments was very high and cytoplasmic organelles were absent with the exception of a few polysomes (fig. 4d). Such accumulations often occupied a large zone, most generally located close to the nucleus.

Effect of Indomethacin

Cells as in controls, but incubated with indomethacin, differed morphologically at 24 h from the control ones (fig. 1e). They were spindle-shaped and their cytoplasm, with the exception of two thin processes, was mostly compacted around the nucleus. However, at 48 h, they were

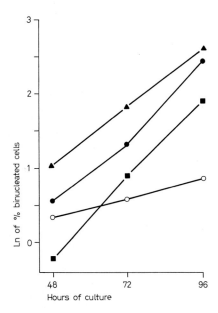

Fig. 2. Effect of progesterone and indomethacin on the percentage of binucle-ated decidual cells *in vitro*. The growth medium was supplemented with: 10% extracted calf serum (■); 10% extracted calf serum and 0.5 μM progesterone (▲); 10% calf serum and 0.5 μM progesterone (●); 10% calf serum, 0.5 μM progeste-rone and 270 μM indomethacin (○). Each value was the mean of more than 1,000 cells.

flattened (fig. 1f) and at 72 h they frequently exhibited triangular or rectangular configurations (fig. 1g). At 96 h (fig. 1h), the cells looked like the controls at 48 h. Under the electron microscope, hollow rod-like structures, 50 nm in diameter and about 250 nm in length, were observed, enclosed in lysosome-like vesicles.

As shown in figure 2, the rate of binucleation was about 4 times less than in controls. This effect of indomethacin on the percentage of bi-nucleated cells is reversible and depends on the continuous presence of the drug. In the experiment illustrated in figure 3, the percentage of binucleated cells was 5% for cultures grown for 4 days with indo-methacin whereas it was 10% when cells were incubated only for 1 day with indomethacin, then 3 days without. In this experiment, the bi-nucleated cells in the control culture grown with 20% calf serum could not be counted due to the packing of cells and the 15% value obtained with only 10% serum is probably an underestimated control level.

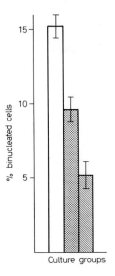

Fig. 3. Effect of the duration of indomethacin treatment on the percentage of binucleated decidual cells at 96 h. Cells isolated in DME containing 20 % heat-inactivated calf serum and progesterone. Cells from control rats were grown in DME with 10 % calf serum (open). Cells from indomethacin-treated rats were isolated in the presence of indomethacin, cultured in medium supplemented with 20 % calf serum containing indomethacin (270 μM) for 24 h (stippled) or 96 h (hatched). Each value was the mean of more than 1,000 cells.

Effect of Progesterone in vivo *and* in vitro

The cells from spayed rats treated by progesterone alone (without oestradiol) differentiated fully *in vitro,* whereas those from spayed rats without hormonal therapy remained fibroblastic [11]. The endometrial cells of rats treated with 2 progesterone injections and cultured for 4 days, although growing less actively than those from the 4-day, progesterone-treated rats, differentiated, but the percentage of binucleated cells was

Fig. 4. The cytoskeleton in decidual cells in culture. *a* Microfilaments and microtubules. *b* Microfilaments radiating from anchorage zones. *c* Dispersed intermediate filaments. *d* Accumulation of intermediate filaments. Electron micrographs of cells fixed in O_sO_4 (b) or glutaraldehyde and O_sO_4 (a, c, d) at the same magnification. Bar = 500 nm. If = Intermediate filaments; L = lipid droplets; M = mitochondria; Mf = microfilaments; Mt = microtubules; p = polysomes.

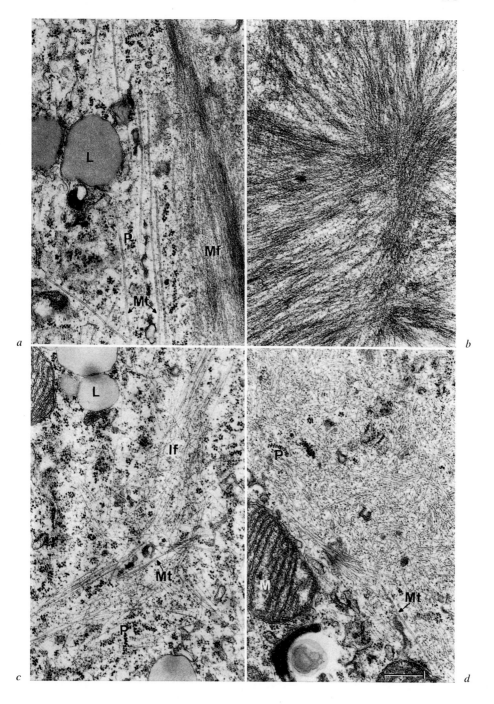

only half that of the control level. We have especially noticed that the survival of cells to the isolation procedure and subsequent anchorage after plating is related to the length of the *in vivo* progesterone treatment.

In order to assess the effect of progesterone *in vitro,* endometrial cells of rats treated by progesterone and oestradiol were grown for 4 days in various conditions. No difference in the rate of binucleation was observed with medium supplemented with 10 % calf serum and progesterone or with medium supplemented with 10 % extracted calf serum with or without progesterone (fig. 2). When plated in serumfree medium, the attached cells exhibited at 24 h a characteristic aspect (fig. 1i) with 2 or 3 slender cytoplasmic processes. Cultured afterwards in DME containing 0.1 % extracted calf serum, these cells remained poorly spread at 48 h (fig. 1j). However, when 10 % calf serum with progesterone was added in the medium from 48 h onwards, the spreading of cells was markedly accentuated (fig. 1k) and was nearly comparable at 96 h (fig. 1l) to the 48-hour control (fig. 1b). The aspect was similar when the medium was supplemented at 48 h with 10 % extracted calf serum with or without progesterone.

Discussion

Although progesterone seems obligatory *in vivo* for maintenance of deciduoma [3], we have observed that progesterone was not necessary for growth and differentiation *in vitro* when rat endometrial cells were grown in medium supplemented with 10 % calf serum (fig. 2); a similar observation was reported by *Vladimirsky et al.* [13]. Since the rate of binucleation was comparable with normal and extracted serum, other factors that are removed during extraction do not seem also indispensable. However, a need for progesterone and/or serum factors has to be considered, owing to the lack of growth in serumfree or low serum concentration medium (fig. 1).

Indomethacin inhibits decidualization *in vivo* [5, 7, 8, 10, 12] and it is noticeable that a similar effect is obtained *in vitro.* The *in vivo* inhibition is dose-dependent [8] and the incomplete effect described in the present study could be attributed to the insufficient availability of indomethacin (low dose, metabolism), unless prostaglandins are not the sole factor involved in triggering the decidual differentiation.

In one experiment (fig. 3), indomethacin was given to the rats and was present in medium during the isolation procedure in order to prevent prostaglandin synthesis. In these conditions, however, there was not more inhibition of the decidual differentiation than when cells were submitted to the drug only at the time of plating (fig. 2). Furthermore, the effect of indomethacin was reversible since the percentage of binucleated cells grown with indomethacin for only 24 h was twice that of those grown with indomethacin for 96 h (fig. 3). The morphology of cells depends on their cytoskeleton. Therefore, the particular aspect at 24 h of cells grown with indomethacin (fig. 1e) suggests an effect on this cellular component.

Microtubules and microfilaments are identical in all cells. However, the intermediate filaments seem to be specific of the cell type [6]. The nature of intermediate filaments in decidual cells (possibly vimentin) has not yet been characterized. It is likely that the large accumulations of these filaments in decidual cells reflect disorders of the cytoskeleton. Such an alteration, impeding the normal functioning of the cell, could be responsible for the polyploidization and the lack of cytoplasmic division at the end of mitosis leading to the formation of binucleated cells. The fact that cells remained flat during mitosis substantiate such an assertion [2].

Acknowledgements

We wish to thank *Monique Gouézou* and *Francine Delahaye* for skillful technical help, *Christiane Barrier* and *Jean-Claude Lambert* for illustrations and *Françoise Boussac* for typing the manuscript.

References

1   Ansell, J. D.; Barlow, P.,W., and McLaren, A.: Binucleate and polyploid cells in the decidua of the mouse. J. Embryol. exp. Morph. *31:* 223–227 (1974).

2   Brecher, S.: The occurrence and possible role of 80–100 Å filaments in PtK1 cells. Expl Cell Res. *96:* 303–310 (1975).

3   De Feo, V.: Decidualization; in Cellular biology of the uterus, pp. 192–290 Appleton, New York 1967).

4   Dupont, H.; Sartor, P.; Dupont, M. A.; Duluc, A. J. et Mayer, G.: Evolution du déciduome expérimental chez la ratte: filiation cellulaire et morphogénèse. Biol. Cell. *32:* 215–222 (1978).

5   Hoffman, L. H.; Strong, G. B.; Davenport, G. R., and Frölich, J. C.: Deciduogenic effect of prostaglandins in the pseudopregnant rabbit. J. Reprod. Fertil. *50:* 321–237 (1977).

6   Lazarides, E.: Intermediate filaments as mechanical integrators of cellular space. Nature, Lond. *283:* 249–256 (1980).

7   Kennedy, T. G.: Evidence for a role for prostaglandins in the initiation of blastocyst implantation in the rat. Biol. Reprod. *16:* 286–291 (1977).

8   Le Goascogne, C.; Sananès, N., and Baulieu, E. E.: Prostaglandins and decidualization; in Prostaglandins and Reproductive Physiology. Série Colloque Inserm 1980 (in press). Int. Symp., Montpellier 1979.

9   Leroy, F.; Bogaert, C.; Hoeck, J. Van, and Delcroix, C.: Cytophotometric and autoradiographic evaluation of cell kinetics in decidual growth in rats. J. Reprod. Fertil. *38:* 441–449 (1974).

10 Sananès, N.; Baulieu, E. E., and Le Goascogne, C.: Prostaglandin(s) as inductive factor of decidualization in the rat uterus. Mol. cell. Endocrinol. *6:* 153–158 (1976).

11 Sananès, N.; Weiller, S.; Baulieu, E. E., and Le Goascogne, C.: *In vitro* decidualization of rat endometrial cells. Endocrinology *103:* 86–95 (1978).

12 Tobert, J. A.: A study of the possible role of prostaglandins in decidualization using a non surgical method for the instillation of fluids into the rat uterine lumen. J. Reprod. Fertil. *47:* 391–393 (1976).

13 Vladimirsky, F.; Chen, L.; Amsterdam, A.; Zor, U., and Lindner, H. R.: Differentiation of decidual cells in cultures of rat endometrium. J. Reprod. Fertil. *49:* 61–68 (1977).

N. Sananès, Ph. D., Unité de recherches sur le métabolisme moléculaire et la physio-pathologie des stéroïdes de l'INSERM, Hôpital de Bicêtre, 78, avenue du Général-Leclerc, F-94270 Bicêtre (France)

Prog. reprod. Biol., vol. 7, pp. 135–142 (Karger, Basel 1980)

Immunological Aspects of the Decidual Cell Reaction[1]

O. Bernard and F. Rachman

Clinique de pédiatrie et unité de recherches d'hépatologie infantile, INSERM U-56, Hôpital de Bicêtre, Kremlin-Bicêtre

Introduction

In rats and mice, the decidua develops from uterine stromal cells and completely surrounds the post-implantation embryo at each implantation site. Although the endocrine and cellular aspects of decidualization have been extensively studied, little is known of its exact role. Several hypotheses have been advanced, including: nutrition of the embryo, protection of the embryo against immunological rejection by the mother, and protection of maternal tissue from excessive invasion by trophoblast. Some data suggest that the decidua could indeed act as a buffer zone contributing to the balance between the embryo and maternal tissues in the post-implantation period.

In this study, the data concerning the possible immunological role of the decidua will be reviewed briefly and results on some immunological aspects of mouse decidua cells presented.

Possible Immunological Role of the Decidua

The mammalian embryo possesses histocompatibility genes of paternal and maternal origin. These genes are expressed as histocompatibility antigens on the surface of all nucleated cells. Thus, in outbred pregnancies, the embryo should be rejected by the pregnant female as an allograft of paternal tissue. Several experimental results and hypotheses have been

[1] Supported by INSERM CRL 78.5.022.4 and UER Kremlin-Bicêtre 743.

presented to date to explain why most embryos can implant in the uterus and successfully develop to term in spite of a known maternal immune response [for review, see 5, 18]. As far as the post-implantation period is concerned, several sets of data demonstrate that, in the mouse, the trophoblast, which completely surrounds the embryo, does not express significant amounts of major histocompatibility antigens [19, 20]. However, these antigens are present on other cells in the peri-implantation embryo [19, 24] and they are later detectable on some trophoblast cells in the placenta [9, 21]. On the other hand, maternal 'blocking' antibodies directed against embryonic or paternal antigens could play a protective role, since maternal immunoglobulins are in close contact with the embryo from the time of implantation [6, 7]. In addition, the uterus, in which a decidual reaction occurs, may possess some features of an immunologically privileged site.

In fact, the immunological role of the decidua in mice and rats can be considered from two points of view: (1) protection of the embryo, and (2) protection of the pregnant female.

Protection of the Embryo

Several experiments suggest that the pregnant uterus may protect the embryo from rejection as an allograft of paternal tissue. Thus, mouse blastocysts transplanted to the uterus of females hyperimmunized against the donor strain can implant and develop, provided a state of pseudopregnancy has been induced in the recipient female. Blastocysts transplanted to the kidney capsule of similarly immunized females, do not develop. Decidua cells could therefore be responsible, at least in part, for the protection afforded to the blastocyst transplanted to the uterus [15].

This was further documented in the rat, using a different model: syngenic skin grafts or suspension of epidermal cells are able to 'implant' and grow when transplanted to the uterus, if the recipient female is treated with estrogens [3]. The rejection of 'implanted' intrauterine skin *allografts* is significantly delayed if the recipient female is pseudopregnant or is in the preimplantation stage of pregnancy and if a decidual reaction develops [4]. If no decidual reaction occurs, the graft is rejected after a few days as would be any allograft in any other site; control experiments excluded the possible role of systemic immunosuppressive hormones released during pregnancy. However, the protection presumably afforded by this decidual reaction cannot be elicited when the recipient females are hyperimmunized against the donor's strain histocompatibility antigens.

From these data it can be concluded that the early post-implantation decidua is probably one among several factors which prevent the embryo from immunological rejection by the mother. However, it is likely that this protection remains limited, since it occurs in immunized females only when the grafted tissue does not express large amounts of major histocompatibility antigens [4, 15]. The mechanism of this protection is still unknown: the decidua could prevent antigens of the embryo from reaching the maternal circulation via the lymphatic route; antigens using the venous route would induce formation of protective 'blocking' antibodies rather than cytotoxic reactions [2]; alternatively, production of immunosuppressive factors by the decidua could contribute to regulation of the immune response of the pregnant female [10].

Protection of Maternal Tissues

It is likely that the decidua plays a role in protecting maternal tissues from excessive invasion by the implanting blastocyst. At the time of implantation, mouse trophoblast cells possess major invasive and phagocytic properties, as has been shown in experiments in which mouse blastocysts are transplanted to the spleen [12]. In one experiment, up to one-third of the spleen was destroyed in a few days by the trophoblast of a single blastocyst [14]. Similarly, when mouse ectoplacental cones are transplanted to the uteri of ovariectomized females in which no decidual reaction can occur, trophoblast cells invade the uterine tissue and sometimes reach the myometrium. In pseudopregnant mice however, transplantation of ectoplacental cones evokes a decidual reaction, and the trophoblast does not invade the uterus but rather fails to develop [14].

Thus, considering the possible immunological roles of the decidua, it should be of interest to define immunologically the cell populations which are present in the early decidua, and to determine whether some are lymphocytes able either to suppress some component of the mother's immune response (suppressor T lymphocytes), or to defend maternal tissue against trophoblast cells.

Immunological Studies of Mouse Decidua Cells

The presence of lymphoid cells among decidual cells derived from the uterine stroma has been observed in the peri-implantation period in the mouse and at later stages in the rat [16, 22] and in human pregnancy

[23]. An analysis of the cell populations present in the mouse decidua in the days after implantation was, therefore, undertaken in an attempt to characterize cells that could be of immunological significance [8].

Mouse decidua cells were obtained following inbred and outbred matings of females pregnant for the first time. After removal of the embryo and remnant of uterine epithelium, 6- to 9-day decidua were treated with collagenase and trypsin to obtain a cell suspension. Deciduomas were induced in pseudopregnant females, treated with the same enzymes, and the properties of deciduoma cells were compared with those of decidua cells. Various membrane markers known to be present on the lymphocyte surface were assessed: (1) major histocompatibility antigens (H-2), Thy-1 antigen and four Lyt antigens were studied using specific antisera in complement-dependent cytotoxicity tests; (2) surface immunoglobulins (sIg) were studied using the direct immunoperoxidase technique and (3) receptors for the Fc portion of immunoglobulin G (FcR) were studied with a rosette technique using antibody-coated sheep erythrocytes (EA). In addition, the distribution of maternal immunoglobulins was studied on sections of decidua or deciduoma from day 6 to day 9 of pregnancy or pseudopregnancy using the direct immunoperoxidase technique.

Cell Surface Antigens on Decidua Cells

H-2 antigens are present on most decidua and deciduoma cells. Thy-1 antigen, known to be present on thymocytes and T lymphocytes, but also on fibroblasts in culture, brain cells epidermis, mammary tissue and early muscle cells in culture [8] can be detected on both decidua and deciduoma cells in similar amounts. On day 6 of pregnancy or pseudopregnancy, 60 % of decidua or deciduoma cells express Thy-1 antigen. Since it could be suggested that at least part of this Thy-1-bearing cell population is made up of T lymphocytes, membrane antigens more specifically characteristic of T lymphocytes were looked for on decidua cells: antisera specific for four different antigens (Lyt 1 to 4) failed to show any reactivity with decidua cells from mice bearing the proper genotype. It is therefore likely that the presence of Thy-1 antigen on decidua and deciduoma cells reflects their fibroblastic origin. Most decidua cells arise from uterine stromal cells in the peri-implantation period. As these fibroblast cells differentiate into decidual cells they may keep some features of their stromal origin and retain Thy-1 antigen.

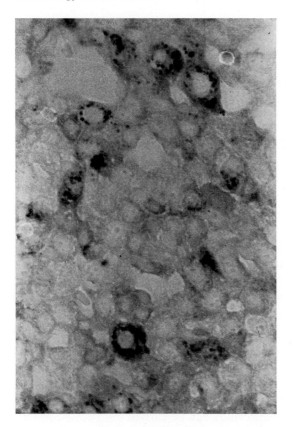

Fig. 1. Immunoglobulin-containing cells in a 7-day mouse decidua. Decidua capsules were fixed in 4 % paraformaldehyde and embedded in polyethylene-glycol 1000. 5-μm sections were incubated with peroxidase-labeled sheep Fab anti-mouse immunoglobulin and peroxidase activity was detected by incubation in diamino-benzidine and H_2O_2. Immunoglobulins appear as dark cytoplasmic granules.

Immunoglobulins

When tested in suspension, using peroxidase-labeled sheep Fab anti-mouse immunoglobulin, no sIg can be detected on decidua or deciduoma cells. However, when the same technique is used on histological sections several cells in the decidua appear to contain immunoglobulin, either as a lining of the cell surface or as cytoplasmic granules (fig. 1). These are found chiefly on the edges of the decidua capsularis but can also be seen in deeper layers. These immunoglobulin-containing cells are also present in deciduoma, but are only detectable on the edges of the capsule.

Table I. Percent EA rosette-forming cells in mouse decidua and deciduoma from day 6 to day 9 of pregnancy or pseudopregnancy (mean ± SD); controls using unsensitized SRBC (E rosettes) were all negative

	Days of pregnancy or pseudopregnancy			
	6	7	8	9
Decidua	6.0 ± 1.2	10.5 ± 1.9	17.8 ± 1.9	17.0 ± 4.8
Deciduoma	5.8 ± 0.6	11.0 ± 0.5	10.7 ± 1.8	n. d.
Deciduoma with progesterone	5.0 ± 2.0	5.2 ± 2.4	11.3 ± 3.6	23.5 ± 8.5

Fc Receptors (FcR)

Studies of FcR-bearing cells were done on decidua, deciduoma cells, and on deciduoma cells from mice treated with progesterone from the time of decidualization on [1]. This treatment prevents regression of deciduoma, which otherwise takes place between day 8 and 9 of pseudopregnancy in the mouse. Table I shows that FcR-bearing cells are indeed present in the post-implantation decidua and in deciduoma. The mean percentage of FcR-bearing cells in decidua increases from 6 to 17 % from day 6 to day 8 and 9. In deciduoma this increase does not continue after day 7 and the number of FcR-bearing cells remains stable on day 8; on day 9 of pseudopregnancy no study can be carried out because of the poor viability of the cells in the deciduoma undergoing regression. However, if regression of deciduoma is prevented by progesterone treatment, the number of FcR-bearing cells on day 9 is similar to that in decidua at the same stage, although in progesterone-treated deciduoma increase of FcR-bearing cells is delayed by about 24 h.

These results suggest the following conclusions: (1) No B or T lymphocytes, as defined by sIg and four Lyt antigens respectively, are detectable in significant amounts in the post-implantation decidua. (2) The presence of Thy-1 antigen on about one-half of decidua and deciduoma cells probably reflects the stromal origin of these cells. Although the role of Thy-1 antigen in the mouse decidua is unknown, experiments showing that heterologous anti-thymocyte antiserum can prevent implantation or alter early embryonic development in mice and rats [13], suggest that the presence of Thy-1-bearing cells in the uterus at the time of implantation could be necessary for normal implantation and subsequent development of the embryo. (3) Detection of an increasing number of FcR-bearing cells in decidua and in deciduoma raises several questions: (a) it is not known whether these cells differentiate *in situ* from 'stem cells' already present in the uterus before implantation, or whether they migrate to the decidua from extrauterine sites. (b) Whatever their origin, they are probably under the control of progesterone. (c) Their role remains a matter of speculation: (i) they could simply be involved in ridding the stroma and decidua of the large amounts of immunoglobulins present in the uterine stroma at the time of implantation; (ii) they could be involved in cell-mediated, antibody-dependent cytotoxicity [17] or 'natural killing' [11] to control the growth of the trophoblast; (iii) alternatively, they could interact with maternal (blocking) antibodies and trophoblast cells to contribute to the protection of the embryo by shedding surface antigens from trophoblast cells in the days after implantation.

References

1   Atkinson, W. B.: The persistence of deciduomata in the mouse. Anat. Rec. *88:* 271–283 (1944).

2   Barker, C. F. and Billingham, R. E.: Immunologically privileged sites. Adv. Immunol. *25:* 1–54 (1977).

3   Beer, A. E. and Billingham, R. E.: Implantation, transplantation and epithelial-mesenchymal relationships in the rat uterus. J. exp. Med. *132:* 721–736 (1970).

4   Beer, A. E. and Billingham, R. E.: Host responses to intrauterine tissue, cellular and fetal allografts. J. Reprod. Fertil. *21:* suppl., pp. 59–88 (1974).

5   Beer, A. E. and Billingham, R. E.: The immunobiology of mammalian reproduction (Prentice Hall, Englewood Cliffs 1976).

6   Bernard, O.: Possible protecting role of maternal immunoglobulins on embryonic development in mammals. Immunogenetics *5:* 1–15 (1977).

7   Bernard, O.; Ripoche, M. A., and Bennett, D.: Distribution of maternal immunoglobulins in the mouse uterus and embryo in the days after implantation. J. exp. Med. *145:* 58–75 (1977).

8   Bernard, O.; Scheid, M. P.; Ripoche, M. A., and Bennett, D.: Immunological studies of mouse decidual cells. I. Membrane markers of decidual cells in the days after implantation. J. exp. Med. *148:* 580–591 (1978).

9   Chatterjee-Hasrouni, S. and Lala, P. K.: Localization of H-2 antigens on mouse trophoblast cells. J. exp. Med. *149:* 1238–1253 (1979).

10 Globerson, A.; Bauminger, S.; Abel, L., and Peleg, S.: Decidual extracts suppress antibody response *in vitro*. Eur. J. Immunol. *6:* 120–122 (1976).

11 Herberman, R. B.; Bartram, S.; Haskill, J. S.; Nunn, M.; Holden, H. T., and West, W. H.: Fc receptors on mouse effector cells mediating natural cytotoxicity against tumor cells. J. Immun. *119:* 322–326 (1977).

12 Kirby, D. R. S.: Development of the mouse blastocyst transplanted to the spleen. J. Reprod. Fertil. *5:* 1–12 (1963).

13 Kirby, D. R. S.: Inhibition of egg implantation and induction of abortion in mice by heterologous immune serum. Nature, Lond. *216:* 1220–1221 (1967).

14 Kirby, D. R. S. and Cowell, T. P.: Trophoblast-host interactions; in Fleischmajer and Billingham, Epithelial-mesenchymal interactions, pp. 64–77 (Williams & Wilkins, Baltimore 1968).

15 Kirby, D. R. S.; Billington, W. D., and James, D. A.: Transplantation of eggs to the kidney and uterus of immunised mice. Transplantation *4:* 713–718 (1966).

16 Peel, S. and Bulmer, D.: The fine structure of the rat metrial gland in relation to the origin of the granulated cells. J. Anat. *123:* 687–696 (1977).

17 Perlmann, P.; Perlmann, H., and Wigzell, H.: Lymphocyte mediated cytotoxicity *in vitro:* induction and inhibition by humoral antibody and nature of effector cells. Transplant. Rev. *13:* 91–114 (1972).

18 Scott, J. S. and Jones, W. R.: Immunology of human reproduction (Academic Press, London 1976).

19 Searle, R. F.; Sellens, M. H.; Elson, J.; Jenkinson, E. J., and Billington, W. D.: Detection of alloantigens during preimplantation development and early trophoblast differentiation in the mouse by immunoperoxidase labeling. J. exp. Med. *143:* 348–359 (1976).

20 Sellens, M. H.: Antigen expression on early mouse trophoblast. Nature, Lond. *269:* 60–61 (1977).

21 Sellens, M. H.; Jenkinson, E. J., and Billington, W. D.: Major histocompatibility complex and non-major histocompatibility complex antigens on mouse ectoplacental cone and placental trophoblastic cells. Transplantation *25:* 173–179 (1978).

22 Smith, L. J.: Metrial gland and other glycogen containing cells in the mouse uterus following mating and through implantation of the embryo. Am. J. Anat. *119:* 15–24 (1966).

23 Tekelioglu-Uysal, M.; Edwards, R. G., and Kisnisci, H. A.: Ultrastructural relationship between decidua, trophoblast and lymphocytes at the beginning of human pregnancy. J. Reprod. Fertil. *42:* 431–438 (1975).

24 Webb, C. G.; Gall, W. E., and Edelman, G. M.: Synthesis and distribution of H-2 antigens in preimplantation mouse embryos. J. exp. Med. *146:* 923–932 (1977).

O. Bernard, MD, Clinique de Pédiatrie, Hôpital d'Enfants,
78, rue du Général-Leclerc, F-94270 Kremlin-Bicêtre (France)

Prog. reprod. Biol., vol. 7, pp. 143–157 (Karger, Basel 1980)

Factors Involved in Uterine Receptivity and Refractoriness

Alexandre Psychoyos and Viviane Casimiri

Laboratoire de Physiologie de la Reproduction, ER 203, CNRS, Hôpital de Bicêtre (Bat. INSERM), Bicêtre

Introduction

Loeb [31], who discovered early in this century the traumatic induction of decidualization, was also the first to observe that this endometrial reaction is possible only during a limited period. In the rat, as shown some 50 years later, the optimal sensitization of the endometrium for a decidual response to stimuli of low intensity is limited to a period of a few hours [13, 27]. In this species the short phase of endometrial responsiveness occurs around noon of the 5th day of pregnancy or pseudopregnancy and is followed by a state of refractoriness which lasts until the end of the luteal period [13, 42, 44, 45].

Thanks to the technique of embryo transfer, these findings were extended and a general rule established for all species studied: beyond the time at which nidation is normally expected to occur, the intra-uterine survival of transferred blastocysts or ova of the pre-blastocyst stages becomes impossible. In the rat, by the end of day 5 of pregnancy or pseudopregnancy, the uterine environment become detrimental to unimplanted embryos [37, 42]. The uterine receptivity manifested on this day, is followed by a state of 'non-receptivity' in terms of deciduogenic sensitivity as well as in terms of egg survival and development [45, 46].

During the last two decades, our knowledge concerning the hormonal regulation of these functional changes of the uterus has increased considerably [13, 18, 29, 44, 45, 67]. However, serious gaps continue to exist in the correlation of these changes with specific factors involved in their induction and manifestation. What makes the endometrium highly sensitive and then insensitive to decidual stimuli? How does the uterine milieu

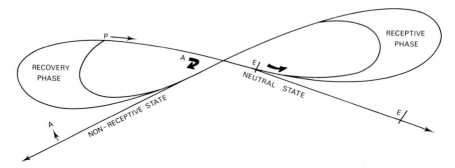

Fig. 1. Diagrammatic representation of the relation between circulating hormones (P = progesterone; E = oestrogen; A = arrest of progesterone effect) and endometrial receptivity. For explanations see text.

influence the metabolic activity of the blastocyst during the conditions of delayed implantation or at the receptive phase? Why and how does this milieu become hostile to blastocysts and ova of the pre-blastocyst stages? Among others, these fundamental questions remain mostly unanswered for the time being.

Hormonal Sequence Which Controls Uterine Receptivity

We are now able to explain the hormonal interplay which makes possible, at least in the rat and mouse, the strictly timed changes in uterine receptivity. In these species it is now well established that the same basic hormonal sequence, i. e. a small amount of oestrogen after 48 h of progesterone, initiates a receptive phase and leads by its biphasic effect, to the establishment of the 'non-receptive' state [42, 44, 45]. The available information and the model proposed for the relationships between hormonally induced events and the changes in endometrial receptivity for nidation are summarized in figure 1. The sequence begins at the point designated as P which represents the initiation of progesterone-dependent events leading after some 48 h to a pre-receptive (neutral) state. The uterus can be maintained indefinitely in this state by a regimen of progesterone alone. If at any time beyond the establishment of this pre-receptive state, oestrogen administration E, is superimposed on the progesterone regimen, a second series of uterine events is initiated which leads to the establishment of the 'non-receptive' state. Midway between E and the occurrence of this latter state, the uterus for a few hours manifests a 're-ceptive' phase. Continuous administration of progesterone permits the maintenance of the 'non-receptive' state permanently. Progesterone with-drawal for a minimal period of 48 h at any time when the uterus is in the pre-receptive or the non-receptive state returns the uterus to a con-dition where the administration of progesterone leads to the re-establish-ment of the pre-receptive (neutral) state. In figure 1, the arrest of the progesterone regimen is designated as A and the interval of return of the uterus to a condition permitting the renewal of its progestational poten-tial is referred to as the recovery phase. The indefinite nature of the duration of the pre-receptive (neutral) and non-receptive states is indi-cated by lines projecting from the cyclic strip [45].

The hormone-dependant molecular events responsible for the uterine changes described above have been the subject of numerous recent studies [19, 28, 29] reviewed by *Dr. Leroy and Dr. Glasser* in this meeting. In terms of practical interest, one should mention that, when the hormonal sequence establishing the 'non-receptive' state can be assured earlier than the normal timing, egg implantation is blocked via an ovo-endometrial asynchrony. In fact, a pre-ovulatory treatment with small amounts of progesterone desynchronizes the ovo-endometrial relationships, and the fertilized ova, when they enter the uterine cavity, face a hostile environ-ment. Such a phenomenon was shown to occur in the rat [41, 45, 67] as well as in the rabbit [4, 11, 34]. There is evidence that it may also be the case in the human, explaining the anti-nidatory effect of the low-dose progestagen oral therapy or of the progesterone releasing IUDs used in contraception [46].

Involvement of Vasoactive Mediators

Attempting to analyze the role of the ovarian hormones upon the onset, magnitude and loss of uterine sensitivity towards deciduogenic stimuli, particular emphasis should be given to uterine vascular changes. It is well known that either the natural decidualization due to the blasto-cyst [40], or the experimentally induced decidual reaction [39] is preceded by a dramatic increase in the permeability of the endometrial capillaries. This vascular change appears to be a *si ne qua non* condition for a de-cidual reaction [39, 43]. Any procedure inhibiting this increase in perme-ability inhibits also the decidual response. Parallely, the *only* period

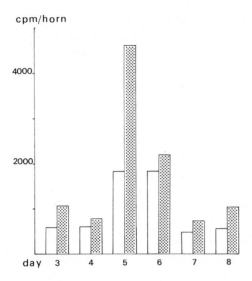

Fig. 2. Comparison of the changes in blood volume and vascular permeability of the uterus, measured by the intravenous injection of ^{131}I-serum albumin, after endometrial scratch trauma applied on different days of pseudo-pregnancy in the rat. Empty bars: [131]I radioactivity corresponding to the blood volume; dotted bars: [131]I radioactivity corresponding to vascular leakage of ^{131}I-serum albumin. From data of *Bitton-Casimiri et al.* [5].

during which the endometrium is able to exhibit a change in vascular permeability in response to a traumatic stimulation is also the *only* period during which endometrial injury leads to deciduogenesis. As we showed several years ago [5, 42], by measuring the vascular leakage of ^{131}I-serum albumin, the degree of increase in capillary permeability, observed after traumatic stimulation, parallels the endometrial sensitivity for a post-trauma decidual reaction (fig. 2). Obviously, information concerning the modulation by hormones or other factors of uterine micro-circulation appears essential for the understanding of the manifestation and loss of uterine receptivity for nidation.

Schayer [53, 54] proposed as an intrinsic regulator system for micro-circulation in general, the induction of histidine decarboxylase in or near the endothelium of small blood vessels. Histamine may thus be formed *in situ* as a consequence of the activation of this enzyme by various agents. As it is known, histamine has been proposed by *Shelesnyak* [55] and *Marcus and Shelesnyak* [32] to be the main tissular mediator responsible

for the decidual reaction. The actual concepts on this possibility are discussed by Dr. *Brandon* in this meeting. Histamine and its receptors appear
in fact involved in the mediation of some oestrogen-dependent phenomena,
as well as on the local vascular response to the presence of a blastocyst.
The systemic administration of a combination of antagonists to histamine
H_1 and histamine H_2 receptors inhibit the oedematous response to oestradiol [6], as well as the stromal oedema characteristic of the receptive
phase in the pregnant rat uterus [8]. This same treatment, by inhibiting
vascular permeability, also reduces the number and intensity of the 'blue'
reaction to blastocysts [8]. Therefore, in the pathway leading to endometrial capillary reactivity, a primary and essential step may be the *de
novo* synthesis of histamine due to the activation of the histidine decarboxylase system via oestrogens and/or other inducers. In the rabbit, the
intra-luminal injection of a specific inhibitor of histidine decarboxylase
on day 5 of pregnancy interrupts the implantation process [14]. Furthermore, in this species, the blastocyst itself exhibits a significant histidine
decarboxylase activity which appears to reflect its own capacity to produce histamine [14]. In any case, in agreement with the involvement of
histidine decarboxylase in endometrial capillary reactivity is also the fact
that the 2–3 h lag time required for the manifestation of a local increase
in capillary permeability after a deciduogenic stimulus, is the same as
that necessary for the induction of this enzymatic activity in vascular
endothelia [45]. A relationship between deciduogenic inducers and activation of the histidine decarboxylase system in small uterine vessels
could possibly be involved in the mechanism of the periarteriolar decidualization of the human endometrium, which occurs in every normal
cycle 'spontaneously', i. e. independently of the presence of a blastocyst.

During the inflammatory response the vascular leakage induced by
mediators such as histamine is potentiated by prostaglandin generation
[61]. The involvement of prostaglandins in the initiation of blastocyst implantation is the subject of extensive studies [24, 25, 60], discussed in
this meeting by Dr. *Kennedy*. The production of prostaglandins is found
to be maximum on the day of optimal endometrial sensitivity [17]. The
agonistic interaction between prostaglandins and histamine in the oestrogen induction of nidation is shown clearly in the results of *Saksena et al.*
[52]: in the progesterone-dominated uteri of mice exhibiting delayed implantation, indomethacin, which inhibits prostaglandin synthesis, also
inhibits the oestrogen induction of nidation; this inhibition can be counter-
acted by the association of prostaglandins and histamine.

In the inflammatory reaction model, another example of the synergistic interactions of prostaglandins with vasoactive mediators is offered by prostaglandins and bradykinin. Prostaglandin generation potentiates the positive effect of bradykinin on vascular permeability, while bradykinin stimulates the synthesis and release of prostaglandins by activating phospholipase A_2 [61]. Bradykinin is also a highly potent deciduogenic inducer [43].

Changes in the generation, storage and release of vasoactive mediators and prostaglandins may certainly be involved in the differences in vascular and decidual response observed in receptive and non-receptive endometria. However, considering the complexity of a common inflammatory vascular reaction, a multitude of additional factors is expected to intervene in response to a deciduogenic stimulus.

Hormone-Dependent Enzymatic Activities and Uterine Receptivity

Alloiteau [1] observed that the presence of eosinophile leucocytes announces the endometrial potentiality for deciduoma formation. Eosinophile leucocytes are abundant in the rat at pro-oestrus and oestrus but also on day 5 of pseudo-pregnancy [30, 58, 59]. They can be considered as target cells for oestrogens since they appear to contain specific receptors for oestradiol [58, 59]. They manifest a strong peroxidase activity which, as shown by *Klebanoff* [26], catalyses the inactivation of oestradiol in the presence of hydrogen peroxide. The activity of peroxidase is enhanced in the receptive rat endometrium [*Rath and Psychoyos,* unpublished data]. This enhancement seems to be due not only to the presence of eosinophile leucocytes but also to a *de novo* synthesis of this enzyme induced by oestradiol in the uterus [2, 3, 9, 23, 35]. Various biochemical processes which may involve hydrogen peroxide may be also linked to this enzymatic activity, and its precise role, in uterine receptivity certainly deserves further investigation.

Collagen disintegration in the inter-cellular stromal space constantly accompanies the edification of the receptive endometrium [15], and lysosomal acid cathepsin has been suggested to be responsible for this change [65]. In fact, an increase in the activity of this enzyme has been shown to occur in the rat endometrium on days 4–7 of pseudo-pregnancy [65] (fig. 3). However, during the nidation process this activity appears to decrease, contrary to other lysosomal enzyme activities which increase [36].

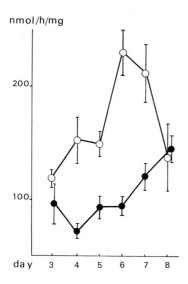

nmol/h/mg

Fig. 3. Changes in activity of cathepsin (◯), and ribonuclease (●), in the rat endometrium during pseudo-pregnancy. The enzymatic activity is expressed in nanomoles of substrate hydrolysed per hour and milligram of endometrial tissue. From data of *Wood and Psychoyos* [65].

Catecholamines and their receptors may also modulate the uterine blood flow and vascular functions. There are marked variations in the catecholamine content of the uterus under the influence of the ovarian hormones [12, 50, 66]. Monoamine oxidase (MAO), the enzyme complex involved in their deactivation, shows an increased activity during the luteal phase in the human endometrium [56]. In the rat [49], the endometrial activity of this enzyme increases sharply on day 3 of pseudo-pregnancy remaining from then on at a plateau of high level (fig. 4). Progesterone is able to increase this activity even *in vitro,* when added to the culture medium of rat uterine epithelium [10].

It has been reported recently [38] that iproniazid, an inhibitor of monoamine oxidase, is able to replace oestrogen and induce implantation in ovariectomized progesterone-treated rats exhibiting delay. However, we were unable to confirm this information by using either this compound or other more specific MAO inhibitors. Several of the effects of catecholamines are attributed to a stimulation of cAMP synthesis. The administration of oestradiol induces an increase in the uterine level of cAMP which could be mediated by catecholamines and their receptors. In general, *α-*

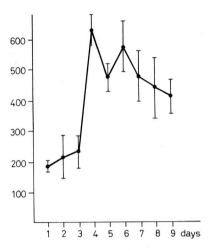

Fig. 4. Changes in activity of monoamine oxidase in the rat endometrium during pseudo-pregnancy. Enzyme activity was measured using tryptamine-^{14}C-bisuccinate as substrate. Data from *Rath et al.* [49].

adrenergic effects are considered to be mediated by a fall, and β-adrenergic effects by an increase in the level of cAMP. Stimulation of α-adrenoreceptors has been found to inhibit the increase in uterine wet weight observed after administration of oestrogen [7]. In any case, cAMP appears to precipitate implantation in mice maintained in experimental delay [62] or to induce the synthesis of specific uterine luminal proteins [57]. These findings favour the possibility that one of the main steps in oestrogen action is mediated by this nucleotide.

Rapidly growing tissues show an enhanced synthesis and accumulation of polyamines such as putrescine and spermidine. In the uterus, following the administration of oestradiol, the polyamine levels rise rapidly [22, 51]. Oestrogens stimulate the activity of the two enzymes involved in the biosynthesis of putrescine and spermidine, namely ornithine decarboxylase and adenosyl-methionine decarboxylase [22, 51]. Ornithine decarboxylase shows a high activity in the pregnant uterus since day 3. In inter-implantation tissue, this activity falls sharply on day 6, remaining at a low plateau from then on [20, 21]. It has also been shown that progesterone-induced accumulation of endometrial glycogen requires the synthesis of spermidine which, in cultured endometrial fragments, is stimulated by this hormone [16]. Polyamines may therefore play an important role in the preparation and manifestation of endometrial receptivity. Among the multiple effects they exert during cellular proliferation and differentiation they appear to be also involved in the stimulation of RNA synthesis, RNA stabilization and inhibition of its degradation [22]. Considering RNA degradation, one should note that in the rat, by day 6 of pseudo-pregnancy, the endometrial activity of ribonuclease rises sharply [65] (fig. 3). Such an enzymatic pattern, increasing during uterine refractoriness, can be considered as an exception. During the refractory period other enzymatic activities fall or remain at a high or a low plateau. Interestingly, in the rat uterus, the oestradiol receptors has been suggested to function as an inhibitor of ribonuclease [68]. It is therefore noteworthy that the activity of ribonuclease is enhanced during the refractory period where the endometrial content in oestrogen receptor plateaus at low values [33].

It is evident that the establishment of a model concerning the factors involved in uterine receptivity and refractoriness becomes more and more complex. Other parts of the puzzle defined as 'receptive uterus' may concern changes of the cell surface, ways in which cells communicate in epithelio-mesenchymal and blastocyst-epithelium relationships. Changes in the components of the inter-cellular and intra-luminal fluids may also play an essential role, which is examined by others in this meeting. Other parts of the same puzzle may concern the existence of factors inhibiting blastocyst metabolic activities and survival.

Uterine Inhibitors of Blastocyst Metabolic Activity and Survival

Direct evidence for the existence in uterine fluid of an inhibitor of blastocyst metabolic activity was offered by our results in the rat [45, 48] and those of *Weitlauf* [63, 64] in the mouse. In our experiments, we have studied the inhibitory activity of uterine flushings on the incorporation *in vitro* of ^{3}H-uridine into blastocyst RNA. Flushings were obtained from animals in various hormonal conditions. Whether they came from uteri of ovariectomized animals, either untreated or treated with progesterone, or from animals sacrificed on one of the first 4 days of pregnancy, similar results were obtained. All culture media containing uterine flushings inhibited both the total uptake of uridine and the incorporation of label into the TCA-insoluble fraction to values about 50% of control. Flushings from the uteri of day 5 pseudo-pregnant rats had a higher inhibitory activity than those of day 4, whereas the flushings from days 6 and 7 of pseudo-

Table I. In vitro development of rat blastocysts cultured for 24 h in medium containing uterine flushings collected on day 2 (F2), or day 6 (F6) of pseudo-pregnancy, or day 6 flushings treated by dialysis or by fractionation on Sephadex G-25/G-10 column[1]

Supplement	Number of blastocysts	% normal	% degenerated
None	117	100	0
F2	38	100	0
F6	45	0	100
F6 dialysed (A)	22	100	0
F6 dialysate (B)	30	0	100
A+B	20	0	100
F6 fraction I	37	100	0
F6 fraction II	30	100	0
F6 fraction III	36	70	0
F6 fraction IV	34	0	100
F6 fraction V	30	90	0
F6 fraction VI	30	100	0

[1] Data from *Psychoyos and Casimiri* [47].

pregnancy inhibited completely the uridine uptake by blastocysts, indicating the presence in these flushings of a toxic factor. In fact, our more recent data [47] show clearly that at least in the rat, uterine flushings collected on day 6 of pseudo-pregnancy exhibit *in vitro* a strong blastotoxic effect. Within 24 h of culture in such flushings all blastocysts degenerate, whereas under the same conditions all blastocysts cultured in flushings collected on day 2 of pseudo-pregnancy exhibit the usual behaviour for the *in vitro* conditions (hatching, attachment and outgrowth). Treatment by dialysis completely abolishes the toxic effect of day 6 flushings. The blastotoxic component is found restricted to the dialysate.

In further experiments, the dialysate of flushings from day 6 uteri was fractionated on a Sephadex G-25/10 column with glass-distilled water as eluant. As shown in table I, the effective substance appears to elute in fraction IV. All blastocysts cultured in medium containing this fraction have degenerated within 24 h of culture. None of the other fractions exhibited such an effect within this time lapse. Further characterization of the blastotoxic component of fraction IV is under way. However, based on the absorbance profile of this fraction and its time of elution, we believe that it might be a polypeptide of a molecular weight of about 700 daltons.

Weitlauf [64], studying in mouse uterine fluid the factors that inhibit the incorporation of ^{3}H-uridine by blastocysts *in vitro,* found after gel filtration on a G-25 column of fluid flushed from the uteri of 'delayed implanting' and 'implanting' mice, several fractions that reduced this incorporation. However, he noticed more inhibitory activity in the void volume fraction of flushings from 'delayed implanting' animals and suggested that this factor may be responsible for the metabolic dormancy of embryos during the diapause associated with delayed implantation. Whether this factor is related to the blastotoxic substance we isolated in the fluid flushed from 'non-receptive' uteri in the rat, remains to be clarified. We proposed [47], until more information is obtained about the chemical structure and biological properties of this toxic substance, to call it *blastocidin.*

The way in which blastocidin affects blastocyst viability *in vitro,* and its possible implication in rendering *in vivo* the 'non-receptive' uterus hostile to unimplanted embryos remains also to be determined.

Conclusions

We do not yet have sufficient information to speculate on the precise events via which the hormonal regulation of uterine receptivity and refractoriness for nidation is exerted. However, some important factors are beginning to appear. They may play a role in the endometrial vascular reactivity by regulating the generation, storage and release of vasoactive mediators. They may involve enzymatic activities related to the metabolism of catecholamines, cyclic nucleotides, and polyamines. They may also control the metabolic activity and survival of the embryo via substances such as blastocidin, which we found in the uterus during the refractory state for nidation.

References

1   Alloiteau, J. J.: Déciduome chez la ratte cyclique ou en début de gestation malgré un traitement œstrogénique dans les jours précédant le traumatisme. Importance de la préparation œstrogénique dans cette résistance à l'œstradiol. C. r. Séanc. Soc. Biol. *157:* 1204–1207 (1963).

2   Anderson, W. A.; Kang, Y., and Desombre, E. R.: Endogenous peroxidase: spe-
    cific marker enzyme for tissues displaying growth dependency on estrogen. J.
    Cell Biol. *64:* 668–681 (1975).
3   Anderson, W. A.; Desombre, E. R., and Kang, Y.: Estrogen-progesterone an-
    tagonism with respect to specific marker protein synthesis and growth by the
    uterine endometrium. Biol. Reprod. *16:* 409–419 (1977).
4   Beier, H. M. and Mootz, U.: Significance of maternal uterine proteins in the
    establishment of pregnancy; in Maternal recognition of pregnancy; Ciba Fdn
    Ser. 64, pp. 111–140 (Excerpta Medica, Amsterdam 1979).
5   Bitton-Casimiri, V.; Vassent, G. et Psychoyos, A.: Réponse vasculaire de l'uté-
    rus au traumatisme au cours de la pseudogestation chez la ratte. C. r. hebd.
    Séanc. Acad. Sci., Paris *261:* 3474–3477 (1965).
6   Brandon, J. M.: Inhibition of the acute oedematous response to oestradiol in
    the immature rat uterus by administration of a combination of mepyramine, a
    histamine H_1-, and burimamide, a histamine H_2-receptor antagonist. J. Endocr.
    73: 42–43 (1977).
7   Brandon, J. M.: Catecholamines and the oedematous reaction to oestradiol in
    the rat uterus. Eur. J. Pharmacol. *47:* 235–238 (1978).
8   Brandon, J. M. and Wallis, R. M.: Effect of mepyramine, a histamine H_1- and
    burimamide, a histamine H_2-receptor antagonist, on ovum implantation in the
    rat. J. Reprod. Fertil. *50:* 251–254 (1977).
9   Brokelmann, J. and Fawcett, D. W.: The localization of endogenous peroxidase
    in the rat uterus and its induction by estradiol. Biol. Reprod. *1:* 59–71 (1969).
10 Casimiri, V.; Rath, N. C.; Parvez, H., and Psychoyos, A.: Effect of sex steroids
    on rat endometrial epithelium and stroma cultured separately. J. Steroid Bio-
    chem. *12:* 293–298 (1980).
11 Chang, M. C.: Fertilization, transportation and degeneration of the egg in
    pseudopregnant or progesterone-treated rabbits. Endocrinology *84:* 356–361
    (1969).
12 Clark, K. E.; Baker, H. A.; Bhatnagar, R.; Van Orden, D. E., and Brody, M. J.:
    Prevention of estrogen-induced uterine hyperemia by α-adrenergic receptor-
    blocking agents. Endocrinology *102:* 903–909 (1978).
13 De Feo, V. J.: Decidualization; in Cellular biology of the uterus, pp. 192–290
    (Appleton Century Crofts, New York 1967).
14 Dey, S. K.; Johnson, D. C., and Santos, J. G.: Is histamine production by the
    blastocyst required for implantation in the rabbit? Biol. Reprod. *21:* 1169–1173
    (1979).
15 Fainstat, T.: Extracellular studies of uterus. I. Disappearance of the discrete
    collagen bundles in endometrial stroma during various reproductive states in
    the rat. Am. J. Anat. *112:* 337–370 (1963).
16 Feil, P. D.; Pegg, A. E.; Demers, L. M., and Bardin, C. W.: Involvement of poly-
    amines in the progestin-induced stimulation of endometrial glycogen synthesis
    during organ culture. Biochem. biophys. Res. Commun. *75:* 1–6 (1977).
17 Fenwick, L.; Jones, R. L.; Naylor, B.; Poyser, N. L., and Wilson, N. H.: Pro-
    duction of prostaglandins by the pseudopregnant rat uterus, *in vitro,* and the
    effect of tamoxifen with the identification of 6-keto-prostaglandin $F_{1\alpha}$ as a
    major product. Br. J. Pharmacol. *59:* 191–199 (1977).

18 Finn, C. A.: The implantation reaction; in Biology of the uterus, pp. 245–308 (Plenum, New York 1977).

19 Glasser, S. R. and McCormack, S. A.: Estrogen-modulated uterine gene transcription in relation to decidualization. Endocrinology 104: 1112–1118 (1979).

20 Heald, P. J.: Biochemical aspects of implantation. J. Reprod. Fertil. 25: suppl., pp. 29–52 (1976).

21 Heald, P. J.: Changes in ornithine decarboxylase during early implantation in the rat. Biol. Reprod. 20: 1195–1199 (1979).

22 Jänne, J.; Pösö, H., and Raina, A.: Polyamines in rapid growth and cancer. Biochim. biophys. Acta 473: 241–293 (1978).

23 Jellinck, P. H. and Lyttle, C. R.: Estrogen-induced uterine enzymes in the control of estradiol action. Adv. Enzyme Reguln 11: 17–33 (1973).

24 Kennedy, T. G.: Evidence for a role for prostaglandins in the initiation of blastocyst implantation in the rat. Biol. Reprod. 16: 286–291 (1977).

25 Kennedy, T. G.: Prostaglandins and increased endometrial vascular permeability resulting from the application of an artificial stimulus to the uterus of the rat sensitized for the decidual cell reaction. Biol. Reprod. 20: 560–566 (1979).

26 Klebanoff, S. J.: Inactivation of estrogen by rat uterine preparations. Endocrinology 76: 301–311 (1965).

27 Kraicer, P. F. et Shelesnyak, M. C.: Détermination de la période de sensibilité de l'endomètre à la décidualisation au moyen de déciduomes provoqués par un traitement empruntant la voie vasculaire. C. r. hebd. Séanc. Acad. Sci., Paris 248: 3213–3215 (1959).

28 Leroy, F.: Aspects moléculaires de la nidation; in L'implantation de l'œuf, pp. 81–92 (Masson, Paris 1978).

29 Leroy, F.; Van Hoeck, J., and Lejeune, B.: Effects of cycloheximide on the uterine refractory state induced by 'nidatory' oestrogen in rats. J. Reprod. Fertil. 56: 187–191 (1979).

30 Lobel, B. L.; Levy, E.; Kisch, E. S., and Shelesnyak, M. C.: Studies on the mechanism of nidation. XXVIII. Experimental investigation on the origin of eosinophilic granulocytes in the uterus of the rat. Acta endocr., Copenh. 55: 451–471 (1967).

31 Loeb, L.: The production of deciduomata and the relation between the ovaries and the formation of the decidua. J. Am. med. Ass. 50: 1897–1901 (1908).

32 Marcus, G J. and Shelesnyak, M. C.: Steroids in nidation; in Advances in steroid biochemistry, vol. II, pp. 373–438 (Academic Press, New York 1970).

33 Martel, D. and Psychoyos, A.: Endometrial content of nuclear oestrogen receptor and receptivity for ovo-implantation in the rat. Endocrinology 99: 470–475 (1976).

34 McCarthy, S. M.; Foote, R. H., and Maurer, R. R.: Embryo mortality and altered uterine luminal proteins in progesterone-treated rabbits. Fert. Steril. 28: 101–107 (1977).

35 McNabb, T. and Jellinck, P. H.: Purification and properties of oestrogen-induced uterine peroxidase. Biochem. J. 151: 275–279 (1975).

36 Moulton, B. C.; Koenig, B. B., and Borkan, S. C.: Uterine lysosomal enzyme activity during ovum implantation and early decidualization. Biol. Reprod. 19: 167–170 (1978).

37   Noyes, R. W.; Dickman, Z.; Doyle, L., and Gates, A. H.: Ovum transfers syn-
     chronous and asynchronous in the study of implantation; in Delayed implan-
     tation, pp. 197–212 (University of Chicago Press, Chicago 1963).
38   Pal, A. K. and Pakrashi, A.: Induction of implantation in the rat by iproniazid.
     J. Reprod. Fertil. 58: 387–388 (1980).
39   Psychoyos, A.: La réaction déciduale est précédée de modifications précoces de
     la perméabilité capillaire de l'utérus. C. r. Séanc. Soc. Biol. 154: 1384–1387
     (1960).
40   Psychoyos, A.: Nouvelle contribution à l'étude de la nidation de l'œuf chez la
     ratte. C. r. hebd. Séanc. Acad. Sci., Paris 251: 3073–3075 (1960).
41   Psychoyos, A.: Nouvelles remarques sur l'état utérin de 'non-réceptivité'. C. r.
     hebd. Séanc. Acad. Sci., Paris 257: 1367–1369 (1963).
42   Psychoyos, A.: Recent research on egg-implantation; in Ciba Foundation Study
     Group on Egg Implantation, pp. 4–28 (Churchill, London 1966).
43   Psychoyos, A.: The hormonal interplay controlling egg-implantation in the rat;
     in Advances in reproductive physiology, pp. 257–277 (Logos Press, London
     1967).
44   Psychoyos, A.: Endocrine control of egg-implantation; in Handbook of physiol-
     ogy: Endocrinology, vol. II, pp. 187–215 (Williams & Wilkins, Baltimore 1973).
45   Psychoyos, A.: Hormonal control of ovoimplantation. Vitams Horm. 31: 201–
     256 (1973).
46   Psychoyos, A.: Hormonal control of uterine receptivity for nidation. J. Reprod.
     Fertil. 25: suppl., pp. 17–28 (1976).
47   Psychoyos, A. and Casimiri, V.: Uterine blastotoxic factors; in Cellular and
     molecular aspects of implantation (Plenum, New York, in Press, 1980).
48   Psychoyos, A.; Bitton-Casimiri, V., and Brun, J. L.: Repression and activation
     of the mammalian blastocysts; in Regulation of growth and differentiated func-
     tion in eukaryotic cells, pp. 509–514 (Raven Press, New York 1975).
49   Rath, N. C.; Olmedo, C.; Casimiri, V.; Parvez, S.; Roche, D.; Parvez, H., and
     Psychoyos, A.: Monoamine metabolism during early pregnancy in the rat; in
     Research on steroids, vol. VIII, pp. 11–15 (Academic Press, New York 1979).
50   Rudzik, A. D. and Miller, J. W.: The effect of altering the catecholamine con-
     tent of the uterus on the rate of contractions and the sensitivity of the myo-
     metrium to relaxin. J. Pharmac. exp. Ther. 138: 88–95 (1962).
51   Russell, D. H. and Taylor, R. L.: Polyamine synthesis and accumulation in the
     castrated rat uterus after estradiol-17β stimulation. Endocrinology 88: 1397–
     1403 (1971).
52   Saksena, S. K.; Lau, I. F., and Chang, M. C.: Relationship between oestrogen,
     prostaglandin F2α and histamine in delayed implantation in the mouse. Acta
     endocr., Copenh. 81: 801–807 (1976).
53   Schayer, R. W.: Evidence that induced histamine is an intrinsic regulator of the
     microcirculatory system. Am. J. Physiol. 202: 66–72 (1962).
54   Schayer, R. W.: Histidine decarboxylase in mast cells. Ann. N. Y. Acad. Sci.
     103: 164–178 (1963).
55   Shelesnyak, M. C.: Aspects of reproduction. Some experimental studies on the
     mechanism of ovo-implantation in the rat. Recent Prog. Horm. Res. 13: 269–
     322 (1957).

56 Southgate, J.; Grant, E. C. G.; Pollard, W.; Prise-Davies, J., and Sandler, M.: Cyclical variation in endometrial monoamine oxidase correlation of histochemical and quantitative biochemical assays. Biochem. Pharmac. *17:* 721–726 (1968).

57 Surani, M. A. H. and Webb, F. T. G.: Effect of dibutyryl cyclic AMP and oestradiol on incorporation of ³H-leucine into proteins of luminal fluid of the rat uterus. J. Endocr. *74:* 431–439 (1977).

58 Tchernitchin, A.: Radioautographic study of the effect of estradiol-17β, estrone, estriol, progesterone, testosterone and corticosterone on the *in vitro* uptake of 2,4,6,7-³H estradiol-17β by uterine eosinophils of the rat. Steroids *19:* 575–586 (1972).

59 Tchernitchin, A.; Roorijck, J.; Tchernitchin, X.; Vandenhende, J., and Galand, P.: Effects of cortisol on uterine eosinophilia and other oestrogenic responses. Mol. cell. Endocrinol. *2:* 331–338 (1975).

60 Tobert, J. A.: A study of the possible role of prostaglandins in decidualization using a non-surgical method for the instillation of fluids into the rat uterine lumen. J. Reprod. Fertil. *47:* 391–393 (1976).

61 Vane, J. R.: Prostaglandins as mediators of inflammation; in Advances in prostaglandin and thromboxane research, vol. 2, pp. 791–801 (Raven Press, New York 1976).

62 Webb, F. T. G.: Implantation in ovariectomized mice treated with dibutyryl adenosine 3'5'monophosphate (dibutyryl cyclic AMP). J. Reprod. Fertil. *42:* 511–517 (1975).

63 Weitlauf, H. M.: Effect of uterine flushings on RNA synthesis by 'implanting' and 'delayed implanting' mouse blastocysts *in vitro*. Biol. Reprod. *14:* 566–571 (1976).

64 Weitlauf, H. M.: Factors in mouse uterine fluid that inhibit the incorporation of ³H-uridine by blastocysts *in vitro*. J. Reprod. Fertil. *52:* 321–325 (1978).

65 Wood, J. C. et Psychoyos, A.: Activité utérine de certains enzymes hydrolytiques au cours de la pseudogestation chez la ratte. C. r. hebd. Séanc. Acad. Sci., Paris *265:* 141–144 (1967).

66 Wurtman, R. J.; Axelrod, J., and Potter, L. T.: The disposition of catecholamines in the rat uterus and the effect of drugs and hormones. J. Pharmac. exp. Ther. *144:* 150–155 (1964).

67 Yoshinaga, K. and Greep, R. O.: Uterine sensitivity with regard to ovo-implantation; in Progress in reproduction research and population control, p. 137 (Publication International, Quebec 1974).

68 Zan-Kowalezewska, M. and Roth, J. S.: On the similarity between ribonuclease inhibitor and β-estradiol receptor protein. Biochem. biophys. Res. Commun. *65:* 833–837 (1975).

A. Psychoyos, MD, Laboratoire de Physiologie de la Reproduction, ER 203, CNRS, Hôpital de Bicêtre (Bat. INSERM), 78, Avenue du Général Leclerc, F-94270 Bicêtre (France)

Prog. reprod. Biol., vol. 7, pp. 158–172 (Karger, Basel 1980)

The Role of Uterine Proteins in the Establishment of Receptivity of the Uterus[1]

Henning M. Beier

Department of Anatomy and Reproductive Biology, RWTH Aachen, Medical Faculty, Aachen

Introduction

Since the question whether the uterus plays an active or a passive role in the process of implantation has induced reproductive biologists to study the basic phenomenon of uterine secretion, we have gained considerable insight into the changes that the uterus undergoes as a consequence of endocrine regulation during the cycle and early pregnancy. The uterus has to prepare a suitable endometrial nest for the embryo. We have evidence that the products of endometrial secretion, which are released into the uterine lumen, play an important role in the preparation of uterine environmental conditions to offer the blastocyst an anchorage ground. The uterus can provide such conditions only during a limited period of the cycle, so that short receptive phases are followed by extended refractory phases. The receptive phase reflects a particular sensitization of the endometrium, the induction of which is possibly best characterized by the development and expression of typical protein patterns of the endometrial secretion. These protein patterns, particularly some dominating protein fractions in the rabbit, can be considered as specific markers for a certain phase of receptivity. Although uterine proteins are very important, they cannot be regarded as the only markers of uterine receptivity for implantation. There are most likely species differences. On the other hand, we have to bear in mind, that merely methodological

[1] The support by grants from the 'Deutsche Forschungsgemeinschaft' in the special programme on 'Biologie und Klinik der Reproduktion' is gratefully acknowledged (Be 524/6; Bonn-Bad Godesberg, FRG).

reasons may have provided the evidence for the involvement of proteins in uterine receptivity, because many modern laboratory methods are easier tools for macromolecular analyses than those available for small molecules (ions, sugars, amino acids, etc.). Nevertheless, uterine proteins may have an essential significance in the establishment of receptivity of the postovulatory uterus, which must interact with the implanting embryo to accomplish placentation and normal pregnancy.

In this article, I will try to assess our state of knowledge on uterine secretion proteins, aspects of their production, release and intraluminal dynamics. On the basis of some egg transfer experiments, I would like to conclude with a few thoughts about the significance of uterine proteins for receptivity.

Endometrial Secretion Proteins

At the time when morphological and cytological transformations preparing the endometrium for implantation are evident, we find even more clearly defined physicochemical alterations of uterine secretions, the characteristics of which can be shown by quantitative and qualitative changes of the secretory components. Volume and viscosity are dependent on the chronological stage of the reproductive cycle and the endocrine background of the maternal system. *Homburger and Tregier* [34] and *Harpel et al.* [33] demonstrated this by data on preovulatory uterine fluid. *Lutwak-Mann et al.* [47] were able to show that the viscosity of uterine fluid increases after ovulation, thereby increasing the protein concentration, which in turn is even more increased by higher secretion rates triggered by postovulatory progesterone.

As our knowledge of the biochemistry of uterine secretion components accumulates, new advances are attained concerning the various ways in which these components (proteins) are released into the lumen of the genital tract, and about the origin of these molecules. Analytical evidence from qualitative and quantitative studies, employing acrylamide gel electrophoresis, Ouchterlony double immunodiffusion, and other immunoelectrophoretical combinations, suggests that whereas many genital tract fluid proteins are identical to those of the blood serum, several are only found in uterine secretion. As shown earlier [10], there is a selective transport of serum proteins into the uterine luminal fluid. Serum-identical proteins are not lined up within a certain category of small molecular sizes, but show, on the contrary, rather diverse molecular weights, disregarding any preference for small protein components that would suggest a simple sieve effect by the blood-uterine secretion barrier. The partial disparity in uterine secretion, compared to their serum proportions, further emphasizes the selectiveness of the transsudation process. The albumin/globulin ratio in blood serum being 1.32, is however, for the day 6 rabbit uterine secretion, only at 0.15, indicating a significant relative albumin decrease in uterine fluid. This feature is even more obvious in the particular case of immunoglobulins. Whereas serum normally contains 15–17 rel%/o of immunoglobulins, uterine preimplantational secretion has not more than 2 rel%/o.

Comparative Animal Data

Changes in the nature of uterine secretion proteins have been extensively studied in several species. A constant finding was that endometrial secretion patterns are hormone dependent and develop time-specific characteristics according to the estrogen/progesterone ratio. The following references give basic information: rabbit [6, 9–12, 14, 16, 18, 25, 31, 35, 39, 42, 61, 62, 71]; mouse [4, 32, 53]; rat [45, 67]; roe deer [2]; cow [30, 46, 58]; pig [55, 64]; sheep [59]; macaque and baboon [37, 49, 50, 57]; man [5, 18, 19, 23, 24, 27, 54, 56, 63, 73, 74].

Selective transsudation of serum proteins evidently occurs into the uterine cavity, irrespectively of the size of protein molecules. This could be shown by immunoelectrophoretic means using available antibodies. More accurate information, however, on the synthesis and transsudation of macromolecules into the uterine secretion could be gained by the use of labelled precursors. Studies of the uterine luminal proteins of the rat have been carried out, using [³H]-leucine and [³H]-fucose [68]. There is little or no protein synthesized and secreted when progesterone alone is dominating the spayed animal; however, when administered together with 17β-estradiol, a substantial increase in the amounts of radiolabelled proteins in uterine secretion was found [68]. The nature of proteins released into the uterine lumen depends on a correct balance between estradiol and progesterone. Similar findings were made in the mouse [4, 32]. By contrast, the most pronounced changes have been found in endometrial secretion of the rabbit where uteroglobin represents the major fraction

(40–50 %) of all proteins, and where this protein is produced in response to progesterone [12, 23]. Interestingly, no clear evidence is available so far that uteroglobin is more than a unique rabbit protein. Although similar protein fractions have been described, no immunological identity could be demonstrated in spite of numerous different methods being used as in research on the human uterine secretions and wash fluids.

Uteroglobin Synthesis and Release

An interesting question since the first observations on progesterone effects on uteroglobin appearance in uterine secretion has been whether this steroid acts by stimulating directly synthesis or only affects release, or both processes together. By now, evidence has been gained showing that progesterone and several other progestagens are capable of largely stimulating synthesis and of controlling release totally. Uteroglobin synthesis is mainly activated by progesterone, although this protein is produced in detectable amounts by the uterus of estrogen-treated castrated does. Under such experimental conditions uteroglobin is present only in epithelial cells, but not extruded into the uterine lumen in amounts detectable by immunohistochemical means, as shown by *Kirchner* [41]. We could never obtain any uteroglobin release within the uterus under the influence of estrogens, even if considerable dosage of 17β-estradiol was used [15].

Progesterone is doubtlessly required to establish the characteristic secretion pattern of uteroglobin during preimplantation *in vivo*. Moreover, *Beato and Arnemann* [7] have shown that this secretion persists when the uterus is removed from the mother, isolated, and perfused under defined experimental conditions.

Production of uteroglobin as the major endometrial secretion protein logically led to a search for identification of true synthesis by demonstration of the corresponding messenger RNA (mRNA). The endometrium of progesterone-stimulated rabbits has been used as a source for the isolation of uteroglobin mRNA. The translation of poly (A)-containing mRNA for uteroglobin has been demonstrated in different systems [8, 26]. Identification of newly *in vitro* synthesized uteroglobin was presented by means of monospecific antibodies. These results confirm that uteroglobin is synthesized *de novo* in the uterus and that progesterone stimulates this synthesis. *Bullock et al.* [26] have demonstrated that the specific mRNA for uteroglobin accounts for an increasing proportion of the poly (A)-rich endometrial mRNA during the preimplantation period, reaching a maximum at day 4 post coitum in normal pregnancy. The pattern of change in uteroglobin mRNA is similar to the pattern of secretion of uteroglobin during normal preimplantation and reflects the changing endocrine control mechanisms of the maternal system.

So far, the process terminating uteroglobin's presence in uterine secretion is unclear. It may well be that the progesterone/estradiol ratio is more important than the level of one of both steroids alone. Since the protein pattern observed during pseudopregnancy suggests that implantation itself terminates uteroglobin release or synthesis, there may be a specific 'message' delivered from the blastocyst or from the decidual tissue, as has been proposed by *Johnson* [36]. The perfusion experiments on isolated uteri and their uteroglobin production [7] indicate that, *in vivo,* the 'switch-off' is a termination of synthesis rather than only of release, since these studies show no long-term accumulation of uteroglobin in the cytosol of endometrial cells.

Studies on Biological Systems, Where Uteroglobin may Act: the in vivo *and the* in vitro *Growing Blastocyst*

Injections of 17β-estradiol benzoate at 6 and 30 h after mating reveal a significant delay in the secretory pattern sequence of the rabbit uterus. Compared to the normal preimplantation patterns, there is a delay of 2–5 days, depending on the stages compared. At earlier stages, the delay is less extended (2–3 days). This feature is not only apparent for the protein patterns, but also for the histology of the endometrium and the enzyme histochemistry of the endometrial epithelia. We have suggested that these asynchronous protein patterns particularly contribute to an unfavorable uterine environment for blastocyst development [13, 15, 23]. Subsequent egg transfer experiments have clearly shown that normally developed blastocysts require a normally developed uterine environment to accomplish implantation and further development [1, 22]. We transferred normal day 4 blastocysts into day 8 pregnant foster-mothers post coitally reated with estrogen, hence showing such delayed uterine secretion. These transferred blastocysts implanted around day 12 (recipient's reproductive stage) and gave about 40 % normal fetuses. Several of these fetuses were allowed to further develop and provided

viable young rabbits. In conclusion, it appears that a strictly synchronized uterine environment particularly as regards the pattern of protein secretion, plays an essential role in early mammalian embryogenesis.
The model of 'delayed secretion' has not been merely studied to demonstrate a questionable growth-inhibition effect on the native blastocysts.
One particular aspect, however, is important: the asynchronous egg transfer in delayed secretory uteri provides biological evidence for the necessity
of a proper uterine environment to blastocyst development. Consequently,
the essential protein environment for normal blastocyst development is
composed of a considerable number of macromolecules. Uteroglobin is
obviously the major component among them, moreover so characteristic
that it represents a specific marker molecule for the ovarian hormonal
status of the animal.

Interestingly, the experimentally induced delay of uterine secretion
is not the only possibility to desynchronize the maternal and embryonic
systems. Prefertilization treatment of estrous rabbits with progesterone
(up to 2 mg/day/animal) for 8 days (day −6 to +1) and induction of
ovulation (by HCG injection on day 0) with consecutive artificial insemination, results in normal egg development during oviducal passage.
However, degeneration of blastocysts occurs after their arrival in the
uterus, mostly within 2 days after being exposed to the uterine secretion
environment [72]. If this treatment is applied, the intrauterine milieu is
modified inasmuch as protein patterns are advanced as compared to the
normal preimplantation state. The exogenous progesterone induces uteroglobin synthesis and release before ovulation and fertilization. There is a
maximum of uteroglobin secretion from day −1 until day +2, designated
consequently as an 'advanced secretion'. Comparable observations were
reported by *Kendle and Telford* [38] and by *McCarthy et al.* [52]. This
phenomenon leads as clearly to failure of implantation as it occurs during
'delayed secretion'.

We have paid particular attention to the possible presence of uterine
secretion proteins in blastocyst fluid. In a recent investigation on the
protein patterns of rabbit blastocyst fluid and blastocyst homogenates
after development *in vivo* and *in vitro,* we have tried to present evidence
for the origin of the blastocyst fluid proteins [20]. Special emphasis was
directed to the protein patterns of *in vitro* grown blastocysts, since these
embryos developed from the 2- and 4-celled stages to expanded blastocysts without any uterine protein in the culture medium, bovine serum
albumin being used as the sole protein source. Patterns from *in vivo* and

in vitro development differ significantly, as judged by acrylamide gel electrophoresis and by several immunochemical test methods. These results demonstrate, that blastocysts grown *in vitro* do not contain uteroglobin or β-glycoprotein in detectable amounts. Compared to the *in vivo* developed blastocysts, this is a striking difference, because uteroglobin and β-glycoprotein have been demonstrated in fluids of *in vivo* expanded blastocysts in considerable amounts, at least in such quantities that our routine immunochemical tests are applicable with positive results. However, our study indicates that the *in vitro* developing blastocyst cannot synthesize uteroglobin. Although it cannot be formally excluded that *in vivo* growing blastocysts synthesize molecules similar to uterine secretion proteins, all available evidence supports the concept that uterine secretion proteins migrate into the blastocyst cells, particularly trophoblast cells on the 7th day post coitum [41], and into the blastocyst coverings [40]. It seems that the blastocyst utilizes these environmental proteins. Our experiments suggest, that under *in vitro* conditions, other proteins supplemented to the culture medium pass into the blastocyst compartments in comparable quantities. This in turn indicates that environmental proteins under *in vitro* conditions do not seem to entail a specific embryotropic activity, but more likely general physicochemical or biochemical effects. If we extrapolate these conclusions from the *in vitro* situation to normal physiology, it appears conceivable that uterine secretion proteins act as integrated parts of a maternal-embryonic molecular system, the function of which is more to protect the genetically unique embryo [15], than a direct embryotropic action as suggested by *Krishnan and Daniel* [42].

In addition, uteroglobin may play an important role in the biochemical reactions and the 'metabolism' of the blastocyst coverings, particularly when the blastocyst enters the uterine cavity on day 4. Striking evidence for the involvement of uteroglobin in the physicochemical conditioning of the blastocyst coverings, the zona pellucida and the mucin coat (mucoprotein layer), has been obtained from the comparison of two experiments on rabbit blastocysts. Blastocysts flushed from the isthmic part of the oviduct or from the upper uterine segment (approximately 70 h post coitum) during 'delayed secretion', show herniations of the trophoblast. The same pictures of trophoblast herniation can be easily obtained by culturing early cleavage stages up to the blastocyst stage *in vitro,* using a defined medium [51] with or without BSA as the only protein source. Both blastocysts are faced to an unfavorable environment, i. e. on one hand to a desynchronized uterine secretion and on the other to a nonnatural medium also 'desynchronized' by a lack of uteroglobin or any other uterine protein.

Probably as a consequence of this disproportion of uterine proteins, in particular of uteroglobin, the mucin coat does not show elasticity and flexibility as usually on day 4, when the normally developing blastocysts start their expansion. Expansion cannot take place when the rigid coverings act under desynchronized conditions as a straitjacket for the blastocyst. We have presented evidence for the influence of the uterine proteins on the coverings and their physicochemical alteration by supplementing *in vitro* protein-free culture media with uterine secretion proteins [51]. Expansion does appear frequently; however, the overall development of the embryonic system occurs more slowly than normally. Since uteroglobin acts like a protease inhibitor [16, 17], it may be regulating proteases which control the structural metabolism of the blastocyst coverings. These proteases can be of uterine or embryonic (trophoblast) origin, and may be controlled by uteroglobin and other uterine secretion components [28, 29].

Proteins of the Human Endometrial Tissue and Secretion

We have subjected human female genital tract secretion to acrylamide disc electrophoresis and several immunochemical tests in order to solve the question whether human uterine secretion follows the same principles as the rabbit model. Follicular and oviductal fluids, uterine and cervical secretions do not contain a predominating postalbumin component comparable to rabbit uteroglobin. However, rather surprising results have been obtained by cross-reactions of human uterine proteins with sera from the goat and the sheep, containing antibodies that had been raised against authentic endometrial uteroglobin from the rabbit. Several patients' uterine secretions samples did contain a uteroglobin-like antigen, which was cross-reacting with the antibodies against rabbit uteroglobin, demonstrated by means of the acrylamide-agar immunoreaction. The appearance of a human uterine protein, that cross-reacts with antibodies to uteroglobin is challenging, since in view of the considerable bulk of data from animal experiments (rabbit), the question whether these results can be extrapolated to human reproductive physiology awaits clarification.

Even samples obtained during the midsecretory phase of the human menstrual cycle, when implantation is normally expected, do not always contain a typical uteroglobin band or a protein with the same electrophoretical mobility that stains well with amido black. To date, we could not isolate by chromatography a predominant postalbumin fraction exhibiting the characteristics of uteroglobin. Such results indicate that the immunological cross-reactions are probably not highly specific, and that even totally nonspecific reactions cannot be excluded. Since there is no immunological cross-reaction of antiuteroglobin sera with any of the uterine flushings from the well-known laboratory animals, such as the mouse, rat, guinea pig, hamster, mink, and also the larger animals, e. g. goat, sheep, pig, cow and even in the baboon, we may conclude that the interspecies cross-reactions with human uterine wash fluid or endometrial tissue homogenate, are of exceptional nature. It is thus an essential problem to biochemically identify uteroglobin-like protein or analogous molecules in the human uterine secretion. Quantitatively, such components may totally differ from the concentrations found in the rabbit uterus, because the type of implantation of the human blastocyst remarkably differs from that in the rabbit (interstitial vs. superficial type). However, there is no reason to deny the presence of uteroglobin-like protein in the human uterus as long as it has not been ruled out by convincing data.

There are characteristic oviducal and uterine proteins in the human being, which we could not as yet directly demonstrate by specific antisera. We have observed several remarkable fractions, among which one prealbumin and one posttransferrin seem to lack any equivalent fraction in blood plasma from the same patient. Comparable posttransferrin bands have been reported by *Mastroianni et al.* [50] in oviductal fluid of the rhesus monkey, by *Moghissi* [54] in the human oviduct and by *Wolf and Mastroianni* [74] in human uterine washings. The prealbumin and the posttransferrin fractions both show a clear PAS-positive reaction indicating their glycoprotein nature.

Sutcliffe et al. [69, 70] have detected a human uterine protein that appears in pregnant uterine tissue, in decidual tissue and in amniotic fluid. This probably tissue-specific protein is characterized as an α_2-uterine globulin, composed of two subunits of approximately 25,000 molecular weight. This molecule was not found by the immunological methods used by *Sutcliffe et al.* [69, 70] in male reproductive tissues, or in blood serum of pregnant or nonpregnant women. Interesting results may be expected from further studies on this protein.

Among the plasma protein components, special attention has been paid to the immunoglobulins in the genital tract secretions. In contrast to other body secretions (milk, saliva, intestinal fluid) where IgA is the major immunoglobulin, the ratio of IgA/IgG in uterine and cervical mucus is approximately the same as in the blood plasma. Both immunoglobulins are present in similar proportions in follicular and oviducal fluid, however, in much lower concentration in uterine and cervical secretions. Interestingly, IgM occurs only in uterine secretion as a permanent component [16, 19].

Receptivity of the Uterus: Maternal Contribution to the Synchronization of the Interacting Systems, Mother and Embryo

The demands for synchronization of the reproductive stages of the mother and embryo are well documented, particularly for small laboratory animals [review in 21], whereas the situation in primates and in the human is much less clear [48, 65, 66]. In our experiments on 'asynchronous' egg transfer in the rabbit, we have observed that advanced (older) embryos in younger uteri develop much better than do the blastocysts in the converse combination, i. e. younger eggs transferred to advanced uteri. The 'beneficial' younger uterine secretion milieu may be more readily susceptible to adaptations and to conversions into receptivity to meet the synchronization demands of the embryo. If the transferred egg is younger (more than 12 h), it may not be able to catch up with the advanced maternal intrauterine environment. The ability to catch up can only be recognized when the advanced stage of the maternal system is changed experimentally by hormone treatment, so that the uterine secretion protein patterns in particular are delayed and do not actually correspond to the overall reproductive stage of the mother. Those uteri are advanced in their general reproductive stage according to the time since mating; however, they are not advanced in the development of their protein secretion pattern. This is, as a matter of fact, what we understand as delayed secretion. If such a delayed secretion uterus serves as the recipient uterus for a transferred egg, and this egg corresponds in its age to the biochemical secretion stage, *identified by the uterine protein pattern,* then it will easily be able to catch up with its new foster environment. The uterus may then be designated 'receptive'. Contrarily, the unsynchronized younger egg, in all cases observed [21], is not able to catch up with the recipient's uterine secretion.

The situation in the human may be different since the birth of a child was obtained after successful transfer of a 2¹/₂-day-old morula from *in vitro* culture to the uterus of its mother at 2¹/₂ days *post ovulationem* [66]. In this case, the morula (about 12 cells) arrived much earlier in the uterine lumen than during normal pregnancy. The human uterus may thus be receptive also for very young embryos, which are normally expected to be in the oviduct. Receptivity for attachment and implantation obviously develops not only into the uterus but also occasionally (or always?) in the oviduct as shown by the frequency of tubal pregnancies in the human. Support to our interpretation of the particular situation in the human can also be found in our analyses of protein patterns of uterine and tubal secretions. There is a striking similarity between both patterns [18, 23], and we may conclude that the 'tubal' embryo feels well in the uterine environment reached earlier than normally. However, from the studies by *Steptoe and Edwards* [65] it is obvious that the human blastocyst in its advanced developmental stage cannot catch up with the receptivity of the luteal phase endometrium when transferred close to the time when implantation normally occurs. The remaining time until maternal receptivity is completed, may be to short for the transferred embryo, which has to overcome considerable 'transfer hazards', to catch up with the maternal system.

Finally, we should bear in mind the possibility that embryonic signals, metabolic or hormonal, may be equally if not more important for the induction and development of receptivity of the uterus, than maternal luminal proteins.

References

1   Adams, C. E.: Asynchronous egg transfer in the rabbit. J. Reprod. Fertil. *35:* 613–614 (1974).
2   Aitken, R. J.: Delayed implantation in roe deer *(Capreolus capreolus)*. J. Reprod. Fertil. *39:* 225–233 (1974).
3   Aitken, R. J.: Embryonic diapause; in Johnson, Development in mammals, vol. 1, pp. 307–359 (North-Holland, Amsterdam 1977).
4   Aitken, R. J.: The protein content of mouse uterine flushings during pseudopregnancy. J. Reprod. Fertil. *50:* 191–192 (1977).
5   Aitken, R. J.: Tubal and uterine secretions; the possibilities for contraceptive attack. J. Reprod. Fertil. *55:* 247–254 (1979).
6   Beato, M.: Physico-chemical characterization of uteroglobin and its interaction

with progesterone; in Johnson, Development in mammals, vol. 1, pp. 173–198 (North-Holland, Amsterdam 1977).

7 Beato, M. and Arnemann, J.: Hormone-dependent synthesis and secretion of uteroglobin in isolated rabbit uterus. FEBS Lett. *58:* 126–129 (1975).

8 Beato, M. and Rungger, D.: Translation of the messenger RNA for rabbit uteroglobin in *Xenopus oocytes*. FEBS Lett. *59:* 305–309 (1975).

9 Beier, H. M.: Das Proteinmilieu in Serum, Uterus und Blastocysten des Kaninchens vor der Nidation; in Beermann, Biochemie der Morphogenese (Deutsche Forschungsgemeinschaft, Konstanz 1966).

10 Beier, H. M.: Veränderungen am Proteinmuster des Uterus bei dessen Ernährungsfunktion für die Blastocyste des Kaninchens. Verh. dt. zool. Ges. *31:* 139–148 (1967).

11 Beier, H. M.: Uteroglobin: a hormone-sensitive endometrial protein involved in blastocyst development. Biochim. biophys. Acta *160:* 289–291 (1968).

12 Beier, H. M.: Biochemisch-entwicklungsphysiologische Untersuchungen am Proteinmilieu für die Blastocystenentwicklung des Kaninchens *(Oryctolagus nuniculus)*. Zool. Jb. Anat. *85:* 72–190 (1968).

13 Beier, H. M.: Hormonal stimulation of protease inhibitor activity in endometrial secretion during early pregnancy. Acta endocr., Copenh. *63:* 141–149 (1970).

14 Beier, H. M.: Die hormonelle Steuerung der Uterussekretion und frühen Embryonalentwicklung des Kaninchens; Habil.-Schrift, Med. Fakult. Univ. Kiel, pp. 1–216 (1973).

15 Beier, H. M.: Oviducal and uterine fluids. J. Reprod. Fertil. *37:* 221–237 (1974).

16 Beier, H. M.: Uteroglobin and related biochemical changes in the reproductive tract during early pregnancy in the rabbit. J. Reprod. Fertil. *25:* suppl., pp. 53–69 (1976).

17 Beier, H. M.: Immunologische und biochemische Analysen am Uteroglobin und dem Uteroglobin-ähnlichen Antigen der Lunge. Medsche Welt *28:* 788–792 (1977).

18 Beier, H. M.: Control of implantation by interference with uteroglobin synthesis, release and utilization; in Ludwig and Tauber, Human fertilization, pp. 191–203 (Thieme, Stuttgart 1978).

19 Beier, H. M. and Beier-Hellwig, K.: Specific secretory protein of the female genital tract. Acta endocr., Copenh. *180:* suppl., pp. 404–425 (1973).

20 Beier, H. M. and Maurer, R. R.: Uteroglobin and other proteins in rabbit blastocyst fluid after development *in vivo* and *in vitro*. Cell Tiss. Res. *159:* 1–10 (1975).

21 Beier, H. M. and Mootz, U.: Significance of maternal uterine proteins in the establishment of pregnancy; in Maternal recognition of pregnancy. Ciba Fdn Ser. 64, pp. 111–140 (Excerpta Medica/North-Holland/Elsevier, Amsterdam 1979).

22 Beier, H. M.; Mootz, U. und Kühnel, W.: Asynchrone Eitransplantation während der verzögerten Uterussekretion beim Kaninchen. 7th Int. Congr. Anim. Reprod. Artif. Insem. München *3:* 1891–1896 (1972).

23 Beier, H. M.; Petry, G., and Kühnel, W.: Endometrial secretion and early mammalian development; in Gibian and Plotz, 21st Coll. Ges. Biol. Chem. Mosbach 1970; Mammalian reproduction, pp. 264–285 (Springer, Berlin 1970).

24 Bernstein, G. S.; Aladjem, F., and Chen, S.: Proteins in human endometrial washings. A preliminary report. Fert. Steril. *22:* 722–726 (1971).

25 Bullock, D. W.: Progesterone induction of messenger RNA and protein synthesis in rabbit uterus. Ann. N. Y. Acad. Sci. *286:* 260–272 (1977).

26 Bullock, D. W.; Woo, S. L. C., and O'Malley, B. W.: Uteroglobin messenger RNA. Translation *in vitro.* Biol. Reprod. *15:* 435–443 (1976).

27 Daniel, J. C., jr.: Uterine proteins and embryonic development. Adv. Biosci. *6:* 191–203 (1971).

28 Denker, H.-W.: Implantation. The role of proteinases, and blockage of implantation by proteinase inhibitors; in Advances in anatomy, embryology and cell biology, vol. 53, fasc. 5 (Springer, Berlin 1977).

29 Denker, H.-W.: The role of trophoblastic factors in implantation; in Spilman and Wilks, Novel aspects of reproductive physiology, pp. 181–212 (Spectrum/Halsted Press/Wiley, New York 1978).

30 Dixon, S. N. and Gibbons, R. A.: Proteins in the uterine secretions of the cow. J. Reprod. Fertil. *56:* 119–127 (1979).

31 Feigelson, M.; Noske, I. G.; Goswami, A. K., and Kay, E.: Reproductive tract fluid proteins and their hormonal control. Ann. N. Y. Acad. Sci. *286:* 273–286 (1977).

32 Fishel, S. B.: Analysis of mouse uterine proteins at pro-oestrus, during early pregnancy and after administration of exogenous steroids. J. Reprod. Fertil. *56:* 91–100 (1979).

33 Harpel, P. C.; Homburger, F., and Tregier, A.: Mouse uterine fluid plasminogen activator, acid phosphatase, and contraceptive hormones. Am. J. Physiol. *215:* 928–931 (1968)

34 Homburger, F and Tregier, A.: Endocrine factors determining the rate of accumulation of endometrial secretions in experimental hydrometra of mice. Endocrinology *61:* 634–639 (1957).

35 Johnson, M. H.: The protein composition of secretions from pregnant and pseudopregnant rabbit uteri with and without a copper intrauterine device. Fert. Steril. *23:* 123–130 (1972).

36 Johnson, M. H.: Studies using antibodies to the macromolecular secretions of the early pregnant uterus; in Centaro and Carretti, Immunology in obstetrics and gynecology. Proc. 1st Int. Congr. Padua 1973, pp. 123–133 (Excerpta Medica, Amsterdam 1974).

37 Joshi, M. S.; Yaron, A., and Lindner, H. R.: Intra-uterine gelation of seminal plasma components in the rat after coitus. J. Reprod. Fertil. *30:* 27–38 (1972).

38 Kendle, K. E. and Telford, J. M.: Investigations into the mechanism of the antifertility action of minimal doses of megestrol acetate in the rabbit. Br. J. Pharmacol. *40:* 759–774 (1970).

39 Kirchner, C.: Untersuchungen an uterusspezifischen Glycoproteinen während der frühen Gravidität des Kaninchens *(Oryctolagus cuniculus).* Wilhelm Roux Arch. EntwMech. Org. *164:* 97–133 (1969).

40 Kirchner, C.: Immune histologic studies on the synthesis of a uterine-specific protein in the rabbit and its passage through the blastocyst coverings. Fert. Steril. *23:* 131–136 (1972).

41 Kirchner, C.: Uteroglobin in the rabbit: I. Intracellular localization in the ovi-

duct, uterus, and the preimplantation blastocyst. Cell Tiss. Res. *170:* 415–424 (1976).

42   Krishnan, R. S. and Daniel, J. C., jr.: 'Blastokinin': inducer and regulator of blastocyst development in the rabbit uterus. Science *158:* 490–492 (1967).

43   Krishnan, R. S. and Daniel, J. C., jr.: Composition of 'blastokinin' from rabbit uterus. Biochim. biophys. Acta *168:* 579–582 (1968).

44   Kulangara, A. C. and Crutchfield, F. L.: Passage of bovine serum albumin from the mother to rabbit blastocysts. II. Passage from uterine lumen to blastocyst fluid. J. Embryol. exp. Morphol. *30:* 471–482 (1973).

45   Kunitake, G. M.; Nakamura, R. M.; Wells, B. G., and Moyer, D. L.: Studies on uterine fluid. I. Disc electrophoretic and disc-gel Ouchterlony analysis of rat uterine fluid. Fert. Steril. *16:* 120–124 (1965).

46   Laster, D. B.: A pregnancy-specific protein in the bovine uterus. Biol. Reprod. *16:* 682–690 (1977).

47   Lutwak-Mann, C.; Boursnell, J. C., and Bennett, J. P.: Blastocyst-uterine relationships: uptake of radioactive ious by the early rabbit embryo and its environment. J. Reprod. Fertil. *1:* 169–185 (1960).

48   Marston, J. H.: Synchrony in embryonic development; in Cellular and molecular aspects of implantation. NICHHD Conf. Houston, Tex. 1979.

49   Mastroianni, L.; Urzua, M.; Avalos, M., and Stambaugh, R.: Some observations on Fallopian tube fluid in the monkey. Am. J. Obstet. Gynec. *103:* 703–709 (1969).

50   Mastroianni, L.; Urzua, M., and Stambaugh, R.: Protein patterns in monkey oviductal fluid before and after ovulation. Fert. Steril. *21:* 817–820 (1970).

51   Maurer, R. R. and Beier, H. M.: Uterine proteins and development *in vitro* of rabbit preimplantation embryos. J. Reprod. Fertil. *48:* 33–41 (1976).

52   McCarthy, S. M.; Foote, R. H., and Maurer, R. R.: Embryo mortality and altered uterine luminal proteins in progesterone-treated rabbits. Fert. Steril. *28:* 101–107 (1977).

53   Mintz, B.: Control of embryo implantation and survival. Adv. Biosci. *6:* 317–342 (1971).

54   Moghissi, K. S.: Human Fallopian tube fluid. I. Protein composition. Fert. Steril. *21:* 821–829 (1970).

55   Murray, F. A.; McGaughey, R. W., and Yarus, M. J.: Blastokinin: its size and shape, and an indication of the existence of subunits. Fert. Steril. *23:* 69–77 (1972).

56   Noske, I. G. and Feigelson, M.: Immunological evidence of uteroglobin (blastokinin) in the male reproductive tract and in non-reproductive ductal tissues and their secretions. Biol. Reprod. *15:* 704–713 (1976).

57   Peplow, V.; Breed, W. G.; Jones, C. M. J., and Eckstein, P.: Studies on uterine flushings in the baboon. Am. J. Obstet. Gynec. *116:* 771–779 (1973).

58   Roberts, G. P. and Parker, J. M.: Macromolecular components of the luminal fluid from the bovine uterus. J. Reprod. Fertil. *40:* 291–303 (1974).

59   Roberts, G. P.; Parker, J. M., and Symonds, H. W.: Macromolecular components of genital tract fluids from the sheep. J. Reprod. Fertil. *48:* 99–107 (1976).

60   Schlafke, S. and Enders, A. C.: Protein uptake by rat preimplantation stages. Anat. Rec. *175:* 539–560 (1973).

61 Schwick, H. G.: Untersuchungen über die Zusammensetzung der Uterus- und Blastocystenflüssigkeit des Kaninchens *(Oryctolagus cuniculus)*; Inaug. Diss., Phil. Fak. Univ. Marburg/L. (1963).

62 Shapiro, A. A.; Jentsch, J. P., and Yards, A. S.: Protein composition of rabbit oviducal fluid. J. Reprod. Fertil. *24:* 403–408 (1971).

63 Shirai, E.; Iizuka, R., and Notake, Y.: Analysis of human uterine fluid protein. Fert. Steril. *23:* 522–528 (1972).

64 Squire, G. D.; Bazer, F. W., and Murray, F. A.: Electrophoretic patterns of porcine uterine protein secretions during the estrous cycle. Biol. Reprod. *7:* 321–325 (1972).

65 Steptoe, P. C. and Edwards, R. G.: Reimplantation of a human embryo with subsequent tubal pregnancy. Lancet *i:* 880–882 (1976).

66 Steptoe, P. C. and Edwards, R. G.: Birth after the reimplantation of a human embryo. Lancet *i:* 366 (1978).

67 Surani, M. A. H.: Uterine luminal proteins at the time of implantation in rats. J. Reprod. Fertil. *48:* 141–145 (1976).

68 Surani, M. A. H.: Cellular and molecular approaches to blastocyst uterine interactions at implantation; in Johnson, Development in mammals, vol. 1, pp. 245–305 (North-Holland, Amsterdam 1977).

69 Sutcliffe, R. G.; Brock, D. J. H.; Nicholson, L. V. B., and Dunn, E.: Fetal- and uterine-specific antigens in human amniotic fluid. J. Reprod. Fertil. *54:* 85–90 (1978).

70 Sutcliffe, R. G.; Bolton, A. E.; Sharp, F.; Nicholson, L. V. B., and MacKinnon, R.: Purification of human alpha uterine protein. J. Reprod. Fertil. *58:* 345–442 (1980).

71 Urzua, M. A.; Stambaugh, R.; Flickinger, G., and Mastroianni, I.: Uterine and oviduct fluid protein patterns in the rabbit before and after ovulation. Fert. Steril. *21:* 860–865 (1970).

72 Visser, J. de: Degeneration of rabbit ova by prefertilization progesterone treatment: effect on endometrium and uteroglobin secretion. ICE-Satellite Symp. Proteins and Steroids in Early Mammalian Development, Aachen 1976; in Beier and Karlson, Reproductive endocrinology (Springer, Berlin in press, 1980).

73 Voss, H.-J. and Beato, M.: Human uterine fluid proteins: gel electrophoretic pattern and progesterone-binding properties. Fert. Steril. *28:* 972–980 (1977).

74 Wolf, D. P. and Mastroianni, L.: Protein composition of human uterine fluid. Fert. Steril. *26:* 240–247 (1975).

H. M. Beier, Professor and Chairman, MD, Department of Anatomy and Reproductive Biology, RWTH Aachen, Medical Faculty, D-5100 Aachen (FRG)

Prog. reprod. Biol., vol. 7, pp. 173–188 (Karger, Basel 1980)

Environmental Factors Involved in Delayed Implantation[1]

M. Bonnin and R. Canivenc

Laboratoire d'Endocrinologie Expérimentale, Université de Bordeaux II, Bordeaux

In 1960, a symposium was organized in Brussels on the theme 'Les fonctions de nidation et leurs troubles'. On this occasion, research into delayed implantation in wild animals [*Canivenc, 1960*] was reviewed. From this date, our knowledge of delayed implantation has increased regularly. The number of species studied has increased greatly as has the number of researchers throughout the world.

The first two major scientists to interest themselves in delayed implantation were *Bischoff* [1854] and *Lataste* [1891]. The first described seasonal delayed implantation in the roe deer, *Capreolus capreolus* (obligatory delayed implantation) and the second the same phenomenon in lactating rodent females (facultative delayed implantation). Even today, these are the two aspects of delayed implantation most studied by researchers in order to find a common denominator and to solve, more than 100 years later, an exciting and unusual problem.

Species with seasonal delayed implantation, the only ones which concern us in this study, have a long pregnancy [*Canivenc, 1960*], except for the mink where pregnancy does not exceed 72 days [*Hansson, 1947; Enders, 1952*]. Generally the period of embryonic diapause lasts for several months. Implantation takes place for some species (such as the marten, the mink) in spring when day length increases, and for others (such as the badger and the black bear) in autumn, when night length increases.

[1] This work was supported by DGRST and CNRS, ERA No. 699. The authors wish to thank Mrs. *I. Aubert* for technical assistance in electron microscopy preparations.

This study will attempt to relate the environmental control of implantation in musteloids with discontinuous development. The two species which will be considered here in detail are the European badger *Meles meles* L. which implants its blastocyst in late autumn or early winter and the European marten *Martes martes* L. which implants its blastocyst in spring. External factors act upon the endocrinological axes in ways still unknown. Internal factors (hormonal or metabolic) then induce resumption of the development of the blastocyst and its possibility of implantation in the endometrium which is in a progestative state at this time.

Species Implanting when Night Length is Increasing. The European Badger Meles meles *L.*

Since the first observation of an unimplanted blastocyst in the badger by *Fries* [1880], several studies have been made of the reproduction of this species. The mating season particularly has been the subject of much disagreement. Coitus takes place in summer (July) for *Fischer* [1931], in February, July or October for *Neal and Harrison* [1958]. The latter authors do not exclude the possibility of renewed ovulations during the delay. For the past 25 years we have systematically observed the genital tracts (ovaries and uterus) of more than 2,000 female badgers throughout the whole year. In February/March, ovulation and fertilization, which occur a few days after parturition, are followed by a prolonged period of embryonic diapause lasting for 300–320 days [*Canivenc and Laffargue,* 1956, 1957a]. Implantation is delayed until mid-December or early January. Parturition takes place 45 days later. Except in February/March we have never observed others signs of ovulation or fertilization such as ovulatory follicles, stigma in ovaries or morula in oviducts. (The mating season for yearlings or nonparous animals remains to be determined but we are convinced that it takes place at the same time as for other females.) Most females litter in February and immediately breed again even though they are still lactating cubs until May or June. 82 % of the females that we have observed had 2–4 blastocysts *in utero,* and of these females, 98 % had the same number of corpora lutea in the ovaries.

Experiments of ligature or sections of oviduct just after normal fertilization in March to prevent the arrival in the uterus of a new generation of eggs together with simultaneous marking of the corpora lutea in ovaries, have demonstrated that the eggs fertilized during the postpartum

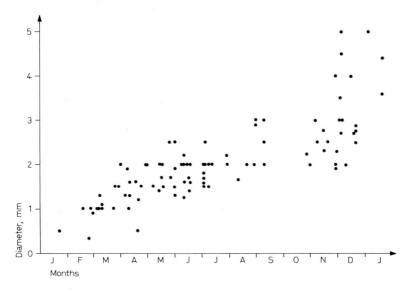

Fig. 1. Blastocyst's diameter during delayed implantation period in the badger.

ovulation were in fact those implanted in December. Corpora lutea marked in March were the only ones present in late December [*Canivenc,* 1958] when females were autopsied.

Contrary to what occurs in the roe deer [*Keibel,* 1902; *Short and Hay,* 1966; *Aitken,* 1974] or in rodents [*Psychoyos,* 1966; *Rumery and Blandau,* 1971], the blastocyst of the badger is surrounded by a zona pellucida during the whole diapause. The zona pellucida is composed of 2 parts: a superficial spongy part increasing in thickness during the diapause and a subjacent lamellar part. The structure of this second part could provide the mechanical resistance necessary for the blastocyst to survive in the uterine environment. The scarcity of microfilaments in the trophoblastic cells supports this hypothesis. On the other hand, in the roe deer, where there is no zona pellucida, the microfilaments are very numerous in the trophoblastic cells. A great quantity of lipidic material and very few granules of glycogen are observed during diapause in the trophoblastic cells of the badger. At the time of implantation, the number of glycogen granules increases and that of lipid droplets decreases considerably.

During the diapause, which lasts for 10 months, the blastocyst grows, slowly during the major part of this period and then very quickly (fig. 1). From March to December, its diameter increases imperceptibly from

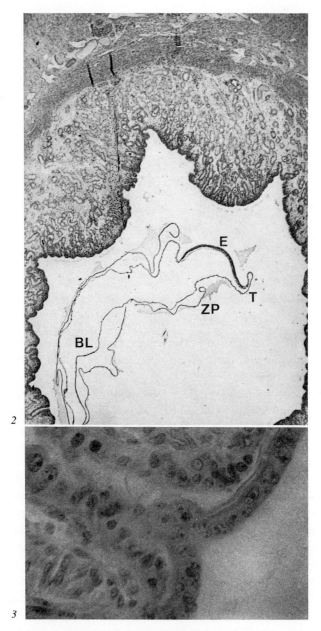

Fig. 2. Hatching of the blastocyst just before implantation. ZP = Zona pellu cida; T = trophoblast; BL = blastocoele; E = embryonic cell mass.

Fig. 3. Trophoblastic penetration of the endometrium.

0.5 to 3 mm. At the end of this period, a sharp increase occurs, from 3 to 5 mm, within a period of 15 days. An increase in the number and size of the trophoblastic cells is responsible for this growth. Just before implantation, the termination of diapause is associated with a slight elongation of the blastocyst which is then not round but oval. However, this enlargement is not comparable to that observed in the roe deer. In this species a sudden rapid elongation greatly increases the size of egg before implantation [*Aitken*, 1974]. These differences between species can be explained by different patterns of anatomical implantation.

The sharp increase in the size of the blastocyst and its enlargement bring it into contact with uterine epithelium. A true hatching occurs, releasing the trophoblast (fig. 2). At this time, the epithelium in the neck of the endometrial glands proliferates and the cells hypertrophy. At regular intervals, areas of trophoblastic invasion are observed (fig. 3). Symplasmal transformation of the hypertrophied endometrial cells occurs and necrotic areas appear. Microvilli of trophoblastic cells establish short desmosome-like junctions with the surface of the endometrial cells (fig. 4). These trophoblastic villi quickly penetrate the mucosa to a considerable degree. During the diapause, regularly spaced uterine contractions are observed. The blastocysts move about the uterus from one uterine horn to the other [*Dutourne and Canivenc*, 1971].

Although it is difficult to think that the badger's blastocyst is autonomous during the whole of this very long period and that it can live without exchanges with uterine environment, we have observed cases where it survived several months after ovariectomy without any damage [*Canivenc and Laffargue*, 1957b]. The quality of the uterine medium is, however, fundamental. The ovarian hormones, progesterone and estrogens, play a role in the maintenance of the myometrial tonicity and in the secretory production of the endometrium. In the badger, the pattern of ovarian secretion during the reproductive cycle and the morphological apparance of corpora lutea are well correlated [*Bonnin et al.*, 1978]. As in other species which exhibit delayed implantation, a peculiar luteal structure during diapause has been described [*Bonnin-Laffargue and Canivenc*, 1962]. The important feature in the life span of corpora lutea is an asthenic phase during which the luteal tissue is relatively inactive [*Canivenc et al.*, 1966]. This phase can be observed during the major part of the diapause (fig. 5). Simultaneously, low levels of progesterone are observed in plasma. Just before the presumed time of implantation in late December or early January, a sharp increase of progesterone is ob-

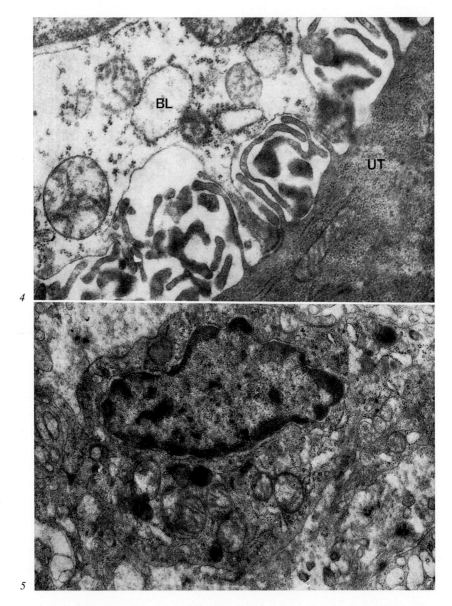

Fig. 4. Interrelations between the trophoblastic cells and the endometrial cells at time of implantation. Desmosome-like junction. BL = blastocyst, UT = uterus. ×16,000.

Fig. 5. Luteal cell during the diapause in the badger. ×29,000. Note the little size, irregular nucleus, lack of lipid droplets and of microvilli in the membrane.

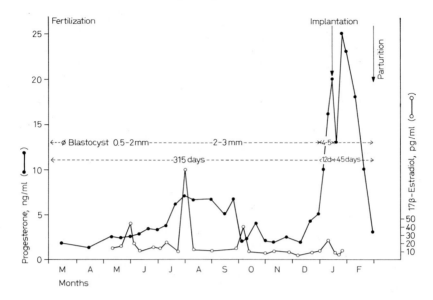

Fig. 6. Progesterone and estrogen profiles during delayed implantation and gestative phase in the badger.

served (fig. 6). It reflects the beginning of full luteal activity (fig. 7) and is associated with implantation. The estrogen levels are low during embryonic diapause [*Mondain-Monval et al.,* 1980]. However, episodic releases of estradiol, regularly spaced over 6–7 weeks, are observed (fig. 6).

Three facts emerge from our observations: (i) the sharp increase in the blastocyst size occurs at the same time as the sharp increase in the progesterone level; (ii) implantation occurs only when the progesterone level is above 15–20 ng/ml; (iii) during the diapause some ovarian activity (estrogen waves) is observed. The nature of the interactions between the endometrium and the blastocyst has not yet been elucidated. During the major part of the delay, the endometrium shows poor hormonal impregnation: the luminal epithelium is prismatic with basally situated nuclei. The endometrial glands are compact, straight and short. The most marked feature revealed by ultrastructural study is the presence of clear vesicles above the nucleus in the endometrial gland cells (fig. 8). Few glycogen granules are observed. Examination of the surface of endometrium by scanning microscopy shows lack of secretion in the

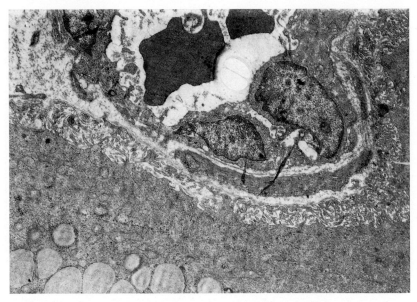

Fig. 7. Luteal cell after luteal reactivation. ×15,000. The size of cell is three times the size. Note lipid droplets and microvilli of cell membrane.

duct openings of the glands (fig. 9). The vascularization of the endometrium is poor. Before implantation, marked changes are observed. The luminal epithelium becomes longer and columnar with round, apically situated nuclei. The uterine glands are dilated. The cells are rich in glycogen granules and the clear vesicles have disappeared (fig. 10). Great quantities of secretion are observed in the duct openings of the glands, which are enlarged (fig. 11). Vascularization is well developed. Analysis of uterine flushings reveals the presence of 7 amino acids. Their concentration is very low during the major part of diapause. Just before implantation, when blastocyst size increases sharply, the total amino acid concentration becomes 12-fold higher. The most important change is observed in the glycine concentration which increases 20-fold. During the diapause the endometrial glands do not have the secretory aspect which is observed when the trophoblastic villi come in contact with them. The lack of uterine secretion could be responsible for the lack of implantation. The secretions are thought to play an embryotrophic role the more so since the trophoblastic implantation is central: in other species implantation is interstitial (rodents, human).

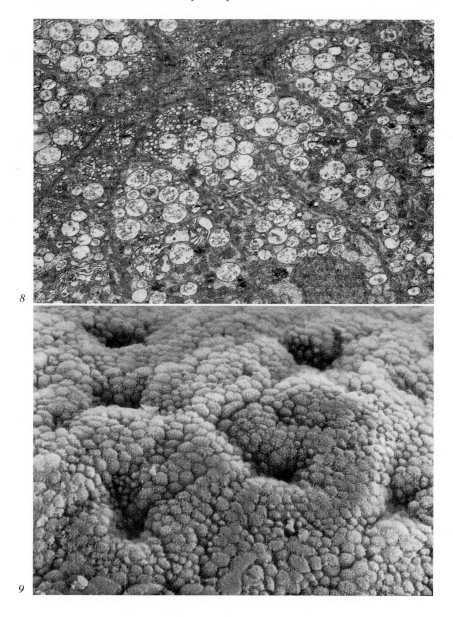

8

9

Fig. 8. Uterine glands during the diapause. ×4,000. Note straight lumen, numerous clear vesicles in the cytoplasm of endometrial cells.

Fig. 9. Surface of endometrium during the diapause. ×1,050.

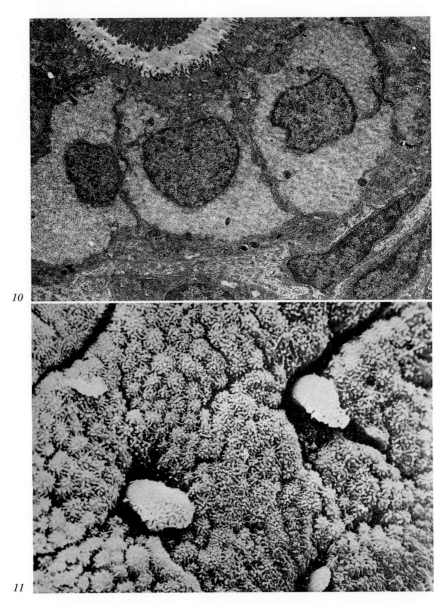

Fig. 10. Uterine glands at time of implantation. ×4,000. Note secretion product in the lumen of gland. The apical pole shows numerous microvilli. In the cytoplasm there is a great quantity of glycogen.

Fig. 11. Surface of the endometrium at time of implantation. ×1,050.

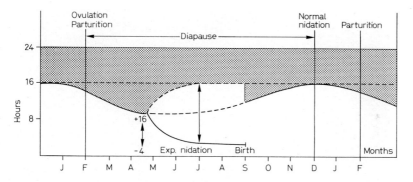

Fig. 12. Ecological program used in the climate chamber for the badger (*M. meles* L.).

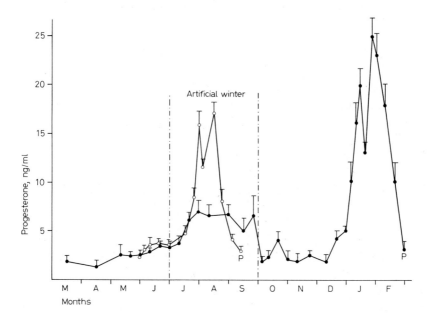

Fig. 13. Progesterone profile during pregnancy in 5 control badgers (●) and in 6 experimental females subjected to 'artificial winter' (○).

Fig. 14. Young badgers: on the right 8-day-old cubs, born in February 1980. On the left, young 6-month-old, born in September 1979 in the climate chamber. Both mothers of these younglings were fertilized in February 1979.

The chronological organization of the ovarian function seems to be controlled with great precision by environmental factors (photoperiod and temperature). In order to verify this hypothesis and to establish the degree of the interaction between ovarian function and environmental factors, animals were submitted during the diapause to the environmental conditions of their normal time of implantation (i. e. December for the badger). Female badgers in delayed implantation were then put into a climate chamber whose photoperiod and temperature can be programmed in order to obtain even during summer low temperature ($+ 4\,°C$) and long nights (8L 16D) (i. e. artificial winter) (fig. 12). These animals modified their behavior and ovarian function in response to this new environment. About 50–60 days after the beginning of experiment, the plasma progesterone level increased and reached the threshold of implantation, which is about 20 ng/ml as we have already found [*Bonnin et al.*, 1978]. The blastocysts resumed their development and implanted in July. The gestative phase lasted for 45 days as in normal conditions (fig. 13). In these experiments, births occurred in summer out of season 6 months be-
these experiments, births occurred in summer out of season 6 months

before the normal time (fig. 14), exposing the cubs to abnormal conditions [*Canivenc and Bonnin,* 1979]. This complete seasonal inversion of reproductive phenomena indicates a direct relationship between external factors, particularly seasonal, and the timing of reproduction in the badger and probably in other species of mammals in the wild.

Species Implanting When Day Length is Increasing.
The European Marten Martes martes L.

The reproductive cycle of this species was demonstrated in detail in our laboratory 13 years ago [*Canivenc et al.,* 1967]. The breeding season takes place once a year, in July. Fertilization is followed by a period of embryonic diapause lasting until the end of February of the next year. Parturition occurs at the end of March, after which females lactate cubs for 2 months. There is no postpartum ovulation in this species. As in other species which exhibit delayed implantation, an asthenic luteal phase was observed during the diapause. Poor hormonal impregnation was observed in the endometrium at this time.

The European marten implants its blastocyst in late February when the day length increases. Gestation is known to become shorter when the photoperiod is increased artificially in the American marten [*Enders and Pearson,* 1943] and in the mink [*Aulerich et al.,* 1963; *Belaiev et al.,* 1963; *Allais and Martinet,* 1978]. By manipulating the photoperiod, we obtained implantation out of season in the European marten [*Canivenc and Bonnin-Laffargue,* 1975].

We have to consider the biological signification of the relationship between reproduction and environment. A species must obviously adapt to this environment. Generally when the young leave their parents, at the end of lactation, they must be able to live autonomously and find sufficient food. This is why for the species with which we are concerned, implantation occurs at the same time of the year, either in autumn or in spring. This concept of a specific programme in the delayed implantation species has been interpreted differently by different authors.

For *Murr* [1929], the external temperature has (as in lower vertebrates) a direct effect in decreasing body temperature in winter and inducing a diapause in the embryonic development. This explanation cannot be valid for all species exhibiting delayed implantation because some animals, such as the armadillo, living in the tropics, exhibit delayed implantation. *Prell* [1930] thinks that delayed implantation is the prerogative of species existing prior to the pleistocene period and then exposed to the rigors of glaciation: during unfavorable ecological conditions, the embryonic development would be stopped and the cubs born in spring. The *Fries* [1880] theory, which is also the oldest, still seems the most valid. For him, delayed implantation in the badger in an adaptive phenomenon for the benefit of the young which are born in spring when conditions are favorable. The bat *(Myotis myotis)* has resolved the same problem by delayed fertilization.

The relationship between environmental factors and reproduction is very strong in species exhibiting delayed implantation. However, birth programming does not seem to be the determining factor. The best the young can hope for is to be able to live autonomously after weaning and find in the wild the best thermic and nutritional conditions. In the wild, female badgers lactate for 3 or 4 months from mid-February to June, after which the young live autonomously. It is a very favorable time as regards temperature and nutritional conditions. In our experiment with the badger using the climate chamber, births out of season occur about mid-September and their weaning in late December. At this time, the young could not live autonomously in the wild. However, other factors are concerned which are perhaps capable of explaining the differences of reproduction pattern in the American badger and European badger. Fertilization occurs in July in the American badger [*Wright,* 1963] but in February in the European badger. However, in both the blastocysts implant at the same time, in December/January.

The last link of the hormonal chain stimulating implantation is the plasma progesterone level. An increase of progesterone always occurs before the resumption of the trophoblastic activity, both in the wild and in the climate chamber. The progestative profile is the same in both cases. The high level of progesterone coinciding with implantation is short lived. After a slight decrease, the progesterone level increases again and this second peak lasts for the major part of the gestative phase (fig. 7). The first peak is probably controlled by the pituitary and results from the stimulation of corpus luteum by external factors (light and/or temperature). These experiments show that to obtain a normal birth the diapause is not obligatory and that its duration can be shortened. Perhaps is it possible to suppress it entirely without altering the egg's chances of survival.

References

Aitken, R. J.: Delayed implantation in the roe deer *(Capreolus capreolus)*. J. Reprod. Fertil. *39:* 225–233 (1974).

Allais, C. and Martinet, L.: Relation between daylight ratio, plasma progesterone levels and timing of nidation in mink *(Mustela vison)*. J. Reprod. Fertil. *54:* 137–143 (1978).

Aulerich, R. J.; Holcomb, L.; Ringer, R. K., and Schaible, P. J.: Influence of photoperiod on reproduction in mink. Q. Bull. Mich. St. Univ. agric. Exp. Stn. *46:* 132–138 (1963).

Belaiev, D. K.; Klotchkov, D. V., and Zhelezova, A. J.: The influence of light condition on the reproductive function and fertility in mink *(Mustela vison* Schr.). Byull. mosk. Obshch. Ispyt. Prir. Otd. Biol. (In Russian.) *78:* 107–125 (1963).

Bischoff, T. L. W.: Entwicklungsgeschichte des Rehes (Ricker'sche Buchhandlung, Giessen 1854).

Bonnin, M.; Canivenc, R., and Ribes, C.: Plasma progesterone levels during delayed implantation in the European badger *(Meles meles* L.). J. Reprod. Fertil. *52:* 55–58 (1978).

Bonnin-Laffargue, M. et Canivenc, R.: Histophysiologie des corps progestatifs et gestatifs chez le Blaireau européen *Meles meles* L. Annls Endocr. *23:* 163–174 (1962).

Canivenc, R.: Destinée des blastocystes et des corps jaunes issus du coït postpartum chez le Blaireau européen *Meles meles* L. C. r. hebd. Séanc. Acad. Sci., Paris *246:* 1914–1917 (1958).

Canivenc, R.: L'ovo-implantation différée des animaux sauvages; in Colloque sur les fonctions de nidation utérine et leurs troubles, pp. 33–86 (Masson, Bruxelles 1960).

Canivenc, R. and Bonnin, M.: Delayed implantation is under environmental control in the Badger *(Meles meles* L.). Nature, Lond. *278:* 849–850 (1979).

Canivenc, R. et Bonnin-Laffargue, M.: Les facteurs écophysiologiques de régulation de la fonction lutéale chez les mammifères à ovoimplantation différée. J. Physiol. Paris *70:* 533–538 (1975).

Canivenc, R.; Bonnin-Laffargue, M. et Relexans, M. C.: Cycles génitaux saisonniers de quelques mustélidés européens. Entretiens de Chizé. *1:* 66–81 (1967).

Canivenc, R. et Laffargue, M.: Présence de blastocystes libres intra-utérins au cours de la lactation chez le Blaireau européen *Meles meles* L. C. r. hebd. Séanc. Soc. Biol., Paris *150:* 1193–1196 (1956).

Canivenc, R. et Laffargue, M.: Relation des corps jaunes et des blastocystes au cours de la nidation différée du Blaireau européen *Meles meles* L. C. r. hebd. Séanc. Soc. Biol., Paris *151:* 561–564 (1957a).

Canivenc, R. et Laffargue, M.: Survive des blastocystes de rat en l'absence d'hormones ovariennes. C. r. hebd. Séanc. Acad. Sci., Paris *245:* 1752–1754 (1957b).

Canivenc, R.; Short, R. V. et Bonnin-Laffargue, M.: Etude histologique et biochimique du corps jaune du Blaireau européen *Meles meles*. Annls Endocr. *27:* 401–413 (1966).

Dutourne, B. et Canivenc, R.: Dynamique de la migration blastocytaire chez le Blaireau européen *(Meles meles* L.). C. r. hebd. Séanc. Acad. Sci., Paris *273:* 2579–2582 (1971).

Enders, R. K.: Reproduction in the mink *(Mustela vison)*. Proc. Am. phil. Soc. *96:* 691–755 (1952).

Enders, R. K. and Pearson, O. P.: Shortening gestation by inducing early implantation with increased light in the marten. Am. Fur Breed. *15:* 18–19 (1943).

Fischer, E.: Die Entwicklungsgeschichte des Dachses und die Frage der Zwillingsbildung. Verh. anat. Ges., Jena *40:* 22–34 (1931).

Fries, S.: Über die Fortpflanzung von *Meles taxus.* Zool. Anz. *3:* 486–492 (1880).

Hansson, A.: Physiology of reproduction in the mink *(Mustela vison* Schreb.) with special reference to delayed implantation. Acta zool. *28:* 1–136 (1947).

Keibel, F.: Die Entwicklung des Rehes bis zur Anlage des Mesoblast. Arch. Anat. *24:* suppl., pp. 293–314 (1902).

Lataste, M. F.: Des variations de durée de la gestation chez les mammifères et des circonstances qui déterminent ces variations: théorie de la gestation retardée. C. r. Séanc. Soc. Biol. *9:* 21–31 (1891).

Mondain-Monval, M.; Bonnin, M.; Canivenc, R., and Scholler, R.: Estrogen levels during delayed implantation in the European badger *Meles meles.* Gen. compar. Endoc. (in press, 1980).

Murr, E.: Zur Erklärung der verlängerten Tragdauer bei Säugetieren. Zool. Anz. *85:* 113 (1929).

Neal, E. G. and Harrison, R. J.: Reproduction in the European badger *(Meles meles).* Trans. zool. Soc. Lond. *29:* 67–130 (1958).

Prell, H.: Die verlängerte Tragzeit der einheimischen Marderarten. Ein Erklärungsversuch. Zool. Anz. *88:* 17–31 (1930).

Psychoyos, A.: Etude des relations de l'œuf et de l'endomètre au cours du retard de la nidation ou des premières phases du processus de nidation chez la ratte. C. r. hebd. Séanc. Acad. Sci., Paris *263:* 1755 (1966).

Rumery, R. E. and Blandau, R. J.: Loss of zona pellucida and prolonged gestation in delayed implantation in mice; in Blandau, The biology of the blastocyst, pp. 115–129 (University of Chicago Press, Chicago 1971).

Short, R. V. and Hay, M.: Delayed implantation in the roe deer *(Capreolus capreolus)*; in Rowlands, Comparative biology of reproduction in mammals, pp. 173–194 (Academic Press, London 1966).

Wright, P. L.: Variations in reproductive cycles in North American mustelids; in Enders, Delayed implantation, pp. 77–98 (University of Chicago Press, Chicago 1963).

M. Bonnin, MD, Laboratoire d'Endocrinologie Expérimentale, Université de Bordeaux II, 146, rue Léo Saignat, F-33076 Bordeaux Cédex (France)

Prog. reprod. Biol., vol. 7, pp. 189–199 (Karger, Basel 1980)

Inhibition of Implantation by Advancement of Uterine Sensitivity and Refractoriness[1]

K. Yoshinaga

Laboratory of Human Reproduction and Reproductive Biology, and Department of Anatomy, Harvard Medical School, Boston, Mass.

Introduction

In the rat, the uterus becomes sensitive to the stimulus of blastocysts on Day 5 of pregnancy (the day when spermatozoa were found in the vaginal smear was designated as Day 1 of pregnancy), and implantation of blastocyst in the uterus takes place on the evening of this day. After Day 6 the uterus loses its sensitivity and does not respond to the stimulus of blastocysts. The uterus at this stage does not even respond to traumatization for induction of decidual response of the endometrium [*Peckam and Greene*, 1947, 1950; *De Feo*, 1958; *Noyes and Dickmann*, 1960]. *Psychoyos* [1963] divided the sequential events of the uterine sensitivity into three successive states: (1) neutral, (2) receptive, and (3) nonreceptive (refractory). The neutral state is attained under the influence of progesterone alone. In rats and mice, the neutral uterus does not respond to the presence of blastocysts and implantation is delayed. When estrogen becomes available after at least 2 days of progesterone influence, the uterus is rendered receptive to blastocysts and implantation takes place. The receptive state lasts for less than 24 h and the uterus automatically loses its receptivity and becomes refractory [*De Feo*, 1963]. Since the series of uterine events take place under the influence of sequential secretions of progesterone and estrogen which occur with special patterns in timing and in magnitude of secretion, measurement of ovarian hormone secretion during

[1] Supported in part by NIH-69-2203, NICHD.

early pregnancy provides important information as to the relationship of the physiological states of the uterus and its sensitivity to blastocysts. In this presentation, therefore, the interaction of ovarian hormone secretion pattern with ovum implantation will be discussed first. Then effects of exogenous hormones will be examined in relation to the uterine sensitivity for ovum implantation and decidual reaction.

Ovarian Hormone Secretion Pattern and Experimental Hormone Treatments

Corpora lutea of the estrous cycle in the rat are 'nonfunctional' in the sense that they do not secrete sufficient amounts of progesterone to prepare the uterus for acceptance of blastocysts. Although they are defined as 'nonfunctional', progesterone secretion does occur on the day of metestrus, soon after the formation of the corpora lutea [*Eto et al.,* 1962; *Telegdy and Endröczi,* 1963; *Hashimoto et al.,* 1968]. The secretion, however, is a temporary phenomenon and the secretion declines on the following day, the day of diestrus.

In pregnant rats, corpora lutea are made 'functional' by the mating stimulus which initiates two daily surges (diurnal and nocturnal) of prolactin [*Butcher et al.,* 1972]. The rate of progesterone output in the ovarian effluent is very low on Day 2 but it increases gradually to a peak on Day 14 [*Hashimoto et al.,* 1968]. According to our recent data [*Yoshinaga,* 1979], progesterone concentration in jugular venous serum measured at around noon of each day during the 1st week of pregnancy increased linearly from Day 1 to Day 5, and it plateaued thereafter. This pattern of progesterone during early pregnancy is in agreement with the results reported by *Fajer and Barraclough* [1967] and *Morishige et al.* [1973]. In pregnant mice, progesterone levels rise sharply during the first 2 days (from Day 1 to Day 3) and then remain approximately on the same level until Day 10 [*Murr et al.,* 1974]. On the other and, *McCormack and Greenwald* [1974] reported that the concentration of progesterone in peripheral plasma was low on Days 1 and 2, and reached a peak on Day 6.

The secretion pattern of estrogen during early pregnancy is reported to increase slightly on the afternoon of Day 4 [*Yoshinaga et al.,* 1969; *Shaikh,* 1971], or to rise rapidly and significantly on the afternoon of Day 2 and to maintain essentially a plateau from Day 3 to Day 8 [*Nimrod et al.,* 1972]. In this presentation, the secretion of estrogen that is essential to induce ovum implantation will be called 'the preimplantation estrogen secretion'. *Zeilmaker* [1963] ovariectomized pregnant rats at various times on Days 4 and 5 and the ovariectomized rats were supplemented with progesterone. When the ovaries were removed before 13.00 h on Day 4, implantation was prevented despite progesterone substitution. Therefore, the preimplantation estrogen secretion is needed to continue until 13.00 h on Day 4 in order to render the uterus receptive on Day 5. The requirement of the ovarian estrogen secretion through the afternoon of Day 4 has been shown by many workers [*Cochrane and Meyer*, 1957; *Psychoyos*, 1960; *Mayer*, 1961].

Pharmacological doses of LH-RH or its potent analogues have been shown to cause a delay in implantation in the rat [*Lin and Yoshinaga*, 1976; *Yoshinaga and Fujino*, 1979]. Recent evidence shows that the peptide hormones act directly on the ovary [*Hsueh and Erickson*, 1979; *Ying and Guillemin*, 1979; *Talbot and Arimura*, 1980] by reducing the secretion of progesterone and estrogen. In order to find out how late in early pregnancy an LH-RH analogue is effective in inhibiting implantation, a potent analogue (des-Gly^{10}-[D-Ala^6]-LHRH-ethylamide) was infused continuously from 12.00 h on Day 3, 12.00 h on Day 4 or 9.00 h on Day 5 until laparatomy on Day 8. Infusion was performed by a subcutaneously implanted osmotic minipump (Alzet) at the rate of 1 μg/24 h. In the rats into which infusion of the LHRH analogue had been started at 12.00 h on Day 3 or Day 4 no implantation swellings were observed in the uterus on Day 8. When their uteri were flushed with physiological saline, free blastocysts were found in the uterine flushings. In those rats into which the analogue infusion had been started at 9.00 h on Day 5, implantation of ova was not prevented, but the implanted embryos were found absorbed in some of the animals at a later stage. Because the time of ovariectomy is so close to the time of initiation of the analogue infusion in suppressing implantation, it is quite likely that the analogue exerted its action directly on the ovary. If the analogue exerted its action through the hypothalamo-pituitary axis, implantation should not have been prevented in the rats into which the analogue infusion was started at 12.00 h on Day 4. As has been shown by *Alloiteau* [1961] and *Zeilmaker* [1963], hypophysectomy or cessation of lactation in the morning of Day 4 results in prevention of implantation despite therapeutic treatment with progesterone.

From the viewpoint of preimplantation 'estrogen surge' [*Shelesnyak*, 1950] different procedures are followed to stimulate the uterus. For example, some investigators will treat ovariectomized rats first with progesterone alone for at least 2 days and then with a combination of estrogen and progesterone to induce implantation [*Psychoyos,* 1961, 1963; *Yoshinaga and Adams,* 1966]. On the other hand, other investigators will treat mated rats ovariectomized on Day 3 with an estrogen-progesterone combination immediately following ovariectomy [*Yochim and De Feo,* 1963; *De Feo and Dickman,* 1966; *Nutting and Meyer,* 1961]. No matter which treatment is used, the uterus responds maximally to the stimulation of blastocysts for implantation, or to chemical or mechanical stimuli for decidualization on Day 5.

Uteri, maximally sensitized on Day 5, will on the following day (Day 6) become automatically refractory [*De Feo,* 1967]. It appears that the chronological aspect of estrogen requirement in preparing implanting blastocysts is better understood than the quantitative aspect. This is due to the fact that a therapeutic regimen that induces implantation in ovariectomized animals does not necessarily reflect the secretory pattern of the ovarian hormones in intact animals. One rather clear pattern of ovarian hormone secretion shown under various physiological conditions is a circadian rhythm. The steroid secretion rate into the ovarian effluent is low in the morning and increases gradually to peak in the late afternoon [*Nimrod et al.,* 1963; *Yoshinaga,* 1974, 1975]. Significance of the late evening peaks in estrogen (or ovarian hormone) secretion remains to be clarified, if the 'estrogen surge' does not occur on Day 4 only.

Advancement of Uterine Sensitization

Since the levels of ovarian venous or peripheral progesterone are low on Days 1 and 2 of pregnancy, or of pseudopregnancy, it was decided to determine whether the administration of progesterone for 2 or more days from Day 1 (or 0) of pregnancy (or pseudopregnancy) and a single injection of estrogen on Day 3 together with progesterone would render the uterus precociously receptive. Since the refractory state automatically follows the receptive state, treatment of pregnant rats with 'the Psychoyos' scheme' (more than 2 days' progesterone and a single injection of estrogen at the end of the progesterone), it should be possible to render the uterus asynchronously refractory when fertilized ova arrive in the uterus. To test this hypothesis, a series of experiments were carried out by using (1) pregnant rats, (2) pseudopregnant rats and (3) lactating-pregnant rats.

Advancement of the Uterine Sensitivity in Pregnant Rats

Progesterone (2 or 5 mg/day) was injected s. c. to pregnant rats from Day 0 or 1 of pregnancy (Day 0 = the day of proestrus, the day before mating) to Day 3 or 8. Various doses of estradiol (10, 50, or 100 ng/4 μl sesame oil) were injected at one spot in the mesometrial adipose tissue of one uterine horn approximately at the middle of the horn. Estrogen thus administered has been shown to exert its action locally near the site of injection [*Yoshinaga*, 1961; *Psychoyos*, 1962; *Yoshinaga and Pincus*, 1963]. The rats were sacrificed on Day 9 and distribution and number of implantation sites relative to the site of locally injected estrogen were examined.

Distribution of implantation sites was abnormal in 19 out of 30 rats (60 %) thus treated. Distribution abnormality was due to lack of implants near the site of the estrogen injection in those rats that received the smallest dose of estrogen. Implants were not observed between the site of estrogen injection and the cervix in the rats which received the next highest dose of estradiol. With the highest dose of estradiol, the portion of the uterine horn that lacked implants spread to the contralateral horn [*Yoshinaga and Greep*, 1971, 1974]. Thus, locally administered estrogen was shown to prevent implantation by acting synergistically with an increased level of progesterone, presumably due to a precocious refractory state in a limited portion of the uterus.

Advancement of the Uterus Sensitivity in Pseudopregnant Rats

As has been already mentioned, the 'neutral' state of the uterus can be obtained by exposing the uterus to progesterone (5 mg/day) from Day 1 of pseudopregnancy. The uterus is rendered 'neutral' on Day 3, as long as the effect of hormone treatment is not interfered with by endogenous hormones. When, in addition to daily progesterone injections, a single injection of estradiol (0.2 μg) was given on Day 3, and on the same day the uterine endometrium was traumatized by scratching with a curved needle along the entire uterine horn, deciduoma was formed in 6 out of 6 rats [*Yoshinaga and Greep*, 1970]. The weight of decidualized horn was, however, approximately 50 % of the decidualized horn usually obtained by traumatizing the endometrium on Day 5 in untreated pseudopregnant rats. Although the decidual response in the progesterone-estrogen treated rats in the above-mentioned study is smaller than normal, advancement of the receptive stage by the hormone treatment is obvious because traumatization of the endometrium on Day 3 of pseudopregnancy did not induce decidual reaction in intact, untreated rats.

Advancement of the Uterine Sensitivity in Lactating-Pregnant Rats

It is well known that implantation is delayed during lactation in the rat and mouse [*Lataste,* 1891] due to reduced secretion of estrogen [*Weichert,* 1941; *Krehbiel,* 1941]. Suckling stimulus increases the release of prolactin from the pituitary. The greater the stimulus, the greater the secretion of prolactin. Thus, the larger the litter size, the more progesterone is secreted [*Yoshinaga et al.,* 1971]. This endocrine condition creates the 'neutral' state of the uterus for ovum implantation.

Using this system, inhibition of ovum implantation and decidualization was demonstrated by advancement (or asynchrony) of the uterine sensitivity at separate portions of the uterus. Female rats were mated at postpartum estrus (Day 1 of pregnancy) and allowed to nurse a litter of more than 8 suckling young in order to cause a delay in implantation. On Day 3 of pregnancy, the uterotubal junction was ligated on one side (sterile horn). On Day 5, 5 ng of estradiol was injected locally in one spot in the mesometrial adipose tissue of both the fertile and sterile horns. In the fertile horn, a blastocyst which happened to be close to the site of the estrogen injection implanted in response to the synergistic action of administered estrogen and the endogenous progesterone, secreted at a high level due to intensive lactation. Other blastocysts which were beyond the estrogen-sensitized portion of the horn remained unimplanted. In the sterile horn, the estrogen-sensitized portion became receptive and then refractory on the following day. The refractoriness of that portion of the uterus was maintained by endogenous progesterone. On Day 9, lactation was terminated by removing the litter from the mother. The cessation of lactation altered the function of the pituitary and increased gonadotropin secretion, which in turn, resulted in an increase in estrogen. Thus, the blastocysts which had remained unattached, implanted soon after the cessation of lactation. On Day 9, laparotomy revealed an implantation site close to the estrogen injection site. The sterile horn was traumatized on this day by scratching the endometrium with a curved needle along the entire horn. Since a limited portion of the sterile horn was made sensitive and then refractory by the locally administered estrogen, the traumatization resulted in decidualization of the entire uterine horn except for the refractory portion of the uterus. On Day 14 of pregnancy, the animals were sacrificed and the uterus examined. As can be seen in figure 1, the fertile horn (left) had 2 sets of embryos at different stages of development (superimplantation). The large swelling seen in the left horn contained an embryo corresponding to Day 14 of age of normal pregnancy which implanted at the

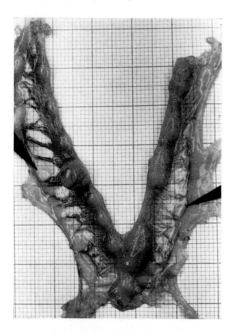

Fig. 1. Rat uterus on day 14 of pregnancy showing asynchrony of uterine sensitivity: the fertile horn (left) and the sterile horn (right).

site of the locally injected estrogen (indicated by a black arrow). The small implantation sites (3 above and 3 below the large implantation site) contained embryos that corresponded to those of day 8–9 of normal pregnancy, which had implanted after cessation of lactation. In the sterile horn (right) deciduomata were observed along the entire horn except in the portion which had been made refractory before the time of traumatization (near the arrow). A similar localized asynchrony in the uterine sensitivity for decidualization was demonstrated in ovariectomized pregnant rats treated with progesterone [*Yoshinaga*, 1977].

Conclusion

In the rat, ovarian hormones secreted during early pregnancy play important roles in preparing the uterine endometrium for implanting blastocysts. When normal secretion is reduced before noon of Day 4, physiologically or artificially, preparation of the uterus remains *incomplete* and implantation is prevented. This prevention of the uterus from appropriate preparation is *one* of the causes that lead to implantation failure. Failure of implantation can also result from excessive stimulation of the ovaries with gonadotrophins [*Yang and Chang,* 1968; *Yoshinaga,* 1979]. In this case preparation of the uterus is altered in such a way that implantation is impossible. This latter case should be distinguished from the first case hormone levels, resulting in implantation, whereas in the latter case implantation is not delayed due to abnormally high levels of estrogen.

There is still another way of preventing implantation. The natural chronological ovarian secretion renders the uterus receptive to blastocysts on Day 5, but the uterus automatically becomes refractory from Day 6 onwards provided the uterus is exposed to progesterone (and estrogen). Advancement of completion of the uterine preparation by exogenous ovarian hormones while fertilized ova are still in the Fallopian tubes, results in the prevention of implantation. This is due to the asynchronous precocious establishment of the refractory state when fertilized ova arrive into the uterus. This last method appears to be the least dangerous to the mother, since interference of implantation is achieved by making use of the physiological principles governing the uterus.

References

Alloiteau, J.-J.: Hypophysectomie au début de la gestation et nidation de l'œuf chez la ratte. C. r. hebd. Séanc. Acad. Sci., Paris *253:* 1348–1350 (1961).

Butcher, R. L.; Fugo, N. W., and Collins, W. E.: Semi-circadian rhythm in plasma levels of prolactin. Endocrinology *90:* 1125–1127 (1972).

Cochrane, R. L. and Meyer, R. K.: Delayed nidation in the rat induced by progesterone. Proc. Soc. exp. Biol. Med. *96:* 155–159 (1957).

De Feo, V. J.: Duration of uterine sensitivity in the pseudopregnant rat. Anat. Rec. *130:* 292 (1958).

De Feo, V. J.: Determination of the sensitive period for the induction of deciduomata in the rat by different inducing procedures. Endocrinology *73:* 488–497 (1963).

De Feo, V. J.: Decidualization; in Wynn, Cellular biology of the uterus, pp. 191–290 (Appleton-Century-Crofts, New York 1967).

De Feo, V. J. and Dickmann, Z.: Hormonal control of embryo-uterine interrelationships from early to late pregnancy in the rat; 2nd Int. Congr. Hormonal Steroids, Milan. Excerpta Med. Inc. Congr. Ser. *III:* 202 (1966).

Eto, T.; Masuda, H.; Suzuki, Y., and Hosi, T.: Progesterone and pregn-4-en-20α-ol-3-one in rat ovarian venous blood at different stages in reproductive cycle. Jap. J. Anim. Reprod. *8:* 34–40 (1962).

Fajer, A. B. and Barraclough, C. A.: Ovarian secretion of progesterone and 20α-hydroxypregn-4-en-3-one during pseudopregnancy and pregnancy in rats. Endocrinology 81: 617–622 (1967).

Hashimoto, I.; Henricks, D. M.; Anderson, L. L., and Melampy, R. M.: Progesterone and pregn-4-en-20α-ol-3-one in ovarian venous blood during various reproductive states in the rat. Endocrinology 82: 333–341 (1968).

Hsueh, A. J. W. and Erickson, G. F.: Extrapituitary action of gonadotropin-releasing hormone: direct inhibition of ovarian steroidogenesis. Science, N. Y. 204: 854–855 (1979).

Krehbiel, R. H.: The effects of theelin on delayed implantation in the pregnant lactating rat. Anat. Rec. 81: 381–392 (1961).

Lataste, F.: Des variations de durée de la gestation retardée. C. r. Séanc. Soc. Biol. 9: 21–31 (1891).

Lin, Y. C. and Yoshinaga, K.: Inhibitory effect of Gn-RH on ovum implantation in the rat. 58th Anna. Meet. Endocr. Soc., San Francisco, Calif., abstr. 174 (1976).

Mayer, G.: Delayed nidation in rats: a method of exploring the mechanisms of ovo-implantation; in Enders, Delayed implantation, pp. 213–228 (University of Chicago Press, Chicago 1961).

McCormack, J. T. and Greenwald, G. S.: Progesterone and oestradiol-17β concentrations in the peripheral plasma during pregnancy in the mouse. J. Endocr. 62: 101–107 (1974).

Morishige, W. K.; Pepe, G. J., and Rothchild, I.: Serum luteinizing hormone, prolactin and progesterone levels during pregnancy in the rat. Endocrinology 92: 1527–1530 (1973).

Murr, S. M.; Stabenfelt, G. H.; Bradford, G. E., and Geschwind, I. I.: Plasma progesterone during pregnancy in the mouse. Endocrinology 94: 1209–1211 (1974).

Nimrod (Zmigrod), A.; Ladany, S., and Lindner, H. R.: Prenidatory ovarian oestrogen secretion in the pregnant rat, determined by gas chromatography with electron capture detection. J. Endocr. 53: 249–260 (1972).

Noyes, R. W. and Dickmann, Z.: Relationship of ovular age to endometrial development. J. Reprod. Fertil. 1: 186–196 (1960).

Nutting, E. F. and Meyer, R. K.: Implantation delay, nidation, and embryonal survival in rats treated with ovarian hormones; in Enders, Delayed implantation, pp. 233–251 (1961).

Peckham, B. M. and Greene, R. R.: Production of secondary deciduomata in the castrated and lactating rat. Endocrinology 41: 277–281 (1947).

Peckham, B. M. and Greene, R. R.: Endocrine influences on implantation and deciduoma formation. Endocrinology 46: 489–493 (1950).

Psychoyos, A.: Nouvelle contribution à l'étude de la nidation de l'œuf chez la ratte. C. r. hebd. Séanc. Acad. Sci., Paris 251: 3073–3075 (1960).

Psychoyos, A.: Nouvelles recherches sur l'ovimplantation. C. r. hebd. Séanc. Acad. Sci., Paris 252: 2306–2307 (1961).

Psychoyos, A.: Production expérimentale des superimplantations ovulaires chez le rat. C. r. hebd. Séanc. Acad. Sci., Paris 254: 1504–1506 (1962).

Psychoyos, A.: Précisions sur l'état de non-receptivité de l'utérus. C. r. hebd. Séanc. Acad. Sci., Paris 257: 1153–1156 (1963).

Shaikh, A. A.: Estrone and estradiol levels in the ovarian venous blood from rats during the estrous cycle and pregnancy. Biol. Reprod. *5:* 297–307 (1971).

Shelesnyak, M. C.: Nidation of the fertilized ovum. Endeavour *19:* 81–86 (1960).

Talbot, S. and Arimura, A.: Luteinizing hormone receptor reduction by (*D*-Trp[6]) LHRH may involve mechanisms other than 'down regulation'. 6th Int. Congr. Endocr., Melbourne, abstr. 736 (1980).

Telegdy, G. and Endröczi, E.: The ovarian secretions of progesterone and 20α-hydroxypregn-4-en-3-one in rats during the estrous cycle. Steroids *2:* 119–123 (1963).

Weichert, C. K.: The effectiveness of estrogen in shortening delayed pregnancy in the rat. Anat. Rec. *81:* 106 (1941).

Yang, W. H. and Chang, M. C.: Interruption of pregnancy in the rat and hamster by administration of PMS or HCG. Endocrinology *83:* 217–233 (1968).

Ying, S.-Y. and Guillemin, R.: (*D*-Trp[6]-Pro[9]-NET)-luteinizing hormone-releasing factor inhibits follicular development in hypophysectomised rats. Nature, Lond. *280:* 593–595 (1979).

Yochim, J. M. and De Feo, V. J.: Hormonal control of the onset, magnitude and duration of uterine sensitivity in the rat by steroid hormones of the ovary. Endocrinology *72:* 317–326 (1963).

Yoshinaga, K.: Effect of local application of ovarian hormones on the delay in implantation in lactating rats. J. Reprod. Fertil. *2:* 35–41 (1961).

Yoshinaga, K.: Daily rhythmic secretion of progestin by the ovary in lactating rats. J. Steroid Biochem. *5:* 386 (1974).

Yoshinaga, K.: Hormonal interplay in the establishment of pregnancy; in Greep, International review of physiology: reproductive physiology II; vol. 13, pp. 201–223 (University Park Press, Baltimore 1977).

Yoshinaga, K.: Suppression of ovarian function by LHRH and its analogues in pregnant rats; in Channing, Marsh and Sadler, Ovarian follicular and corpus luteum functions, pp. 729–734 (Plenum, New York 1979).

Yoshinaga, K. and Adams, C. E.: Endocrine aspects of egg implantation in the rat. J. Reprod. Fertil. *12:* 583–586 (1966).

Yoshinaga, K. and Fujino, M.: Hormonal control of implantation in the rat: inhibition by luteinizing hormone-releasing hormone and its analogues. Ciba Fdn Symp. 64 (new series), pp. 85–105 (1979).

Yoshinaga, K. and Greep, R. O.: Precocious sensitization of the uterus in pseudopregnant rats. Proc. Soc. exp. Biol. Med. *134:* 725–727 (1970).

Yoshinaga, K. and Greep, R. O.: Local inhibition of ovo-implantation in the rat. Endocrinology *88:* 627–632 (1971).

Yoshinaga, K. and Greep, R. O.: Uterine sensitivity with regard to ovo-implantation; in Husain and Guttmacher, Progress in reproduction research and population control, pp. 137–146 (Publications International, Quebec 1974).

Yoshinaga, K.; Hawkins, R. A., and Stocker, J. F.: Estrogen secretion by the rat ovary *in vivo* during the estrous cycle and pregnancy. Endocrinology *85:* 103–112 (1969).

Yoshinaga, K. and Lin, Y. C.: LH involvement in the circadian rhythm of ovarian progestin secretion in pseudopregnant rats. Fed. Proc. *34:* 323 (1975).

Yoshinaga, K.; Moudgal, N. R., and Greep, R. O.: Progestin secretion by the ovary in lactating rats: effect of LH-antiserum, LH and prolactin. Endocrinology *88:* 1126–1130 (1971).

Yoshinaga, K. and Pincus, G.: Local effect of estrogen on cholesterol synthesis in the uterus of ovariectomized rats. Steroids *1:* 656–663 (1963).

Zeilmaker, G. H.: Experimental studies on the effects of ovariectomy and hypophysectomy on blastocyst implantation in the rat. Acta endocr., Copenh. *44:* 355–366 (1963).

Zeilmaker, G. H.: Quantitative studies on the effect of the suckling stimulus on blastocyst implantation in the rat. Acta endocr., Copenh. *46:* 483–492 (1964).

K. Yoshinaga, MD, Reproductive Sciences Branch, Center for Population Research, NICHD, NIH, Bethesda, MD 20205 (USA)

Prog. reprod. Biol., vol. 7, pp. 200–215 (Karger, Basel 1980)

Initiation of Implantation at the Subcellular Level

F. Leroy, G. Schetgen and M. Camus

Laboratory of Gynaecology and Research on Human Reproduction,
Saint-Pierre Hospital, Free University Brussels, Brussels

A model indicating how implantation might be triggered has been previously proposed [30] and the present review will attempt to further analyze this concept. Again, we shall mainly rely on data obtained from rats and mice which have been the most extensively studied species. Although the decidual cell reaction will be used as a criterion for the initiation of nidatory changes, its intimate nature is left to be dealt with elsewhere in these proceedings.

On the endometrial side, our main concern will be some of the events occurring in principle at the level of surface epithelial cells that surround the blastocyst. The role of these cells is considered central to implantation and can be visualized as threefold: (i) regulating the blastocyst's metabolism through the control of its humoral environment [13, 37, 43, 55]; (ii) providing an appropriate matrix for changes to occur at the interface between throphoblast and epithelium [12]; (iii) transmitting information towards the underlying stroma [17, 43].

After these few remarks it will be recalled how our working hypothesis was conceived. Some results of its experimental probing will be given and we shall try to correlate our conclusions with relevant information from the recent literature.

The concept which was arrived at stemmed from an apparent contradiction between two sets of published data concerning the role of estrogen-dependent RNA synthesis in the onset of implantation. On one side were the data presented by *Segal et al.* [47] indicating that uterine RNA obtained from estrogen-treated ovariectomized rats would induce implantation by delayed blastocysts. These results were obtained by topical injection into the parametrial fat of castrated pregnant animals that were maintained

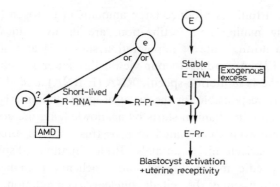

Fig. 1. Schematic hypothesis for chain of events involved in the onset of implantation. E = Proestrus estrogen; P = luteal progesterone; e = nidatory estrogen; Pr = proteins; R = repressive; AMD = actinomycin D.

on progesterone treatment. Conversely, it was found by *Finn* [11] and *Finn and Downie* [16], using the same type of recipients in mice, that implantation could be initiated by a single systemic injection of actinomycin D. In each case the experimental procedure thus mimics the action of trace amounts of estrogen which are known to be requisite for implantation in both species [17, 43].

As evidence has accumulated indicating that stimulation of RNA synthesis is essential to estrogen action on uterine cells [47], it was surprising that a potent inhibitor of transcription [7] should be capable of initiating nidatory changes. The results, however, were clear-cut and independently confirmed [1]. Against the interpretation of results obtained with exogenous RNA, it could be argued that such extracts might be contaminated with enough estrogen to elicit implantation since the dose required locally is very small [43, 47, 59]. However, because control experiments had shown that RNA submitted to specific enzymic degradation or nonuterine RNA were ineffective [47], this objection appeared untenable. Also, from data obtained with highly labelled estradiol, we were able to calculate that possible contamination of RNA topically applied in *Segal's* experiments, was far below the minimal dose of estrogen capable of inducing implantation [30].

The working hypothesis by which we have tried to resolve this conflict proposes that implantation basically depends on two paths of transcriptional and translational events of which the interaction has to be disrupted by the nidatory stimulus (fig. 1).

At proestrus, ovarian follicles produce large amounts of estrogen (E) which, besides allowing mating and fertilization, are involved in the further production of a normal rate of implantation sites [23] and adequate decidualization [17]. According to our views, this latter secretion would stimulate the synthesis of uterine-specific RNA (E-RNA), coded for proteins (E-Pr) which are responsible for the onset of implantation. Argumentation in favor of such a mechanism starts by acknowledging the close similarity between uterine events correlated to proestrus and the effects of estradiol given to ovariectomized animals. Both situations display a common series of specific morphological and biochemical changes. Among those modifications may be cited: nuclear accumulation of estrogen-receptor complex [6, 58], estrogen receptors' replenishment [27, 58], general protein and IP synthesis [6, 25], epithelial cells' hypertrophy [12, 47] and proliferation [12, 29]. Also the shift from predominantly glandular towards stromal mitosis which requires estrogen priming before artificial uterine sensitization in ovariectomized mice [17], closely mimics the normal course of events during early pregnancy in this species. Therefore, it appears that proestrus estrogen, being released when the uterus is under no other hormonal influence, exerts its effects through the classical mechanism of steroid action, validated for 17β-estradiol in immature or ovariectomized animals (i. e. binding to cytoplasmic receptors entailing nuclear translocation and subsequent effect on transcription and genetic expression). Its becomes logical, then, to postulate that during the normal preovulatory period, the synthesis of RNA is stimulated, which is endowed with the same nidatory properties as that extracted by *Segal* and his associates from estrogen-stimulated uteri.

However, since blastocysts do not attach before several days after proestrus, this specific E-RNA has to remain inactive and stable. Variability of mRNAs half-lives is currently admitted to be involved in the regulation of gene expression and there are well-documented examples of delayed translation, namely in relation to development and differentiation [20, 24, 46].

Induction of implantation by estrogen-dependent exogenous RNA remains to be confirmed. Some of our work, however, bears on the role that may be assigned to proestrus estrogen. Ovariectomized mice were given hormonal treatment mimicking progestational ovarian secretions in order to provide maximal uterine sensitivity (fig. 2). In this sequence, priming with estrogen, equivalent to proestrus secretion, is requisite to obtain a full-blown decidual response [17]. This pretreatment with estradiol

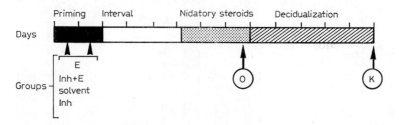

Fig. 2. Effect of transcriptional and translational inhibitors and estrogen priming on decidual cell reaction in ovariectomized mice. Basic experimental program.
Priming treatments: E = 150 ng 17β-estradiol daily; Inh = inhibitor (see legend to
figures 3 and 4); interval = no treatment; nidatory steroids = 500 μg progesterone +
10 ng 17β-estradiol daily. Treatment during decidualization: 400 μg progesterone +
40 ng 17β-estradiol daily; 0 = intrauterine injection of oil; K = killing for assessment of decidual reaction.

was combined with different inhibitors of which the interference could be
assessed later on in terms of amplification of the decidual reaction which
was induced by oil at 6 days after priming. Under such conditions (fig. 3,
4) inhibitors of mRNA processing, such as α-amanitin and cordycepin,
while having no effect in nonprimed controls, were found responsible
for a significant reduction of the decidual reaction in animals pretreated
with estradiol. This was not the case with cycloheximide which, given in
a similar manner, had no effect whatsoever on decidualization. As far as
they go, these results are compatible with the hypothesis that the action
of proestrus estrogen in implantation requires transcriptional stimulation but does not necessitate protein synthesis to *immediately* ensue (fig. 1).

From experiments on ovariectomized animals, it appears that estrogen
priming is not quite necessary for implantation to occur, but is needed to
normalize its yield [17, 23]. It is possible that under nonpriming conditions
extraovarian sources of estrogen might act as a weak surrogate allowing
low-rate implantation. Another explanation would be that, in this respect, the role of proestrus secretion is to *amplify* some specific RNA
synthesis maintained at basic level in the uterus, rather than to inaugurate
transcription of original genomic sequences.

In rats and mice, without the action during progestation of small
amounts of estrogen, implantation is prevented and blastocysts are kept
into metabolic dormancy as long as high progesterone levels are
maintained [37, 43]. This latter state appears to be due to liberation of

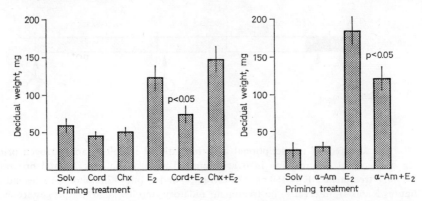

Fig. 3, 4. Two experiments on the effect of inhibitors and estrogen priming on decidualization (animals basically treated as indicated in fig. 2). Daily priming treatments: Solv = solvent controls; α-Am = 10 μg α-amanitin; E_2 = 150 ng 17β-estradiol; Cord = 2 mg 3'-deoxyadenosine (cordycepin); Chx = 60 μg cycloheximide.

inhibitory proteins into the uterus only dominated by progesterone. First suggested by the finding that diapausal embryos resume proliferative activity and invade surrounding tissues when transplanted at extrauterine sites [28, 36, 42], this conclusion was confirmed by recent work analyzing the metabolic behavior of blastocysts incubated *in vitro,* in the presence of uterine flushings obtained under various hormonal conditions [43, 44, 55, 56]. We were led to assume that synthesis of inhibitory proteins (R-Pr) depends on a sustained output of short-lived RNA, thereby enabling actinomycin D to induce implantation through unblocking translation of stable E-RNA. The same final effect would be obtained with topical application of uterine RNA from estrogenized-spayed animals, by mere quantitative overcoming of the inhibitory process (fig. 1). In such species as rats and mice, release from inhibition allowing implantation to proceed, would normally be effected by luteal estrogen secretion [13, 43].

Although the mechanism by which progesterone affects uterine cells is far from elucidated it is clear that this hormone profoundly modifies their response to estrogen. As opposed to pure estrogenic effects, the capacity of inducing epithelial cell division by estradiol is blocked by progesterone, while stromal mitosis becomes strongly stimulated [35]. Still another difference is that a relatively small amount of estradiol can elicit the complete sequence of nidatory modifications in the progesterone-treated uterus [13, 43], whereas about 20 times as much is required to obtain a full biosynthetic response in ovariectomized animals [18]. These features hint at a fundamental difference in the mechanism of estrogen action, depending on whether progesterone is involved or not.

Additional evidence was available from the work by *Finn and Martin* [17] who showed that actinomycin D given before final sensitization of the progestational uterus by estradiol, could not inhibit the sequence of early nidatory changes depending on this hormone, i. e., the 'second stage' of uterine closure, increased vascular permeability and edema of the stroma as well as ultimate decidualization in response to an oil stimulus. Also blastocysts transferred to actinomycin D-treated foster mothers were not prevented from implanting [15]. Quite to the contrary, this drug allowed delayed blastocysts to attach [11, 16].

Several questions arise in relation to the nidatory properties of actinomycin D: (i) Is implantation obtained through a direct action on uterine cells or is it somehow mediated through metabolic changes occurring outside the uterus? (ii) Is this effect truly attributable to antitranscriptional activity rather than to an unknown ancillary action of the drug? (iii) If so: can it be obtained by direct inhibition of translation as well as through preventing RNA synthesis? (iiii) Do the nidatory capabilities of actinomycin D apply to other species than the mouse?

Our results regarding these problems are presented in terms of implantation rates by delayed blastocysts in tables I and II [5, 10]. Because of higher general toxicity in the rat, it was not possible to obtain a significant number of animals with implantation sites in this species when actinomycin D was given by intraperitoneal route. However, when the drug was applied locally at a relatively low dose, implants were found in 70 % of cases, only in the homolateral horn. Actinomycin D would thus precipitate implantation through direct action on the uterus. Results with α-amanitin were inconclusive, while cycloheximide remained largely inactive. The model (fig. 1) would predict this latter inefficiency since direct inhibition of translation would prevent all protein synthesis, including that of implantation-inducing factors (E-Pr).

Table II shows that cordycepin injections induced implantation in nearly 50 % of mice when repeated at 3-hour intervals. In the total of animals treated with this substance a lower rate of pregnancy was found than in control groups ($p < 0.001$). This points to the possibility of direct toxicity to delayed blastocysts. Therefore, only positive results can be interpreted with some confidence. It appears, however, that another inhibitor of RNA processing than actinomycin D, is capable of initiating

Table I. Effect of systemic administration of actinomycin D and of intraparametrial injections of actinomycin D, α-amanitin and cycloheximide on the implantation rate of delayed blastocysts in the rat

Groups	Treatment	Total number of rats treated	Number of rats with implantation sites	Number of implantation sites/rat
1	actinomycin D 1 × 25 μg intraperitoneally	10 (2)	0	0
2	actinomycin D 1 × 50 μg intraperitoneally	6 (1)	0	0
3	actinomycin D 1 × 75 μg intraperitoneally	15 (9)	2	5.5
4	actinomycin D 3 × 10 or 15 μg into parametrium	14	10	2.5
5	α-amanitin 3 × 5 or 10 μg into parametrium	21	4	7.5
6	cycloheximide 3 × 40 or 60 μg into parametrium	15	1	4
7	solvent (H$_2$O) 3 × 0.02 ml into parametrium	8	0	0
8	17β-estradiol 3 × 10 ng into parametrium	15	13	3.6

Figures in parentheses indicate animals that died before the end of the experiment.
Statistical significance of differences in implantation rate:
Group 8 is significantly different from all other groups but group 4 ($p \leq 0.001$ for groups 1, 2, 5, 6 and 7; $p < 0.03$ for group 3).
Group 4 is significantly different from groups 1, 2, 5, 6 and 7 at $p < 0.002$, $p = 0.01$, $p < 0.001$, $p < 0.01$, $p < 0.001$ and $p < 0.002$, respectively.
Other differences are not significant.
Animals were ovariectomized without damaging Fallopian tubes at day 3 of pregnancy (day 1: spermatozoa in vaginal smear) and further treated with 5 mg progesterone/day. Injections into parametrium were evenly spaced along one uterine horn.

Table II. Effect of 3′-deoxyadenosine on the rate of implantation of delayed blastocysts in mice

Groups	Treatment	Number of mice treated[1]	Number of mice pregnant[2] at autopsy	Number of free blasto-cysts/mouse	Number of mice with im-plantation sites	Number of implantation sites/mouse
1	20 ng 17β-estradiol	13	13	–	13	4 ±
2	1 × 2 mg cordycepin	12	6	3.5	0	–
3	2 × 1 mg cordycepin (I = 24 h)	9	6	2.5	0	–
4	3 × 1 mg cordycepin (I = 3 h)	16	12	3	5	2.8
5	4 × 1 mg cordycepin (I = 12 h)	15	8	4	0	–
6	3 × 1 mg adenosine (I = 3 h)	15	14	2.4	0	–
7	3 × 0.3 ml NaCl 0.9% (I = 3 h)	15	13	3	0	–

[1] Animals with a vaginal plug at the beginning of the experiment (day 1).

[2] Animals with free intrauterine blastocysts or implantation sites.

Animals were ovariectomized without damaging Fallopian tubes at day 3 of pregnancy and further treated with 1 mg progesterone/day. Treatments were given intraperitoneally in 0.3 ml 0.9% NaCl solution, except 17β-estradiol which was given in propylene glycol. I = Interval between injections.

Statistical evaluation of group differences in numbers of mice with implantation:

Group 4 vs. 6 and 7: in terms of number of mice initially treated (¹): p = 0.026 and p = 0.026, respectively.

Group 4 vs. 6 and 7: in terms of number of mice pregnant at autopsy (²): p = 0.013 and p = 0.015, respectively.

Group 1 vs. others: all differences significant (p < 0.01).

implantation. In view of the alleged specific action of 3′-deoxyadenosine [33], the RNA involved might be of messenger nature. Also the hypothesis attributing an inhibitory role to RNA endowed with a rapid turnover (fig. 1), is consistent with the finding that cordycepin injections have to be repeated at short intervals to enable implantation to occur.

Glasser and McCormack [19] have recently shown in the rat that pretreatment with progesterone considerably amplifies transcriptive activity of uterine chromatin, but that a superimposed dose of estradiol entails within 4 h an abrupt fall of the number of sites available for RNA chain initiation. This decrease thus occurs while the uterus is becoming fully sensitized, particulary at the level of the epithelium [17]. Together with data on implantation effected by transcriptional inhibitors, this latter information allows to conclude that nidatory estrogen quantitatively operate by inhibition of RNA synthesis rather than by its stimulation. It remains to be defined to which tissular component(s) this restraining mechanism applies and also if it is triggered by direct action on chromatin or at some other level of the cell machinery.

It has been repeatedly suggested in the case of the uterus [3, 21, 38, 39], as well as for other target systems [48, 52] that steroid hormones might control RNA translation directly and extensive reviews have emphasized that posttranscriptional control applies to a large variety of endocrine and nonendocrine systems [46, 53]. In experiments where implantation was obtained with actinomycin D as the sole inducing agent, embryonic development and uterine response were less advanced than in animals receiving estradiol in addition to the drug [16]. According to the molecular pattern suggested in figure 1, this delay could be related to the turnover time of repressive proteins (R-Pr) or RNA (R-RNA), while estrogen would be expected to alleviate E-RNA repression in a more direct manner and therefore more rapidly.

The uterus, rendered receptive by luteal estrogen only for a limited period, becomes thereafter hostile to unimplanted blastocysts and incapable of decidualizing [43]. Since uterine refractoriness occurs as the end effect of nidatory estrogen and provides therefore a good criteron for their action, we have investigated this state in relation to transcriptional and translational events. The comparison between groups 1 and 2 in table III confirms that under chronic progesterone treatment, it is impossible to resensitize the uterus to decidual stimuli [43], once it has already gone through a previous receptivity-nonreceptivity sequence induced by a first dose of estradiol. Actinomycin D given instead of (group

Table III. Effect of actinomycin D and cycloheximide on the uterine refractory state induced by nidatory estrogen in ovariectomized rats

Treatment groups	Day of treatment before decidual induction							Number of rats with decidual response	Weight (mg + SEM) of cornua treated by	
	5	6	7	8	9	10	11		oil	scratching
1	P	P	P	P	P	P+e	Ind.	8/8	696 ± 103	1,276 ± 75
2	P	P	P+e	P	P	P+e	Ind.	0/8	95 ± 6	93 ± 5
3	P	P	P+Act.D	P	P	P+e	Ind.	8/8	556 ± 110	836 ± 106
4	P	P	P+Act.D+e	P	P	P+e	Ind.	1/6	174 ± 77	235 ± 134
5	P	P	P+Chx	P	P	P+e	Ind.	7/7	539 ± 97	1,024 ± 214
6	P	P	P+Chx+e	P	P	P+e	Ind.	8/8	401 ± 95	755 ± 88

All groups were given 500 ng 17β-estradiol on days 1 and 2 and no treatment on days 3 and 4. P = 3 mg progesterone; e = 50 ng 17β-estradiol; Act.D = 50 μg actinomycin D; Chx = 500 μg cycloheximide. After induction (Ind.) all animals received daily 2 mg progesterone + 200 ng 17β-estradiol for 5 days after which they were killed.

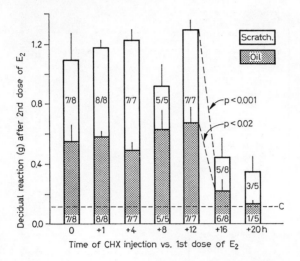

Fig. 5. Effect of delaying cycloheximide treatment on the uterine refractory state induced by estrogen in rats. Basic treatment as in table III, group 6. $E_2 = 17\beta$-estradiol; CHX = cycloheximide. Ratios in bars indicate number of rats with decidual reaction per group. C = Nondecidualized control level.

3) or prior to (group 4) the first estradiol injection, can neither elicit nor prevent uterine refractoriness [31]. On the contrary, under similar experimental conditions, 500 μg cycloheximide proved able to fend off the effect of estradiol, the uterus being maintained in its 'neutral' state (group 6). Using the same basic schedule, it was found through delaying cycloheximide administration after giving estradiol, that protein synthesis which is involved, takes place between 12 and 16 h after estrogen treatment [32]. Beyond that interval, uterine desensitization can no longer be totally avoided (fig. 5). Therefore, it would seem that nidatory estrogen somehow stimulates protein synthesis by acting at a posttranscriptional level.

The role played by steroid receptors being discussed in a following paper by *Martel et al.,* we shall only briefly allude to an aspect of estrogen binding wich might also be consistent with an extratranscriptional mechanism for triggering implantation. Autoradiographic data have indicated that in contrast to observations in nonprimed ovariectomized animals, tritiated estradiol no longer accumulates in the nuclei of surface epithelial cells while it still heavily labels those of glands and stroma, in the progesterone-dominated uterus [49, 51]. In contradiction to these findings, some biochemical results would suggest that progesterone priming does not modifiy estrogen receptors in epithelial nuclei [26, 45]. However, by using the labelled estradiol exchange assay, *Vass and Green* [54] have recently shown that progesterone pretreatment entails about a 50 % reduction of nuclear estradiol receptors in isolated uterine epithelium. In such studies, there is room, moreover, for significant contamination between tissular fractions. Also artefactual alteration and relocation of estrogen-receptor complexes during tissue processing cannot be excluded How luteal estrogen induces implantation thus remains largely open to controversy and further investigation.

Removal of the inhibition exerted through specific gene expression in the uterine milieu provides an acceptable hypothesis for a fundamental triggering mechanism of implantation. Such a mode of control, eventually combined with production of stimulatory factors [37, 50, 57] as suggested by our model, might apply to species endowed with compulsory delayed implantation [2] as well as to laboratory rodents [5, 13]. Data in rats and mice indicate the disappearance at implantation of a low molecular weight substance which is secreted in the progesterone-dominated uterus [2, 40, 43]. It remains doubtful, however, whether the synthesis of such an inhibitor is related to luteal function, for blastocysts refuse to implant in the uterus of normally cycling recipients [8] and become metabolically inactive in uterine flushings from ovariectomized animals [44]. Also during obligatory delay, the overall luteal activity is generally considered to be low [2]. In favor, however, of hormonally mediated inhibition is the case of the armadillo in which ovariectomy during delay, precipitates implantation [4].

So far, the role of luteal estrogen in removing intrauterine inhibition has been demonstrated only in rats and mice [13, 43]. In a number of other cases it would seem that progesterone is the only hormonal requisite to implantation [2, 14]. An alternative source of estrogen might be the blastocyst itself [9, 22], but the general applicability of this concept and its relevance to the onset of implantation have yet to be worked out. We are thus left with the need to look for more complex triggering mechanisms than estrogen action alone. When involved, this latter would seem to be paramount at the level of the uterine surface epithelium for enabling the induction of nidatory changes [17]. Therefore, it may be that species differences as regards hormonal requirements for implantation are to some extent related to the variety of uterine epithelium-trophoblast relationships displayed by eutherian mammals [14].

It has been long known that in spite of being the privileged receptacle of gestation, the uterus is endowed with the inherent property shared with no other tissue, to *prevent* implantation, except at transient hormonally defined moments [34]. This fundamental paradox still largely escapes our understanding, for we have no general hormonal pattern to rely on, let alone a basic explanation in metabolic and molecular terms.

Acknowledgments

Personal results given in this paper were obtained thanks to a grant from the Belgian FRSM. *L. F.* and *C. M.* are 'Chercheur Qualifié' and 'Aspirant' at the Belgian FNRS, respectively. We also thank Mrs. *Van Hoeck* and Mrs. *Bogaert* for technical assistance.

References

1   Aitken, R. J.: The influence of actinomycin D on the protein composition of mouse uterine flushings. J. Reprod. Fertil. *50:* 193–195 (1977).

2   Aitken, R. J.; Bonnin, M. et Canivenc, R.: Léthargie et activation du blastocyste; in du Mesnil du Buisson, Psychoyos and Thomas, L'implantation de l'œuf, pp. 255–272 (Masson, Paris 1978).

3   Barker, K. L.; Adams, D. J., and Donohue, T. M., jr.: Regulation of the levels of mRNA for glucose-6-phosphate dehydrogenase and its rate of translation in the uterus by oestradiol; in Glasser and Bullock, Cellular and molecular mechanism of implantation (Plenum Press, New York, in press 1980).

4   Buchanan, G. D.; Enders, A. C., and Talmage, R. V.: Implantation in armadillos ovariectomized during the period of delayed implantation. J. Endocr. *14:* 121–128 (1956).

5   Camus, M.; Lejeune, B., and Leroy, F.: Induction of implantation in the rat by intraparametrial injection of actinomycin D. Biol. Reprod. *20:* 1115–1118 (1979).

6   Clark, J. H.; Anderson, J., and Peck, E. J., jr.: Receptor-estrogen complex in the nuclear fraction of rat uterine cells during the estrus cycle. Science *176:* 528–530 (1972).

7   Cooper, H. L. and Braverman, R.: The mechanism by which actinomycin D inhibits protein synthesis in animal cells. Nature, Lond. *269:* 527–529 (1977).

8   Cowell, T. P.: Implantation and development of the mouse eggs transferred to uteri of non-progestational mice. J. Endocr. *49:* 345–346 (1969).

9   Dickmann, Z.; Dey, S. K., and Sen Gupta, J.: A new concept: control of early pregnancy by steroid hormones originating in the preimplantation embryo. Vitams Horm. *34:* 215–242 (1976).

10 Fernandez-Noval, A. and Leroy, F.: 3'-Deoxyadenosine and implantation of delayed blastocysts in mice. J. Endocr. *81:* 351–354 (1979).

11 Finn, C. A.: The induction of implantation in mice by actinomycin D. J. Endocr. *60:* 199–200 (1974).

12 Finn, C. A.: The endometrium; in Finn and Porter, The uterus, pp. 17–104 (Elek Science, London 1975).

13 Finn, C. A.: The implantation reaction; in Wynn, Cellular biology of the uterus (Plenum Press, New York 1977).

14 Finn, C. A.: Species variation in implantation. Prog. reprod. Biol., vol. 7, pp. 253–261 (Karger, Basel 1980).

15 Finn, C. A. and Bredl, J. C. S.: Studies on the development of the implantation reaction in the mouse uterus: influence of actinomycin D. J. Reprod. Fertil. *34*: 247–253 (1973).

16 Finn, C. A. and Downie, J. M.: Changes in the endometrium of mice after the induction of implantation by actinomycin D. J. Endocr. *65*: 259–264 (1975).

17 Finn, C. A. and Martin, L.: The control of implantation. J. Reprod. Fertil. *39*: 195–206 (1974).

18 Galand, P.; Flandroy, L., and Mairesse, N.: Relationship between the estrogen induced protein IP and other parameters of estrogenic stimulation. A hypothesis. Life Sci. *22*: 217–238 (1978).

19 Glasser, S. R. and McCormack, S. A.: Estrogen-modulated gene transcription in relation to decidualization. Endocrinology *104*: 1112–1118 (1979).

20 Gross, P. R.: Protein synthesis during cleavage; in Sussman, Molecular genetics and developmental biology, pp. 102–127 (Prentice Hall, Englewood Cliffs 1972).

21 Hamilton, T. H.: Steroid hormones, ribonucleic acid synthesis and transport, and the regulation of cytoplasmic translation; in Smellie, The biochemistry of steroid hormone action, pp. 49–84 (Academic Press, London 1971).

22 Heap, R. B.; Flint, A. P., and Gadsby, J. E.: Embryonic signals that establish pregnancy. Br. med. Bull. *35*: 129–135 (1979).

23 Humphrey, K. W.: Induction of implantation of blastocysts transferred to ovariectomized mice. J. Endocr. *44*: 299–305 (1969).

24 Kafatos, F. C.: mRNA stability and cellular differentiation; in Diczfaluzy, Gene transcription in reproductive tissue, pp. 319–341 (Bogtrykkeriet Forum, Copenhagen 1972).

25 Katzenellenbogen, B. S.: Synthesis and inducibility of the uterine estrogen-induced protein, IP, during the rat estrous cycle: clues to estrogen sensitivity. Endocrinology *96*: 289–297 (1975).

26 Kelly, P. A. T.; Morrison, C., and Green, B.: Effect of progesterone treatment on the uptake of estradiol-17β by uterine tissues of ovariectomized rats. Mol. cell. Endocrinol. *10*: 319–325 (1978).

27 King, R. J. B. and Mainwaring, W. I. P.: Oestrogens. Steroid-cell interactions, pp. 190–263 (Butterworths, London 1974).

28 Kirby, D. R. S.: Ectopic autografts of blastocysts in mice maintained in delayed implantation. J. Reprod. Fertil. *14*: 515–517 (1967).

29 Leroy, F.: Etude fonctionnelle de l'acide désoxyribonucléique dans l'endomètre du rongeur. Analyse morphologique semi-quantitative, pp. 116–125 (Arscia, Brussels 1974).

30 Leroy, F.: Aspects moléculaires de la nidation; in du Mesnil du Buisson, Psychoyos and Thomas, L'implantation de l'œuf, pp. 81–92 (Masson, Paris 1978).

31 Leroy, F.: Van Hoeck, J., and Bogaert, C.: Inability of actinomycin D to act

upon the uterine refractory state resulting from nidatory oestrogen action in rats. J. Endocr. *68:* 137–140 (1976).

32   Leroy, F.; Van Hoeck, J., and Lejeune, B.: Effects of cycloheximide on the refractory state induced by 'nidatory' oestrogen in rats. J. Reprod. Fertil. *56:* 187–191 (1979).

33   Maale, G.; Stein, G., and Mans, R.: Effects of cordycepin and cordycepin triphosphate on polyadenylic and ribonucleic acid-synthesising enzymes from eukaryotes. Nature *255:* 80–82 (1975).

34   Mallonee, R. C. and Yochim, J. M.: The uterus during progestation: hormonal modulation of pyridine nucleotide activity in relation to decidual sensitivity. J. Steroid Biochem. *11:* 745–755 (1979).

35   Martin, L.; Das, R. M., and Finn, C. A.: The inhibition by progesterone of uterine epithelial proliferation in the mouse. J. Endocr. *57:* 544–559 (1973).

36   Mayer, G. et Duluc, A. J.: Transplantation intratesticulaire chez le rat de blastocystes en léthargie et de fragments de corne utérine en phase de non-réceptivité. C. r. hebd. Séanc. Acad. Sci., Paris *267:* 509–512 (1968).

37   McLaren, A.: Blastocyst activation; in Crozier, Corfman, Condliffe and Segal, The regulation of mammalian reproduction, pp. 321–328 (Thomas, Springfield 1973).

38   Müller, G. L.: The role of RNA and protein synthesis in oestrogen action; in Karlson, Mechanisms of hormone action, pp. 228–245 (Academic Press, London 1965).

39   Notides, A. and Gorski, J.: Estrogen-induced synthesis of specific uterine protein. Proc. natn. Acad. Sci. USA *56:* 230–235 (1966).

40   Pollard, J. W.; Finn, C. A., and Martin, L.: Actinomycin D and uterine epithelial protein synthesis. J. Endocr. *69:* 161–162 (1976).

41   Prasad, M. R. N.; Sar, M., and Stumpf, W. E.: Autoradiographic studies on (^{3}H)-oestradiol localization in the blastocysts and uterus of rats during delayed implantation. J. Reprod. Fertil. *36:* 1–7 (1974).

42   Psychoyos, A.: Mécanismes de la nidation. Archs Anat. microsc. Morph. exp. suppl. *56:* 3–4, pp. 616–623 (1967).

43   Psychoyos, A.: Hormonal control of ovoimplantation. Vitams Horm. *32:* 210–256 (1973).

44   Psychoyos, A.; Bitton-Casimiri, V., and Brun, J. L.: Repression and activation of the mammalian blastocyst; in Talwar, Regulation of growth and differentiated function in eukaryote cells, pp. 509–514 (Raven Press, New York 1975).

45   Quarmby, V.: Thesis, University of London (1980).

46   Revel, M. and Groner, Y.: Posttranscriptional and translational controls of gene expression in eukaryotes. A. Rev. Biochem. *47:* 1079–1126 (1978).

47   Segal, S. J.; Scher, W., and Koide, S. S.: Estrogens, nucleic acids and protein synthesis in uterine metabolism; in Wynn, Biology of the uterus, pp. 139–201 (Plenum Press, New York 1977).

48   Smith, L. D. and Ecker, R. R.: Regulatory processes in the maturation and early cleavage of amphibian eggs. Curr. Top. devl. Biol. *5:* 1–38 (1970).

49   Stumpf, W. E. and Sar, M.: Autoradiographic localization of estrogen, androgen, progestin and glucocorticosteroid in 'target tissues' and 'non-target tissues'; in Pasqualini, Receptors and mechanism of action of steroid hormones, pp. 41–84 (Dekker, New York 1976).

50   Surani, M. A. H.: Radiolabelled rat uterine luminal proteins and their regula-
     tion by oestradiol and progesterone. J. Reprod. Fertil. *50:* 289–296 (1977).
51   Tachi, C.; Tachi, S., and Lindner, H. R.: Modification by progesterone of oestra-
     diol-induced cell proliferation, RNA synthesis and oestradiol distribution in the
     rat uterus. J. Reprod. Fertil. *31:* 59–76 (1972).
52   Tomkins, G. M.; Gelehrter, T. D.; Granner, D.; Martin, D., jr.; Samuels, H. J.,
     and Thompson, B.: Control of specific gene expression in higher organisms.
     Science *166:* 1474–1480 (1969).
53   Tomkins, G. M.; Levinson, B. B.; Baxter, J. D., and Dethlefsen, L.: Further
     evidence for posttranscriptional control of inducible tyrosine aminotransferase
     synthesis in cultured hepatoma cells. Nature new Biol. *239:* 9–14 (1972).
54   Vass, M. A. and Green, B.: Effect of progesterone pretreatment on the uptake
     of estradiol-17β by the uterine epithelium of the rat. Cell tiss. Res. *202:* 171–
     175 (1979).
55   Weitlauf, H. M.: Effect of uterine flushings on RNA synthesis by 'implanting'
     and 'delayed implanting' mouse blastocysts *in vitro*. Biol. Reprod. *14:* 566–571
     (1976).
56   Weitlauf, H. M.: Factors in mouse uterine fluid that inhibit the incorporation
     of (^{3}H)-uridine by blastocysts *in vitro*. J. Reprod. Fertil. *52:* 321–325 (1978).
57   Webb, F. T. G. and Surani, M. A. H.: Influence of environment on blastocysts
     proliferation, differentiation and implantation; in Talwar, Regulation of growth
     and differentiated function in eukaryote cells, pp. 519–522 (Raven Press, New
     York 1975).
58   White, J. D.; Thrower, S., and Lim, L.: Intracellular relationships of the oestro-
     gen receptor in the rat uterus and hypothalamus during the oestrous cycle. Bio-
     chem. J. *172:* 37–47 (1978).
59   Yoshinaga, K.: Effect of local application of ovarian hormones on the delay in
     implantation in lactating rats. J. Reprod. Fertil. *2:* 35–41 (1961).

F. Leroy, MD, Ph. D., Laboratoire de gynécologie, Hôpital Saint-Pierre,
322, rue Haute, B-1000 Bruxelles (Belgium)

Prog. reprod. Biol., vol. 7, pp. 216–233 (Karger, Basel 1980)

Behavior of Uterine Steroid Receptors at Implantation[1]

Dominique Martel and Alexandre Psychoyos

Laboratoire de Physiologie de la Reproduction, ER 203, CNRS,
Hôpital de Bicêtre (Bat. INSERM), Bicêtre

Introduction

The preparation of the uterus for egg implantation varies considerably from one species to the other. Most of the investigations concerning this process have been carried out on small rodents, and little information is available for other mammals, in particular primates. In all species, ovo-implantation requires a strict synchronism between egg development and endometrial preparation. In the rat and mouse, the timing of the endo-metrial evolution is controlled by steroid hormones [1, 21, 63]. It has been shown that a precise hormonal sequence is required to induce an endo-metrial receptive phase; thus only after a progesterone priming of at least 48 h, can the estradiol lead, 18 h later, to endometrial receptivity which lasts a few hours. This period is followed by a nonreceptive state during which the endometrium becomes hostile to nonimplanted blastocysts.

Progesterone priming is essential, but implantation occurs in some species (rabbit, hamster, sheep) without the apparent intervention of estrogen [1, 5, 35, 56, 75]. Whether this intervention is generally required in mammalian species remains controversial. Recently the hypothesis has been advanced that the blastocyst itself secretes estrogen necessary for implantation [16, 17, 58].

In spite of considerable information concerning steroid receptors, the precise molecular mechanism by which the hormones interfere with gene

[1] This work was supported by the Centre National de la Recherche Scientifique, the Délégation Générale à la Recherche Scientifique et Technique and the Ford Foundation.

expression in target cells is still unknown. However, it is accepted that the steroids, once they have entered the uterine cell, bind with high affinity and specificity to a protein macromolecule, the receptor, present in the cytoplasm. The binding of the steroid to its receptor is followed by translocation of the hormone receptor complexes to the nucleus, where probably by combining with acceptor sites they interfere with the gene expression to trigger the cellular response [3, 7, 29, 31].

According to this model the modulation of the amount of receptor molecules available in the cytoplasm, and the occupancy of nuclear sites could control the response of the target cell. Such a correlation between binding sites and biological response has been reported for the estrogen receptor and the uterine growth response [13].

Since progesterone and estradiol are highly implicated in the uterine timing for ovoimplantation, it is of interest to correlate this process with the intracellular movement of receptors for these hormones.

Estradiol Receptor

Characteristics of the Estrogen Receptor in the Uterus

The characteristics of the estrogen receptor have been largely documented during the last decade [28, 30, 31, 68, 76] and no major differences exist between mammalian species. The estrogen is noncovalently bound to its receptor with a high affinity varying from 0.1 to 1 nM. The binding affinity of estrogens for this receptor is: estradiol = DES > 17β-estradiol > estrone > estriol; progesterone, testosterone or corticosterone at physiological doses are not bound.

The receptor is a protein macromolecule, present in the cytoplasm at very low concentrations. In the cytosol, the native form of the receptor sediments in a sucrose gradient as an 8S component; however, the receptor tends to form aggregates. Under definite conditions, i. e. exposure to high ionic strength, calcium action or mild proteolysis, the receptor is converted to smaller subunits with a sedimentation coefficient of 4–4.5S. In primates, because of the presence of uterine proteolytic enzymes, a sedimentation coefficient of about 3S is found for cytosol estrogen receptor complexes [18, 47, 53]; in the presence of protease inhibitors, the classical 8S component is observed.

Several groups are working on the purification of the receptor, the molecular characteristics of which appear to be slightly different depending upon the degree of purity and the form purified [cf. 19, 54, 64, 70]. The incubation of the receptor with estradiol at 20–37 °C results in receptor activation. The active form of the receptor sediments as a 5S component indistinguishable from the soluble nuclear receptor.

Regulation of the Cytosol Estrogen Receptor Concentration

In various species, fluctuations in the uterine estrogen receptor content were detected during the estrous or menstrual cycle, the maximal receptor levels coinciding with the increase in plasmatic estrogen in proestrus for rodents [11, 20, 37] or during the proliferative phase for primates [4, 47, 59].

The regulation of the cytosol receptor concentration is controlled by the two ovarian hormones. The effect of estradiol has been clearly established: it induces a *de novo* synthesis of its own receptor. The cytosol receptor concentration increases markedly some 24 h after estradiol injection [65]. This process has been repeatedly described in the literature and seems to be common to all mammalian species.

Progesterone is generally considered to antagonize the action of estrogen on a majority of uterine responses. However, at least in the rat, the species most commonly studied for receptor regulation, progesterone exhibits a stimulatory or inhibitory action on the estrogen receptor concentration, depending on the uterine tissular component examined. In the endometrium of ovariectomized rats (fig. 1), progesterone induces an increase in the estrogen receptor content, well defined after at least 48 h of treatment. In the myometrium (fig. 1, 2) progesterone alone does not augment the receptor content, and if associated with estradiol, it blocks the receptor increase induced by estradiol alone (fig. 2). Other workers studying the whole uterus of immature [24] or ovariectomized rats [57] failed to observe the stimulatory effect of progesterone on the estrogen receptor production. This is probably due to the fact that myometrium represents quantitatively the major uterine cell type (more than 90 %) and results obtained for the whole uterus mainly concern the myometrium.

The increase in estrogen receptor production in the rat endometrium under the influence of progesterone is in good agreement with the preparative role of progesterone during the preimplantation period in the rat. It appears that the lag period required to induce a significant increase in the estrogen receptor concentration by progesterone (48 h), is similar to that required for certain specific responses of the progesterone primed endometrium to the estrogen, i. e. a receptive phase for nidation [44,

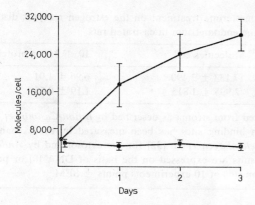

Fig. 1. Effect of progesterone (4 mg/rat/day) on the cytosol estrogen receptor concentration in the endometrium (●) and myometrium (■) of ovariectomized rats. Cytosol preparations were incubated with increasing amounts of (^{3}H)-estradiol (0.2–8 nM), unbound and nonspecifically bound steroids were removed by differential dissociation in the presence of a charcoal dextran suspension. The number of estrogen-binding sites were calculated according to the graphical representation of *Scatchard* [67]. The results are expressed on the basis of DNA [8], points are the means ± SEM of 4 experimental points.

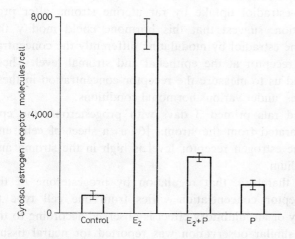

Fig. 2. Ovariectomized rats treated daily with estradiol (E₂), progesterone (P) or both for 3 days were sacrificed 24 h after the last injection. The number of estrogen-binding sites in the cytoplasmic fraction of the myometrium was determined using the graphical representation of *Scatchard* [67] as described in figure 1. The results are expressed on the basis of DNA [8], points are the means ± SEM of 8 experimental points.

Table I. Effect of 3 days of progesterone treatment on the estrogen receptor distribution in different cell types of the endometrium in castrated rats

	Molecules/cell	10^{-13} M/mg protein
Stroma	17,133 ± 3,490	6.96 ± 1.01
Epithelium	2,995 ± 1,815	1.19 ± 0.65

The epithelium has been separated from stroma as described by *Bitton-Casimiri et al.* [6] and the number of estrogen binding sites has been measured by the exchange techniques described by *Katzenellenbogen et al.* [32] for the cytosol and by *Anderson et al.* [2] for the nuclei. Results are expressed on the basis of DNA [8] or protein [40] content. Points are the means of 10 experiments points ± SEM.

63] and the redirection of the mitogenic effect from the epithelium to the stroma cells [48, 74].

During progestation, in the rat or mouse, progesterone priming leads to opposite effects in the different cell types of the endometrium. The epithelial cell responses to estradiol in terms of mitosis, (^{3}H)-thymidine or uridine incorporation or protein synthesis, are reduced [48, 72, 74], while those responses are enhanced in the stromal cells. These investigations and the autohistoradiographic studies of *Tachi et al.* [74], who showed an increase of the (^{3}H)-estradiol uptake by rat uterine stroma after progesterone administration, suggest that this hormone could modify the cellular response to the estradiol by modulating differently the concentration of the estrogen receptor at the epithelial and stromal level. These observations prompted us to measure the receptor concentration in these different uterine tissues, under various hormonal conditions.

In ovariectomized rats primed 3 days with progesterone, the epithelium is easily separated from the stroma [6] as a sheet of cells and table I shows that the estrogen receptor level is high in the stroma and very low in the epithelium.

It can be seen, therefore, that regulation by progesterone of the cytosol estrogen receptor concentration varies from one cell type to another; a stimulatory or an inhibitory effect is observed according to the cellular function. A similar observation was reported for neural tissue, where progesterone enhanced (^{3}H)-estradiol retention in the median eminence [39].

One can speculate that in the uterus this phenomenon is valid not only in the rat but in all species where the nidation process is programmed by the sequential action of progesterone and estradiol.

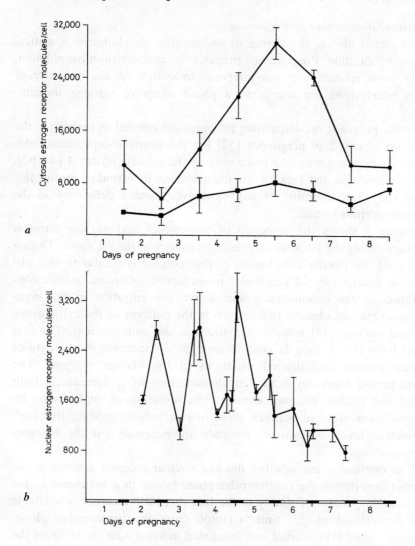

Fig. 3. Estrogen receptor in the rat endometrium during early pregnancy. *A* The cytosol estrogen-binding sites in the endometrium (●) and myometrium (■) were determined using the graphical representation of *Scatchard* [67] as described in figure 1. *B* The nuclear binding sites (●) were measured by an (³H)-estradiol exchange technique [2]. The results are expressed on the basis of DNA [8]. Vertical bars show the standard error of the mean.

Estradiol Receptor and Nidation

As stated above, the timing of endometrial development is critical for ovoimplantation. Progesterone prepares the endometrium for nidation, but in some species, estrogen intervene to switch on the sequence of events which lead the uterus to a phase receptive for egg implantation.

In the pregnant rat, circulating estrogens are present as early as in the afternoon of day 2 of pregnancy [52] but the estrogen-dependent endometrial changes begin to take place only by the evening of day 4 [44, 62]. It is possible that the lack of uterine response to estradiol during this period (day 2 to morning of day 4) derives from a deficiency in the estrogen-receptor system.

Figure 3 shows the evolution of the cytosol and nuclear estrogen receptors during the 1st week of pseudopregnancy in the rat uterus. During this period the tissular distribution of the receptor is similar to that observed in ovariectomized rats under progesterone treatment: a high concentration in the endometrium and a low concentration in the myometrium. One can observe two periods in the patterns of the cytoplasmic [49] and nuclear [45] receptor variations in the endometrium. The first period from day 1 to 5 is characterized by an increase of the cytosol receptor content and diurnal variations of the nuclear receptor. The second period from day 5 to 7 is characterized by a decrease in both cytosol and nuclear receptor content. The initiation of implantation by estrogen takes place in between these two periods, suggesting that high endometrial levels of estrogen receptor are necessary for the nidatory estrogen action.

The question arose whether the low nuclear receptor content in the endometrium during the postreceptive phase is due to a deficiency in the translocation process or simply to the fact that little receptor is available from the cytosol at this time. Cytosol, from the prereceptive phase, saturated with (^{3}H)-estradiol was incubated *in vitro* with nuclei from the postreceptive phase and vice versa. The results show that the amount of receptor that enters the nuclei depends only on the concentration of the hormone receptor complexes present in the cytosol [*Malet et al.,* to be published; 42].

As shown above, one of the actions of progesterone during progestation could be to increase the level of estrogen receptor in the stroma cells which could thus respond to estradiol. However, it is obvious from our experimental results that the absence of such a response before the evening of day 4 is neither due to the absence of cytoplasmic receptor, nor to a block in the nuclear translocation process. As soon as day 3, there is an adequate amount of receptor located in the nucleus to promote a cellular response. As defined by *Clark et al.* [12], 1,000–3,000 molecules/ nucleus are sufficient for this response. *O'Farell and Daniel* [55] reached similar conclusions, concerning the cytosol receptor, when studying a mammal with an obligate delayed implantation [northern fur seal]. In these species also the block in the uterine response to estrogen until the end of the delay period, is not explainable by an apparent failure in the estrogen binding system. Thus the turning on of the endometrial response to estradiol on the evening of day 4 or at the end of an obligate delay, suggests either the existence of a control mechanism acting on a transcriptional or posttranscriptional step [23, 38] or the implication of other unknown factors.

Recently, the hypothesis has been proposed that the blastocyst itself could secrete estrogens inducing its own implantation. A steroidogenic activity of the blastocyst has been shown in rabbits [16, 17], pigs [58] and cows [69]. It was also hypothesized that the oocyte could have absorbed estrogen when in the follicule, and released it at the time of implantation [17].

There is no evidence for estrogen production by the rat blastocyst, although 3-hydroxysteroid deshydrogenase activity has been shown to increase in day 4 and 5 embryos [16]. However, *Sartor* [66] and *Ward et al.* [81] observed that (^{3}H)-estradiol uptake was reduced in the uterine implantation sites. One explanation was that the uterine tissue adjacent to the embryo was saturated by estrogen released from the blastocyst.

In order to test this possibility, pregnant rats were ovariectomized according to the procedure of *Psychoyos and Alloiteau* [62] on day 3 of pregnancy and received a daily injection of progesterone (4 mg/rat/day). The implantation was induced on day 7 or 8 postcoitum, by a single injection of estradiol (0.25 μg/rat) and the animals were sacrificed 24 h after the estrogen injection. The implantation sites were revealed by injecting 1 ml of Evans blue at 0.5 % in the femoral vein, 30 min before sacrifice [61]. The uteri were removed, opened longitudinally and cut into nonblue and blue-stained segments. Since we showed previously that the myometrium is not concerned by the estrogen effect during implantation, only the nonblue and blue endometrium were collected and assayed for receptor content. Control animals in which the oviducts containing the eggs were removed simultaneously to the ovariectomy were treated in a similar way.

Table II. Estrogen receptor in the endometrium was measured by a (^{3}H)-estradiol exchange technique [2, 32] in cytosol and nuclear fraction prepared from implantation (blue) or interimplantation sites (nonblue): the results are expressed on the basis of DNA [8] or protein [40] content; points are the means of 3 experimental points ± SEM

	Control sterile horn	Nonblue area	Blue area
Progesterone delayed *+ estradiol 24 h*[1]			
Cytosol	29,472 ± 2,128	29,253 ± 2,470	13,705 ± 1,735
Nuclei	245 ± 143	227 ± 400	130 ± 45
Day 6 of pregnancy[2]			
Cytosol		11,532 ± 1,451	10,754 ± 1,542
Nuclei		287 ± 332	838 ± 573

[1] Animals were in delayed implantation after ovariectomy and progesterone treatment. Egg implantation was induced by estradiol 24 h prior to sacrifice.
[2] Animals were sacrificed on day 6 of normal pregnancy.

The estrogen receptor concentration in the control sterile horns is the same as that in the nonstained area, whereas a statistically significant decrease of the cytosol receptor concentration is observed in the blue area. This decrease is of 42 % when the results are expressed per DNA, and 63 % when expressed per protein (table II).

However, no receptor was found located in the nucleus, either in the blue or in the nonblue parts, indicating the absence of estrogen in both endometrial areas. The decrease of the estrogen receptor concentration we observed in the cytosol of the blue region may therefore reflect rapid modifications, such as increased DNA synthesis and protein accumulation in the endometrium area surrounding the blastocyst, rather than a translocation to the nucleus.

The estrogen receptor was also measured (table II) in the implantation-blue and control-nonblue sites on day 6 of pregnancy (10–12 a.m.), i.e. some 36 h after the normal intervention of the nidatory estrogen. The nuclear estrogen receptor concentration was very low in both regions. At the time of the initiation of egg implantation, therefore, the rat blastocyst does not seem to release estrogens that could trigger a local change in the intracellular distribution of estrogen receptor and could influence locally

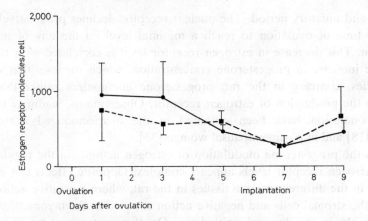

Fig. 4. Baboon endometrium was obtained by biopsy. The estrogen receptor concentrations in the cytoplasm (■) and the nuclei (●) were measured using an (^{3}H)-estradiol exchange technique [46, 47]. Results are expressed on the basis of DNA content [8]. The samples were dated from the day of ovulation which was determined according to two parameters: swelling of the sex skin and menstrual bleeding.

the target cell functions, as postulated by others [66, 81]. However, our results concerning the rat, do not rule out the existence of such a mechanism in other species, in which it will be of interest to perform similar studies. It cannot be excluded either that the rat trophoblast can secrete estrogens later (days 7–8), when more advanced in its development.

To our knowledge, information is missing concerning the endometrial estrogen receptor concentration during the nidatory period in species which appear not to require estrogen for the induction of the egg implantation process. In the human and the laboratory primates, indirect indications are offered by data concerning the luteal phase of the menstrual cycle, which can be considered as equivalent to pseudopregnancy in rodents. All the authors agree that the concentration of the estrogen receptor in endometrial cytosol is lower after than before ovulation. However, the highest concentration in the nuclear fraction is reported to occur before [9, 59] or around the ovulatory period [4].

The results of a study we carried out on our baboon colony are shown in figure 4. The cytosol receptor in the baboon endometrium remains at a base level which is very low (1,000 molecules/cell) during the prenidatory and nidatory period. The nuclear receptor declines progressively from the time of ovulation to reach a minimal level on the day of implantation. This decrease in estrogen receptor level is correlated with the plasmatic increase in progesterone concentration, which suggests that in this species, contrary to the rat, progesterone antagonizes rather than promotes the production of estrogen receptor. Observations, leading to a similar conclusion have been obtained in the ovariectomized rhesus monkey [18] and in postmenopausal women [33].

Thus the progesterone modulation of estrogen action, via the regulation of estrogen receptor levels appears complex. Opposite effects can be observed in the different uterine tissues in the rat, where a positive action exists in the stromal cells and negative action is noted in the myometrium and probably in the luminal epithelium. On the contrary, there is no evidence of a positive action of progesterone on the estrogen receptor content in the endometrium of primates. Whether this reflects differences in the progesterone-estradiol interactions and hormonal requirements during the prenidatory and nidatory period between these species, remains to be defined.

Progesterone Receptors

Characteristics of the Progesterone Receptor in the Uterus
Characterization of the progesterone receptor has been more difficult than for estrogen receptor because of its lability and the rapid dissociation rate of the (^3H)-progesterone receptor complexes. Investigations have also been complicated by the presence in the cytosol of cortico-binding globulin (CBG) that binds progesterone with high affinity. The synthesis of new progestins that do not bind to the CBG and that dissociate slowly from the receptor (R 5020 and ORG 2028) facilitated these studies. The characteristics of the progesterone receptor appear to be roughly the same in many mammals, although minor differences exist in binding kinetics and specificity between some species. Progesterone binds to its receptor with high affinity Kd: $1-5 \cdot 10^{-9}\,M$ [14, 36, 50, 79], the relative affinity of binding for other steroids is in the order: R 5020, desoxycorticosterone, testosterone, estradiol, cortisol; in some species, namely in the guinea pig, differences are noted for the binding of receptor to progestins substituted in the D ring [34].

As was observed for the estrogen receptor, two forms of the progesterone receptor are found in the uterine cytosol preparation, with sedimentation coefficients of 4–4.5S and 6–7S. According to the species studied and also to the hormonal regimen, one or the other form is predominant [4, 14, 25, 27, 36, 43, 51, 60]. Molecular characteristics of the human progesterone receptor has been determined in crude or purified preparations [73, 77].

Regulation of the Progesterone Receptor Concentration

As for the estrogen receptor, the evolution of the progesterone receptor content in the uterus during the ovarian cycle gives evidence that the progesterone receptor level is under the double control of estradiol and progesterone. The highest levels are found at proestrus or, in primates. during the proliferative phase. The regulation process is simple and appears to be the same in the different uterine tissues [41] and for all mammals [10, 22, 26, 36, 51, 80].

Studies in castrated animals have provided evidence that estrogens increase the cytosol concentration of progesterone receptor within 6 to 24 h; this increase is dependent on RNA and protein synthesis. Progesterone decreases the cytosol content of its own receptor as a consequence of two processes: first the nuclear translocation of the receptor, and second probably by increasing the degradation rate of the cytosol receptor. The negative effect of progesterone on its receptor production is dependent on protein synthesis, and can be overcome by a reinjection of estradiol.

Progesterone Receptors and Implantation

The evolution of cytosol progesterone receptors during progestation has been studied in the myometrium of the pregnant rabbit [14] and in the whole uterus of the guinea pig and the rat [78]. In these species, the receptor concentrations decrease after estrous and were at a low level at the time of implantation. Nuclear receptor concentrations, measure only in the whole uteri of the rat, are low but a small increase is observed on the day of implantation.

The few data obtained in primates give evidence of the following situation: in the human endometrium, the lowest level of cytosol progesterone receptors was detected during the 1st week after ovulation, while the receptor concentration was high in the nuclei [4, 59]. In the baboon endometrium our results show (fig. 5) that progesterone receptor content in the cytosol remains at a baseline value during all the prenidatory and nidatory period. Although the nuclear receptor content sharply decreases

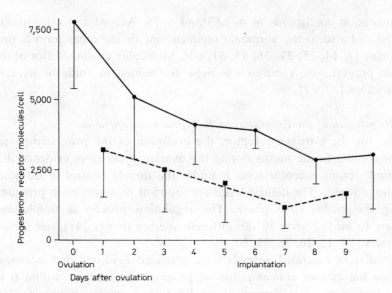

Fig. 5. Baboon endometrium was obtained by biopsy. The progesterone receptor concentrations in the cytoplasm (■) and the nuclei (●) were measured using a (³H)-progesterone exchange technique [43]. Results are expressed on the basis of DNA content [8]. The samples were dated from the day of ovulation which was determined according to two parameters: swelling of the sex skin and menstrual bleeding.

from the ovulation to the 10th day after ovulation, the amount of the receptor present in the nuclei during the nidatory period is relatively high (4,000 molecules/cell) when compared to the other periods of the menstrual cycle. As yet, investigations concerning the progesterone receptor around the implantation period in the uterus of primates and laboratory rodents are limited. Some studies concern the whole uterus while others are performed on the endometrium, thus it is premature to draw any definitive conclusions. However, the absence of the receptor in the cytoplasm, coinciding with a nonnegligible level of nuclear receptor, appears to be common to the nidatory period in all species studied.

Studies concerning progesterone receptor in the uteri strongly suggest that hormonal regulation of its concentration and evolution of uterine receptor content during definite physiological conditions may be roughly the same in the various species. This situation is completely different from that described for estrogen receptor.

References

1 Aitken, R. J.: The hormonal control of implantation; in Maternal recognition of pregnancy. Ciba Fdn Symp. 64, pp. 53–83 (Excerpta Medica, Amsterdam. 1979)

2 Anderson, J.; Clark, J. H., and Peck, E. J., jr.: Oestrogen and nuclear binding sites. Biochem. J. *126:* 561–567 (1972).

3 Baulieu, E. E.; Alberga, A.; Jung, I.; Lebeau, M. C.; Bodard, C.; Milgrom, E.; Raynaud, J. P.; Raynaud-Jammet, C.; Rochefort, H.; Truong, H., and Robel, P.: Metabolism and protein binding of sex steroids in target organs: an approach to the mechanism of action. Recent Prog. Horm. Res. *27:* 351–419 (1971).

4 Bayard, F.; Damilano, S.; Robel, P., and Baulieu, E. E.: Cytoplasmic and nuclear oestradiol and progesterone receptors in human endometrium. J. clin. Endocr. Metab. *46:* 635–648 (1978).

5 Bindon, B. M.: The role of progesterone in implantation in the sheep. Aus. J. biol. Sci. *24:* 149–158 (1971).

6 Bitton-Casimiri, V.; Rath, N. C., and Psychoyos, A.: A simple method for separation and culture of rat uterine epithelial cells. J. Endocr. *73:* 537–538 (1977).

7 Buller, R. E. and O'Malley, B. W.: The biology and mechanism of steroid hormone receptor interaction with the eukaryotic nucleus. Biochem. Pharmac. *25:* 1–12 (1976).

8 Burton, K.: A study of the conditions and mechanism of the diphenylamine reaction for the colorimetric estimation of desoxyribonucleic acid. Biochem. J. *62:* 315–322 (1956).

9 Brenner, R. M.; West, N. B.; Norman, R. L., and Sandow, B. A.: Progesterone suppression of the estradiol receptor in the reproductive tract of macaques, cats and hamsters; in Leavitt and Clark, Steroid hormone receptor systems, pp. 173–196 (Plenum Publishing, New York 1979).

10 Chen, T. J. and Leavitt, W. W.: Nuclear progesterone receptor in hamster uterus, measurement by (^{3}H)-progesterone exchange during the estrous cycle. Endocrinology *79:* 1588–1597 (1979).

11 Clark, J. H.; Anderson, J., and Peck, E. J., jr.: Receptor oestrogen complex in the nuclear fraction of rat uterine cells during the oestrous cycle. Sciences *176:* 528–530 (1972).

12 Clark, J. H.; Eriksson, H. A., and Bardin, J. H.: Uterine receptor estradiol complexes and their interaction with nuclear binding sites. J. Steroid Biochem. *7:* 1039–1043 (1976).

13 Clark, J. H.; Paszko, Z., and Peck, E., jr.: Nuclear binding and retention of the receptor estrogen complex relation to the agonistic and antagonistic properties of estriol. Endocrinology *100:* 91–96 (1977).

14 Davies, J. I.; Challis, J. R., and Ryan, K.: Progesterone receptors in the myometrium of pregnant rabbits. Endocrinology *95:* 165–173 (1974).

15 De Feo, V. J.: Decidualization; in Wynn, Cellular biology of the uterus, pp. 191–290 (Appleton Century Crofts, New York 1967).

16 Dickmann, Z.; Dey, S. K., and Sen Gupta, J.: A new concept: control of early

pregnancy by steroid hormones originating in the preimplantation embryo. Vitams Horm. *34:* 215–242 (1976).

17   Dickmann, Z.: Blastocyst oestrogen: an essential factor for the control of implantation. J. Steroid Biochem. *11:* 771–773 (1979).

18   Elsner, C. W.; Illingworth, D. V.; de Groot, K.; Flickinger, G., and Mikhail, G.: Cytosol and nuclear oestrogen receptor in the genital tract of the rhesus monkey. J. Steroid Biochem. *8:* 151–155 (1977).

19   Erdos, T. and Fries, J.: The endometrial nuclear oestradiol receptor of the pregnant cow has a molecular weight of 53,000 in 0.6 *M* guanidine-KCl. Mol. cell. Endocrinol. *13:* 203–209 (1979).

20   Feherty, P.; Robertson, D. M.; Waynforth, H. B., and Kellie, A. E.: Changes in the concentration of high affinity oestradiol receptors in rat uterine supernatant preparations during the oestrus cycle, pseudopregnancy, pregnancy, maturation and after ovariectomy. Biochem. J. *120:* 837–844 (1970).

21   Finn, C. A.: The implantation reaction; in Wynn, Biology of the uterus, pp. 245–308 (Plenum Press, London 1977).

22   Freifeld, M. L.; Feil, P., and Bardin, W.: The *in vivo* regulation of the progesterone 'receptor' in guinea pig uterus: dependance on oestrogen and progesterone. Steroids *23:* 93–103 (1974).

23   Glasser, S. R. and McCormack, S.: Estrogen modulated uterine gene transcription in relation to decidualization. Endocrinology *104:* 1112–1118 (1979).

24   Hsueh, A. J.; Peck, E. J., and Clark, J. M.: Control of uterine estrogen receptor levels by progesterone. Endocrinology *98:* 438–444 (1976).

25   Illingworth, D. V.; Elsner, C.; de Groot de la Cruz, K.; Flickinger, G. L., and Mikhail, G.: A specific progesterone receptor of myometrial cytosol from the rhesus monkey. J. Steroid Biochem. *8:* 157–160 (1977).

26   Isomaa, V.; Isotalo, H.; Orava, M., and Jänne, O.: Regulation of cytosol and nuclear progesterone receptors in rabbit uterus by oestrogen, antioestrogen and progesterone administration. Biochim. biophys. Acta *585:* 24–33 (1979).

27   Jänne, O.; Kontula, F.; Luukkainen, T., and Vihko, R.: Oestrogen induced progesterone receptor in human uterus. J. Steroid Biochem. *6:* 501–509 (1975).

28   Jänne, O.; Isomaa, V.; Isotalo, H.; Hokko, E., and Vierikko, P.: Uterine estrogen and progestin receptors and their regulation. Uppsala J. med. Sci., suppl. 22, pp. 62–70 (1978).

29   Jensen, E. V. and Jacobson, H. I.: Basic guides to the mechanism of oestrogen action. Recent Prog. Horm. Res. *18:* 387–414 (1962).

30   Jensen, E. V. and De Sombre, E. R.: Mechanism of action of the female sex hormone. A. Rev. Biochem. *41:* 203–230 (1972).

31   Katzenellenbogen, B. S. and Gorski, J.: Oestrogen action on biosynthesis of macromolecules in target cells; in Litwack, Biochemical action of hormones, pp. 188–243 (1975).

32   Katzenellenbogen, J.; Johnson, H. J., and Carlson, K. E.: Studies on the uterine cytoplasmic estrogen binding protein. Thermal stability and ligand dissociation rate. An assay of empty and filled sites by exchange. Biochem. *12:* 4092–4099 (1973).

33   King, R. J. B.; Whitehead, M. I.; Campbell, S., and Minardi, J.: Effect of estrogen and progestin treatments on endometria from postmenopausal women. Cancer Res. *39:* 1094–1101 (1979).

34   Kontula, K.; Jänne, O.; Vihko, R.; de Jager, E.; de Visser, J., and Zeelen, F.:
      Progesterone binding proteins: *in vitro* binding and biological activity of dif-
      ferent steroidal ligands. Acta endocr., Copenh. *78:* 574–592 (1975).
35   Kwun, J. K. and Emmens, C. W.: Hormonal requirements for implantation and
      pregnancy in the ovariectomized rabbit. Aust. J. biol. Sci. *27:* 275–283 (1974).
36   Leawitt, W. W.; Toft, D. O.; Strott, C. A., and O'Malley, B. W.: A specific pro-
      gesterone receptor in the hamster uterus, physiologic properties and regulation
      during the oestrus cycle. Endocrinology *94:* 1041–1953 (1974).
37   Lee, C. and Jacobson, H. I.: Uterine estrogen receptor in rat during pubescence
      and the estrous cycle. Endocrinology *88:* 596–601 (1971).
38   Leroy, F.: Aspects moléculaires de la nidation; in l'Implantation de l'œuf, pp.
      81–92 (Masson, Paris 1978).
39   Lisk, R. D. and Reuter, L. A.: *In vivo* progesterone treatment enhances (³H)-
      estradiol retention by neural tissue of the female rat. Endocrinology *100:* 1652–
      1658 (1977).
40   Lowry, O. H.; Rosebrough, N. J.; Farr, A. L., and Randall, R. J.: Protein mea-
      surement with the Folin phenol reagent. J. biol. Chem. *193:* 265–275 (1951).
41   Luu Thi, M.; Baulieu, E. E., and Milgrom, E.: Comparison of the characteristics
      and of the hormonal control of endometrial and myometrial progesterone re-
      ceptors. J. Endocr. *66:* 349–356 (1975).
42   Malet, C.: Etude de la migration et de la rétention nucléaire du complexe œs-
      tradiol-récepteur, sous diverses conditions hormonales chez la ratte. Diplôme
      Etudes Approfondies Université Paris VI (1976).
43   Malet, C.; Martel, D.; Monier, M. N., and Psychoyos, A.: Cytosolic and nu-
      clear progesterone receptor in the baboon endometrium throughout the men-
      strual cycle. J. Endocr. (to be published, 1980).
44   Marcus, G. J. and Shelesnyak, M. C.: Steroids in nidation. Adv. steroid bio-
      chem. pharmacol. *2:* 273–440 (1970).
45   Martel, D. and Psychoyos, A.: Endometrial content of nuclear estrogen receptor
      and receptivity for ovoimplantation in the rat. Endocrinology *99:* 470–475 (1976).
46   Martel, D.; Malet, C.; Monier, M. N.; Dubouch, P., and Psychoyos, A.: Nu-
      clear receptor for oestrogen in the baboon endometrium, detection, characteri-
      zation and variation in its concentration during the menstrual cycle. J. Endocr.
      84: 273–280 (1980).
47   Martel, D.; Malet, C.; Olmedo, C.; Monier, M. N.; Dubouch, P., and Psy-
      choyos, A.: Oestrogen receptor in the baboon endometrium: cytosolic receptor,
      detection, characterization and variation of its concentration during the men-
      strual cycle. J. Endocr. *84:* 261–272 (1980).
48   Martin, L. and Finn, C. A.: Hormonal regulation of cell division in epithelial
      and connective tissues of the mouse uterus. J. Endocr. *41:* 363–371 (1968).
49   Mester, J.; Martel, D.; Psychoyos, A., and Baulieu, E. E.: Hormonal control of
      oestrogen receptor in uterus and receptivity for ovoimplantation in the rat.
      Nature, Lond. *250:* 776–778 (1974).
50   Milgrom, E.; Atger, M., and Baulieu, E. E.: Progesterone in uterus and plasma.
      IV. Progesterone receptors in guinea pig uterus cytosol. Steroids *16:* 741–754
      (1970).
51   Milgrom, E.; Luu Thi, M.; Atger, M., and Baulieu, E. E.: Mechanisms regulat-

ing the concentration and conformation of progesterone receptor(s) in the uterus. J. biol. Chem. *248:* 6366–6374 (1973).

52 Nimrod, A.; Ladany, S., and Lindner, M. R.: Prenidatory ovarian oestrogen secretion in the pregnant rat, determined by gas chromatography with electron capture detection. J. Endocr. *53* 249–260 (1972).

53 Notides, A. C.; Hamilton, D. E., and Rudolph, J. H.: The action of human uterine protease on the estrogen receptor. Endocrinology *93:* 210–216 (1973).

54 Notides, A. C.; Hamilton, D. E., and Muechler, E. K.: A molecular analysis of the human estrogen receptor. J. Steroid Biochem. *7:* 1025–1030 (1976).

55 O'Farrell, P. H. and Daniel, J. C., jr.: Oestrogen binding in the uteri of mammals with delayed implantation. Endocrinology *88:* 1104–1106 (1971).

56 Orsini, M. W. and Meyer, R. K.: Implantation of the castrate hamster in the absence of exogenous estrogen. Anat. Rec. *134:* 619–620 (1959).

57 Pavlik, E. J. and Coulson, P. B.: Modulation of oestrogen receptors in four different target tissues: differential effects of oestrogen versus progesterone. J. Steroid Biochem. *7:* 369–376 (1976).

58 Perry, J. S.; Heap, R. B., and Amoroso, E. C.: Steroid hormone production by pig blastocyst. Nature, Lond. *245:* 45–47 (1973).

59 Pollow, K.; Boquoi, E.; Schmidt-Gollwitzer, M., and Pollow, B.: The nuclear estradiol and progesterone receptors of human endometrium and endometrial carcinoma. J. mol. Med. *1976:* 325–342.

60 Pollow, K.; Schmidt-Gollwitzer, M., and Pollow, B.: Characterization of a cytoplasmic receptor for progesterone in normal and neoplastic human endometrial tissue. J. mol. Med. *3:* 55–69 (1978).

61 Psychoyos, A.: Nouvelle contribution à l'étude de la nidation de l'œuf chez la rate. C. r. hebd. Séanc. Acad. Sci., Paris *251:* 3073–3075 (1960).

62 Psychoyos, A. et Alloiteau, J. J.: Castration précoce et nidation de l'œuf chez la rate. C. r. Séanc. Soc. Biol. *254:* 46–49 (1962).

63 Psychoyos, A.: Hormonal control of ovoimplantation. Vitams Horm. *31:* 201–255 (1973).

64 Puca, G. A.; Nola, E.; Sica, V., and Bresciani, F.: Estrogen binding proteins of calf uterus. Partial purification and preliminary characterization of two cytoplasmic proteins. Biochemistry *10:* 3769–3780 (1971).

65 Sarff, M. and Gorski, J.: Control of oestrogen binding protein concentration under basal conditions and after estrogen administration. Biochemistry *10:* 2557–2563 (1971).

66 Sartor, P.: Exogenous hormone uptake and retention in the rat uterus at the time of ovoimplantation. Acta endocr., Copenh. *84:* 804–812 (1977).

67 Scatchard, G.: The attraction of proteins for molecules and ions. Ann. N. Y. Acad. Sci. *51:* 660–672 (1949).

68 Segal, S. J.; Scher, W., and Koide, S. S.: Estrogens, nucleic acids and protein synthesis in uterine metabolism; in Wynn, Biology of the uterus, pp. 139–201 (Plenum Press, London 1977).

69 Shemesh, M.; Melaguir, F.; Lair, S., and Ayalon, N.: Steroidogenesis and prostaglandin synthesis by cultured bovine blastocysts. IIth Annu. Meet. Soc. Study of Reprod., abstr. 61 (1978).

70 Sica, V.; Nola, E.; Puca, A., and Bresciani, F.: Estrogen binding proteins of

calf uterus. Inhibition of aggregation and dissociation of receptor by chemical perturbation with NaSCN. Biochemistry *15:* 1915–1923 (1976).

71 Smith, J. A.; Martin, L.; King, R. J. B., and Vertes, M.: Effects of oestradiol-17β and progesterone on total and nuclear protein synthesis in epithelial and stromal tissues of the mouse uterus and of progesterone on the ability of these tissues to bind oestradiol-17β. Biochem. J. *119:* 773–784 (1970).

72 Smith, J. A.; Martin, L., and King, R. J. B. Effects of steroid hormones on protein synthesis in uterine epithelium and stromal tissues; in Hubinont, Leroy and Galand, Basic actions of sex steroids on target organs, pp. 221–226 (Karger, Basel 1971).

73 Smith, R. G.; Iramain, C. A.; Buttram, V. C., and O'Malley, B. W.: Purification of human uterine progesterone receptor. Nature, Lond. *253:* 271–272 (1975).

74 Tachi, G.; Tachi, S., and Lindner, H. R.: Modification by progesterone of oestradiol induced cell proliferation, RNA synthesis and oestradiol distribution in the rat uterus. J. Reprod. Fertil. *31:* 59–76 (1972).

75 Thibault, C.: L'implantation: sa programmation; in l'Implantation de l'œuf, pp. 1–19 (Masson, Paris 1978).

76 Truong, H. and Baulieu, E. E.: :Interaction of uterus cytosol receptor with estradiol equilibrium and kinetic studies. Biochim. biophys. Acta *237:* 167–172 (1971).

77 Verma, U. and Laumas, K. R.: *In vitro* binding of progesterone to receptors in the human endometrium and the myometrium. Biochim. biophys. Acta *317:* 403–419 (1973).

78 Vu Hai, M. T.; Logeat, F.; Warembourg, M., and Milgrom, E.: Hormonal control of progesterone receptors. Ann. N. Y. Acad. Sci. *286:* 199–209 (1977).

79 Walters, R. M. and Clark, J. H.: Cytosol progesterone receptors of the rat uterus assay and receptor characteristics. J. Steroid Biochem. *8:* 1137–1144 (1977).

80 Walters, M. R. and Clark, J. H.: Cytosol and nuclear compartmentalization of progesterone receptors of the rat uterus. Endocrinology *103:* 601–609 (1978).

81 Ward, W. F.; Frost, A. G., and Ward, M. O.: Estrogen binding by embryonic and interembryonic segments of the rat uterus prior to implantation. Biol. Reprod. *18:* 598–601 (1978).

D. Martel, Ph. D., Laboratoire de Physiologie de la Reproduction, ER 203, CNRS, Hôpital de Bicêtre (Bat. Inserm), 78 avenue du Général Leclerc, F-94270 Bicêtre (France)

Prog. reprod. Biol., vol. 7, pp. 234–243 (Karger, Basel 1980)

Prostaglandins and the Endometrial Vascular Permeability Changes Preceding Blastocyst Implantation and Decidualization[1]

T. G. Kennedy

MRC Group in Reproductive Biology, Departments of Obstetrics and Gynaecology and of Physiology, The University of Western Ontario, London, Ont.

In all species which have been investigated, one of the earliest signs of blastocyst implantation is an increase in endometrial vascular permeability in areas adjacent to the blastocysts [*Psychoyos, 1973*]. Increased endometrial vascular permeability also precedes the decidualization induced by the application of artificial stimuli to properly sensitized uteri. *Psychoyos* [1973] has proposed that an increase in endometrial vascular permeability is essential if decidualization is to occur.

At present, the identity of the factor(s) responsible for the increase in endometrial vascular permeability is uncertain. Recently, considerable evidence has accumulated which suggests that prostaglandins have an essential role in mediating this increase in permeability; this evidence will be reviewed here.

Increased Endometrial Vascular Permeability in Response to Blastocysts

Studies in which inhibitors of prostaglandin biosynthesis were used provided the first indication that prostaglandins are involved in implantation. *Gavin et al.* [1974] and *Saksena et al.* [1976] reported that indomethacin, an inhibitor of prostaglandin synthesis [*Vane, 1971*], prevented implantation in the rat and mouse, respectively, as indicated by the ab-

[1] The research reported in this review was supported by grants from the Medical Research Council (Canada) and the World Health Organization.

sence of implantation swellings. However, since this inhibitor of prostaglandin synthesis also inhibits decidualization in response to artificial stimuli [*Castracane et al.*, 1974; *Tobert*, 1976; *Sananes et al.*, 1976; *Rankin et al.*, 1979a], it was not clear whether it was the initiation of implantation or the subsequent decidualization which had been affected.

The possibility that prostaglandins are involved in the initiation of implantation was examined by *Kennedy* [1977]. Indomethacin, given to rats on day 5 of pregnancy (day 1 = first day of vaginal sperm), inhibited the increase in endometrial vascular permeability on the evening of day 5, as indicated by the absence of uterine dye sites 15 min after the intravenous injection of Evans blue. This inhibition was transitory; on day 6, when indomethacin was presumably no longer present in concentrations sufficient to inhibit prostaglandin synthesis, increased endometrial vascular permeability was found in both treated and control rats. Since indomethacin also effectively inhibited the increase in endometrial vascular permeability when exogenous steroids, adequate for the initiation of implantation, were given, the inhibitor affected implantation by a mechanism which was independent of an effect, if any, on ovarian stereoidogenesis. Additional evidence for a role of prostaglandins at the uterine level was obtained by comparing prostaglandin concentrations in the areas of increased permeability with those elsewhere in the uterus. The concentrations of prostaglandins of the E and F series, as well as prostaglandin I_2 (measured as its stable breakdown product, 6-oxo-prostaglandin $F_{1\alpha}$) were elevated in uterine dye sites [*Kennedy*, 1977; *Kennedy and Zamecnik*, 1978].

Indomethacin has also been shown to inhibit the initiation of implantation in the hamster [*Evans and Kennedy*, 1978] and rabbit [*Hoffman et al.*, 1978].

Increased Endometrial Vascular Permeability in Response to Artifical Deciduogenic Stimuli

Increased endometrial vascular permeability precedes the decidualization induced by the application of artificial stimuli to the properly sensitized uterus. Experiments were therefore conducted to determine if, as in the pregnant animal, prostaglandins mediate the change in permeability. The artifical deciduogenic stimulus in our studies [*Kennedy*,

1979] was the unilateral injection into the uterine lumen of 50 μl phosphate buffered saline containing gelatin (PBS-G). Changes in endometrial vascular permeability were quantified using ^{125}I-labelled bovine serum albumin (^{125}I-BSA) [*Psychoyos*, 1961]. Based on the findings that first, indomethacin inhibited the permeability response to intrauterine PBS-G, second, uterine concentrations of prostaglandins E were elevated following the intrauterine injection (preceding detectable changes in endometrial vascular permeability), and finally, the injection into the uterine lumen of small amounts of prostaglandin E_2 resulted in increased endometrial vascular permeability in rats in which endogenous prostaglandin production had been inhibited, it was suggested that prostaglandins, probably of the E series, mediate the increase in permeability in response to intraluminal PBS-G.

The proposal that prostaglandins mediate the changes in endometrial vascular permeability following artificial stimuli is supported by other work. The uterine contents of prostaglandins E and F in mice are elevated by deciduogenic stimuli [*Jonsson et al.*, 1978; *Rankin et al.*, 1979a]. In addition, indomethacin treatment inhibits decidualization induced by artificial stimuli in rats [*Castracane et al.*, 1974; *Tobert*, 1976; *Sananes et al.*, 1976] and mice [*Rankin et al.*, 1979a]. Since increased endometrial vascular permeability is thought to be an essential prerequisite for the decidual cell reaction [*Psychoyos*, 1973], it is possible that indomethacin inhibited decidualization by preventing the normal increase in endometrial vascular permeability. However, since *Tobert* [1976] observed inhibition of decidualization even when indomethacin administration was delayed until 8 h after the stimulus, by which time the increase in permeability had presumably already occurred [*Psychoyos*, 1973; *Kennedy*, 1979], it seems likely that inhibition of prostaglandin synthesis affects not only endometrial vascular permeability, but also the later stages of the decidual cell reaction. Finally, prostaglandins administred into the uterine lumen of the rat [*Sananes et al.*, 1976] and rabbit [*Hoffman et al.*, 1977] induce decidualization.

Control of Uterine Sensitivity

Blastocyst implantation requires strict synchronization between the development of the embryo and endometrium [*Psychoyos*, 1973]. Moreover, decidualization in response to artificial stimuli can only be obtained during a limited period of pregnancy, pseudopregnancy, or when the uterus has been sensitized by an appropriate regimen of hormone treatments [*Psychoyos, 1973; Finn and Porter, 1975*]. In addition, it is well established that estrogens, in low dosages, act synergistically with progesterone to sensitize the rat and mouse uterus for the decidual cell reaction [*Yochim and De Feo, 1963; Armstrong and King, 1971; Finn and Porter, 1975*]. That these changes in uterine sensitivity might be related to the ability of the uterus to produce prostaglandins is suggested first, by the report of *Fenwick et al.* [1977] that the production of prostaglandins by uterine homogenates from pseudopregnant rats is maximal on day 5, corresponding to when uterus is most sensitive to deciduogenic stimuli [*De Feo, 1963*], and secondly by the observations that estrogens affect uterine prostaglandin production [*Castracane and Jordan, 1975; Kuehl et al., 1976*]. However, investigations of the timing of uterine sensitivity [*Kennedy, 1980a*] and its modification with estrogens [*Kennedy, 1980b*] indicated that uterine prostaglandin levels in response to a standardised artificial stimulus did not provide a ready explanation for the changes in sensitivity. In both cases, maximum uterine sensitivity for the decidual cell reaction corresponded with the maximum ability of the endometrium to respond to intrauterine prostaglandin E_2 with increased endometrial vascular permeability.

There are at least two possible explanations for these findings. Endometrial responsiveness may be related to the properties of endometrial receptors for prostaglandins which may, for example, be present in the greatest concentrations in the maximally sensitized uterus. Alternatively, changes in endometrial vascular permeability may require mediators in addition to prostaglandins and it is the production, release or action of these other mediators which determines maximum uterine sensitivity.

Which Prostaglandins?

Based on measurements of prostaglandins at the site of implantation, prostaglandins of the E and I series are possible mediators of the changes in endometrial vascular permeability. Levels of E-series prostaglandins are elevated in uterine dye sites, relative to other areas of the uterus, in rats [*Kennedy, 1977*] and hamsters [*Evans and Kennedy, 1978*]. As indicated by levels of 6-oxo-prostaglandin $F_{1\alpha}$, the stable breakdown product of prostaglandin I_2, levels of prostaglandin I_2 are also elevated in uterine dye sites of rats [*Kennedy and Zamecnik*, 1978]. By contrast, levels of prostaglandins F are elevated in implantation sites of rats [*Kennedy*, 1977] but not hamsters [*Evans and Kennedy*, 1978].

After the application of an artificial deciduogenic stimulus to the sensitized rat uterus, the levels of prostaglandins E, F and I_2 (measured as 6-oxo-prostaglandin $F_{1\alpha}$) are elevated within the uterus [*Kennedy*, 1979; *Kennedy et al.*, 1980]. Intrauterine administration of prostaglandin E_2, but not prostaglandin $F_{2\alpha}$ [*Kennedy*, 1979], prostaglandin I_2 [*Kennedy et al.*, 1980], or thio-prostacyclin, a stable analogue of prostaglandin I_2 [*Kennedy*, unpubl. observations], to rats in which endogenous prostaglandin production had been inhibited, increased endometrial vascular permeability.

While the above data suggest that prostaglandin E_2 is the mediator, there is indirect evidence that prostaglandin $F_{2\alpha}$ may also be involved. *Saksena et al.* [1976] and *Oettel et al.* [1979] have reported the induction of implantation with prostaglandin $F_{2\alpha}$ in mice and rats, respectively. In addition, intrauterine administration of prostaglandin $F_{2\alpha}$ results in decidualization in rats [*Sananes et al.*, 1976] and rabbits [*Hoffman et al.*, 1977], although in the latter study, prostaglandin E_2 was considerably more effective than prostaglandin $F_{2\alpha}$. If the usual changes in vascular permeability accompanied implantation and decidualization, these studies suggest that prostaglandin $F_{2\alpha}$ may be a mediator. *Oettel et al.* [1979] did report increased endometrial vascular permeability in response to intrauterine prostaglandin $F_{2\alpha}$, but this may have been mediated by endogenously produced prostaglandins rather than the exogenous prostaglandin $F_{2\alpha}$ as no inhibitor of prostaglandin synthesis was used.

The suggestion that prostaglandin I_2 may be the mediator of decidualization has come from *Rankin et al.* [1979b], who found that tranylcypromine, a selective inhibitor of prostaglandin I_2 synthesis [*Gryglewski et al.*, 1976] inhibits decidualization in mice. However, the selectivity of this inhibitory action of tranylcypromine has been questioned [*Rajtar and de Gaetano*, 1979], and the inhibition has not been overridden with prostaglandin I_2.

Given the uncertainty about which prostaglandins are involved in implantation and given the similarities between the early stages of implantation and the inflammatory response, it is of interest to note that it has been prostaglandins of the E, and more recently of the I, series which have been implicated in the inflammatory response [*Williams and*

Morley, 1973; *Kuehl et al.,* 1977; *Williams and Peck,* 1977; *Williams,* 1979].

Source of Prostaglandins

The two most likely sources of the prostaglandins which are involved in the initiation of implantation are the blastocysts themselves and the endometrium.

The observation that rabbit blastocysts contain prostaglandins [*Dickmann and Spilman,* 1975] raised the possibility that the blastocyst is the source of the prostaglandins [*Kennedy,* 1977]. *Kennedy and Armstrong* [1980], however, were unable to find evidence of significant prostaglandin production by rat blastocysts *in vitro.* More recently, *Biggers* [1980] reported that rabbit blastocysts converted radioactive arachidonic acid to prostaglandins. This, however, does not necessarily mean that the prostaglandins which mediate the endometrial vascular response are of blastocyst origin; blastocyst-produced prostaglandins may have functions within the blastocyst [*Biggers et al.,* 1978].

The endometrial cells as a source of the prostaglandins regulating endometrial vascular permeability is an attractive hypothesis since it is capable of explaining the increases in permeability brought about by both blastocysts and artificial stimuli. It is possible that blastocysts, by virtue of their close contact with the endometrial epithelium [*Psychoyos,* 1973], and artificial stimuli have the common property of damaging the endometrium, thereby stimulating prostaglandin production. There is evidence that artificial deciduogenic stimuli cause tissue damage [*Finn,* 1977; *Lundkvist et al.,* 1977] and in other tissues, injury is known to stimulate prostaglandin biosynthesis [*Ramwell and Shaw,*1970;*Piper and Vane,*1971].

Mode of Action of Prostaglandins

Little is known about the mechanisms by which prostaglandins bring about increased endometrial vascular permeability and the subsequent decidualization. Arguing by analogy with the inflammatory response, *Kennedy and Armstrong* [1980] suggested that there may be two mediators of the endometrial vascular permeability response: one, a prostaglandin possibly of the E or I series, may cause vasodilation; the other, possibly

histamine, may increase vascular permeability. In support of this, there
is evidence that vasodilation accompanies the increased endometrial
vascular permeability induced by an artificial stimulus [*Bitton et al.*, 1965].
The involvement of histamine in the endometrial vascular permeability
response is suggested by the demonstration of *Brandon and Wallis*
[1977] that histamine H_1 and H_2 receptor antagonists reduce the number
and intensity of uterine dye sites.

At the cellular level, the effects of prostaglandins may be mediated by
alterations in intracellular cyclic AMP levels. Prostaglandins of the E and
I series are stimulators of cyclic AMP synthesis in a variety of cell types
[*Kuehl et al.*, 1976; *Goff et al.*, 1978; *Omini et al.*, 1979] and artificial
deciduogenic stimuli bring about a rapid increase in uterine cyclic AMP
levels [*Leroy et al.*, 1974; *Rankin et al.*, 1977, 1979a]. The increase in
cyclic AMP levels in response to intrauterine oil is inhibited by indomethacin, indicating that the response is prostaglandin-mediated [*Rankin
et al.*, 1979a]. The cells within the endometrium which respond to prosta-
gladins with increased cyclic AMP synthesis are unknown. It would be of
endothelial cells within the endometrium, as this may alter interest to
know if prostaglandins stimulate cyclic AMP synthesis in their function.
Alternatively, cyclic AMP synthesis may be stimulated by prostaglandins
in stromal cells, and this may be of importance for the differentiation of
these cells into decidual cells.

Summary and Conclusions

Evidence suggesting the involvement of prostaglandins in the endometrial vascular permeability changes associated with implantation and
the decidual cell reaction includes the observation that inhibition of
prostaglandin synthesis delays or prevents the increase in permeability in
response to blastocysts or artificial stimuli. The concentrations of prostaglandins are higher in areas of increased endometrial vascular permeability
than elsewhere in the uterus. When administered into the uterine lumen,
prostaglandin E_2 increases endometrial vascular permeability.

It is suggested that blastocysts and artificial stimuli have the common
property of injuring the endometrium, thereby stimulating prostaglandin
biosynthesis, particularly of the E and/or I series. The prostaglandins may
then act to cause the vasodilation necessary for the increase in endometrial
vascular permeability.

References

Armstrong, D. T. and King, E. R.: Uterine progesterone metabolism and progestational response. Effects of estrogens and prolactin. Endocrinology *89:* 191–197 (1971).

Biggers, J. D.: Discussion of paper by *Kennedy and Armstrong* (1980).

Biggers, J. D.; Leonov, B. V.; Baskar, J. F., and Fried, J.: Inhibition of hatching of mouse blastocysts *in vitro* by prostaglandin antagonists. Biol. Reprod. *19:* 519–533 (1978).

Bitton, V.; Vassent, G. et Psychoyos, A.: Réponse vasculaire de l'utérus au traumatisme, au cours de la pseudogestation chez la ratte. C. r. hebd. Séanc. Acad. Sci., Paris *261* 3474–3477 (1965).

Brandon, J. M. and Wallis, R. M.: Effect of mepyramine, a histamine H_1-, and burimamide, a histamine H_2-receptor antagonist, on ovum implantation in the rat. J. Reprod. Fertil. *50:* 251–254 (1977).

Castracane, V. D. and Jordan, V. C.: The effect of estrogen and progesterone on uterine prostaglandin biosynthesis in the ovariectomized rat. Biol. Reprod. *13:* 587–596 (1975).

Castracane, V. D.; Saksena, S. K., and Shaikh, A. A.: Effect of IUDs, prostaglandins and indomethacin on decidual cell reaction in the rat. Prostaglandins *6:* 397–404 (1974).

De Feo, V. J.: Determination of the sensitive period for the induction of deciduomata in the rat by different inducing procedures. Endocrinology *73:* 488–497 (1963).

Dickmann, Z. and Spilman, C. H.: Prostaglandins in rabbit blastocysts. Science, N. Y. *190:* 997–998 (1975).

Evans, C. A. and Kennedy, T. G.: The importance of prostaglandin synthesis for the initiation of blastocyst implantation in the hamster. J. Reprod. Fertil. *54:* 255–261 (1978).

Fenwick, L.; Jones, R. L.; Naylor, B.; Poyser, N. L., and Wilson, N. H.: Production of prostaglandins by the pseudopregnant rat uterus, *in vitro,* and the effect of tamoxifen with the identification of 6-keto-prostaglandin $F_{1\alpha}$ as a major product. Br. J. Pharmacol. *59:* 191–199 (1977).

Finn, C. A.: The implantation reaction; in Wynn, Biology of the uterus, pp. 245–308 (Plenum Press, New York 1977).

Finn, C. A. and Porter, D. G.: The uterus, pp. 86–95 (Publishing Sciences Group, Acton 1975).

Gavin, M. A.; Dominguez Fernandez-Tejerina, J. C.; Montañes de las Heras, M. F., and Vijil Maeso, E.: Efectos de un inhibidor de la biosíntesis de las prostaglandinas (indometacina) sobre la implacion en la rata. Reproduccion *1:* 177–183 (1974).

Goff, A. K.; Zamecnik, J.; Ali, M., and Armstrong, D. T.: Prostaglandin I_2 stimulation of granulosa cell cyclic AMP production. Prostaglandins *15:* 875–879 (1978).

Gryglewski, R. J.; Bunting, S.; Moncada, S.; Flower, R. J., and Vane, J. R.: Arterial walls are protected against deposition of platelet thrombi by a substance (prostaglandin X) which they make from prostaglandin endoperoxides. Prostaglandins *12:* 685–713 (1976).

Hoffman, L. H.; Di Pietro, D. L., and McKenna, T. J.: Effects of indomethacin on uterine capillary permeability and blastocyst development in rabbits. Prostaglandins *15:* 823–828 (1978).

Hoffman, L. H.; Strong, G. B.; Davenport, G. R., and Fröhlich, J. C.: Deciduogenic effect of prostaglandins in the pseudopregnant rabbit. J. Reprod. Fertil. *50:* 231–237 (1977).

Jonsson, H. T.; Rankin, J. C.; Ledford, B. E., and Baggett, B.: Prostaglandin levels following stimulation of the decidual cell reaction in the mouse uterus. Endocrine Soc., 60th Ann. Meet., 1978, abstract No. 502.

Kennedy, T. G.: Evidence for a role for prostaglandins in the initiation of blastocyst implantation in the rat. Biol. Reprod. *16:* 286–291 (1977).

Kennedy, T. G.: Prostaglandins and increased endometrial vascular permeability resulting from the application of an artificial stimulus to the uterus of the rat sensitized for the decidual cell reaction. Biol. Reprod. *20:* 560–566 (1979).

Kennedy, T. G.: Timing of uterine sensitivity for the decidual cell reaction. Role of prostaglandins. Biol. Reprod. (in press, 1980a).

Kennedy, T. G.: Estrogen and uterine sensitization for the decidual cell reaction (DCR). Role of PGEs. 6th Int. Congr. Endocr., 1980b, abstract.

Kennedy, T. G. and Armstrong, D. T.: Role of prostaglandins in endometrial vascular changes at implantation; in Bullock and Glasser, Cellular and molecular aspects of implantation (Plenum Press, New York, in press, 1980).

Kennedy, T. G. and Zamecnik, J.: The concentration of 6-keto-prostaglandin $F_{1\alpha}$ is markedly elevated at the site of blastocyst implantation in the rat. Prostaglandins *16:* 599–605 (1978).

Kennedy, T. G.; Barbe, G. J., and Evans, C. A.: Prostaglandin I₂ and increased endometrial vascular permeability preceding the decidual cell reaction; in Kimball, The endometrium (Spectrum, in press, 1980).

Kuehl, F. A.; Cirillo, V. J.; Zanetti, M. E.; Beveridge, G. C., and Ham, E. A.: The effect of estrogen upon cyclic nucleotide and prostaglandin levels in the rat uterus. Adv. Prostaglandin Thromboxane Res. *1:* 313–323 (1976).

Kuehl, F. A.; Humes, J. L.; Egan, R. W.; Ham, E. A.; Beveridge, G. C., and Arman, C. G. van: Role of prostaglandin endoperoxide PGG₂ in inflammatory processes. Nature, Lond. *265:* 170–173 (1977).

Leroy, F.; Vansande, J.; Schetgen, G., and Brasseur, D.: Cyclic AMP and the triggering of the decidual reaction. J. Reprod. Fertil. *39:* 207–211 (1974).

Lundkvist, Ö.; Ljungkvist, I., and Nilsson, O.: Early effects of oil on rat uterine epithelium sensitized for decidual induction. J. Reprod. Fertil. *51:* 507–509 (1977).

Oettel, M.; Koch, M.; Kurischko, A., and Schubert, K.: A direct evidence for the involvement of prostaglandin $F_{2\alpha}$ in the first step of estrone-induced blastocyst implantation in the spayed rat. Steroids *33:* 1–8 (1979).

Omini, C.; Folco, G. C.; Pasargiklian, R.; Fano, M., and Berti, F.: Prostacyclin (PGI₂) in pregnant human uterus. Prostaglandins *17:* 113–120 (1979).

Piper, P. and Vane, J.: The release of prostaglandins from lung and other tissues. Ann. N. Y. Acad. Sci. *180:* 363–385 (1971).

Psychoyos, A.: Perméabilité capillaire et décidualisation utérine. C. r. Acad. hebd. Séanc. Sci., Paris *252:* 1515–1517 (1961).

Psychoyos, A.: Endocrine control of egg implantation; in Greep, Astwood and Geiger, Handbook of physiology, section 7, vol. II, part 2, pp. 187–215 (American Physiological Society, Washington 1973).

Rajtar, G. and Gaetano, G. de: Tranylcypromine is not a selective inhibitor of prostacyclin in rats. Thromb. Res. *14:* 245–248 (1979).

Ramwell, P. W. and Shaw, J. E.: Biological significance of the prostaglandins. Recent Prog. Horm. Res. *26:* 139–173 (1970).

Rankin, J. C.; Ledford, B. E., and Baggett, B.: Early involvement of cyclic nucleotides in the artificially stimulated decidual cell reaction of the mouse uterus. Biol. Reprod. *17:* 549–554 (1977).

Rankin, J. C.; Ledford, B. E.; Jonsson, H. T., and Baggett, B.: Prostaglandins, indomethacin and the decidual cell reaction in the mouse uterus. Biol. Reprod. *20:* 399–404 (1979a).

Rankin, J. C.; Ledford, B. E., and Baggett, B.: The effect of tranylcypromine on the artificially stimulated decidual cell reaction in the mouse uterus. Soc. Study Reprod., 12th Ann. Meet., 1979b, abstract No. 174.

Saksena, S. K.; Lau, I. F., and Chang, M. C.: Relationship between oestrogen, prostaglandin $F_{2\alpha}$ and histamine in delayed implantation in the mouse. Acta endocr., Copenh. *81:* 801–807 (1976).

Sananes, N.; Baulieu, E.-E., and Le Goascogne, C.: Prostaglandin(s) as inductive factor of decidualization in the rat uterus. Molec. Cell. Endocr. *6:* 153–158 (1976).

Tobert, J. A.: A study of the possible role of prostaglandins in decidualization using a nonsurgical method for the instillation of fluids into the rat uterine lumen. J. Reprod. Fertil. *47:* 391–393 (1976).

Vane, J. R.: Inhibition of prostaglandin synthesis as a mechanism of action of aspirin-like drugs. Nature New Biol. *231:* 232–235 (1971).

Williams, T. J.: Prostaglandin E₂, prostaglandin I₂ and the vascular changes of inflammation. Br. J. Pharmacol. *65:* 517–524 (1979).

Williams, T. J. and Morley, J.: Prostaglandins as potentiators of increased vascular permeability in inflammation. Nature, Lond. *246:* 215–217 (1973).

Williams, T. J. and Peck, M. J.: Role of prostaglandin-mediated vasodilatation in inflammation. Nature, Lond. *270:* 530–532 (1977).

Yochim, J. M. and De Feo, V. J.: Hormonal control of the onset, magnitude and duration of uterine sensitivity in the rat by steroid hormones of the ovary. Endocrinology *72:* 317–326 (1963).

T. G. Kennedy, MD, MRC Group in Reproductive Biology, Departments of Obstetrics and Gynaecology and of Physiology, The University of Western Ontario, London, Ont. (Canada)

Prog. reprod. Biol., vol. 7, pp. 244–252 (Karger, Basel 1980)

Some Recent Work on the Role of Histamine in Ovum Implantation

Janet M. Brandon

Department of Agricultural Science, Parks Road, Oxford

Introduction

Histamine is found in most mammalian tissues studied. Preformed histamine may be stored in, and released from, tissue mast cells, blood components such as platelets and other tissue-specific sites [18]. Histamine may also be formed *de novo* in response to local stimuli by induction of the enzyme histidine decarboxylase, a process which is believed to be involved in the control of the microcirculation [27]. The cell receptors at which histamine acts to produce its physiological effects have not been defined physically or chemically but have been defined pharmacologically be means of specific competitive antagonists. The classical antihistamines, such as diphenhydramine and mepyramine, which were introduced in the 1940s, behave as competitive antagonists to some actions of histamine, such as stimulation of smooth muscle contraction in the gut and bronchi, but have no effect on other actions of histamine, such as stimulation of gastric acid secretion and increased heart rate *in vitro* [1]. The receptors at which the classical antihistamines act have been defined as histamine H_1-receptors [1] and these drugs are referred to as H_1-antagonists. The existence of at least one other population of histamine receptors was inferred from studies of the effects of H_1-antagonists before any other receptors were defined. Novel antihistamines which inhibited actions of histamine refractory to inhibition by H_1-antagonists, that is, burimamide [2], metiamide [3] and cimetidine [9], were introduced in the 1970s and the receptor at which these antagonists act was defined as the histamine H_2-receptor [2]. Histamine is a mixed agonist activating both H_1- and H_2-receptors. Some actions of histamine, such as vasodilatation, usually involve both receptors and a combination of H_1- and H_2-antagonists is required for inhibition of such actions [25]. Compounds have been developed which stimulate either H_1- or H_2-receptors specifically [2, 15, 26]. The development of H_2-antagonists and of H_1- and H_2-agonists advanced our understanding of the physiology and pharmacology of histamine and provided an opportunity to re-examine the significance of histamine in various physiological processes including those involved in ovum implantation.

Materials and Methods

Animals. Rats from the Wistar-derived SPF colony maintained by the Smith Kline & French Research Institute (Welwyn Garden City, UK) were used throughout. Animals were maintained in a lighting cycle of 14 h white light and 10 h darkness or low-wattage red light with food and tap water available *ad libitum.*

Dating and Timing of Events in Pregnancy. Virgin females were paired with males of proven fertility and mating was confirmed by observation of vaginal plugs or sperm. The dark period during which mating occurred was designated night 0 of pregnancy, followed by day 1 (plug or sperm observed), night 1, day 2 and so on. The time of ovulation was investigated by examining animals during the dark period during which a pro-oestrous vaginal smear was observed. Swelling of the ampullary portion of the oviduct was taken to indicate recent ovulation. The earliest time at which all animals examined had ovulated was the time at which the next light period began, i. e. the start of day 1 in mated animals [4]. The time of 100% ovulation was designated 0 h of pregnancy. The time of blastocyst attachment was investigated by examination of fixed, cleared uteri for Pontamine Sky Blue (PSB) sites [24] and the earliest time at which the mean number of PSB sites/rat was 10 or more was found to be 113 h of pregnancy, i. e. the third hour of night 5 [4]. This time was considered to represent the earliest time of 100% attachment.

Blastocyst Attachment Experiments. This experimental system has been described in detail elsewhere [8]. Animals were maintained on a lighting cycle with white light from 21.00 to 11.00 h. Histamine antagonists were given intraperitoneally (i. p.) at 09.00, 12.30 and 16.00 h on day-to-night 2, 3 and 4 and/or at 09.00 and 12.30 h on day-to-night 5, and the animals were killed at 113 h of pregnancy, i. e. 14.00 h on night 5.

Drug Uptake Studies. Pregnant rats were given [³H]-burimamide or [³H]-metiamide i. p. and killed 5 min to 4 h thereafter. All animals were killed between 113 and 115 h of pregnancy. The uterus was removed and washed in 4 changes of saline before being weighed and then oxidised in a Packard Tri-Carb Sample Oxidiser using 1 min combustion time. The products of oxidation were dissolved in scintillant and the radioactivity present assessed in a scintillation counter.

Ovariectomized Pregnancy Experiments. This experimental system has been described in detail elsewhere [7]. Mated rats were ovariectomized on day or night 3 of pregnancy taking care to leave the oviducts intact. Estradiol (E_2) was given i. v. and the number of PSB sites observed in fresh uteri was assessed 24 h later. Histamine antagonists were given i. p. 30 min before the E_2. Histamine (i. p. or i. v.) and the histamine agonists (i. v.) were given immediately after the E_2.

Agents Administered. PSB 6BX (Searle Diagnostic) was given as a 2 % (w/v) solution in saline. 1 ml of the solution was given i. v. 15–20 min before the animals were killed. Estradiol (BDH) was given i. v. as a solution in 1 % (v/v) ethanol in saline. Histamine (BDH), 2-(2-aminoethyl)thiazole (SK&F), dimaprit (SK&F), S-[3-(N,N-dimethylamino)propyl]isothiourea) and mepyramine (May & Baker) were dissolved in saline. Burimamide (SK&F) and metiamide (SK&F) were dissolved in HCl and the solution brought to pH 6–7 using NaOH. All agents were given in 1 ml kg^{-1}.

Results

In early experiments pregnant rats were treated with the H_1-antagonist mepyramine and/or the H_2-antagonist burimamide during the pre-implantation and implantation phases of pregnancy (days 2–7). Treatment with mepyramine plus burimamide resulted in reduced uterine weight assessed on days 10, 14 or 18. A detailed study of this effect was undertaken in which mepyramine 30 μmol kg^{-1} and/or burimamide 300 μmol kg^{-1} were given i. p. once every 6 h from the morning of day 2 to the end of night 5 and the animals were killed on day 18. This treatment schedule was adopted because of the known short half-life of H_2-antagonists [21]. The results are summarised in table I. The placental weights (not shown) and number of fetuses/pregnant animal were not affected by any of the treatments. Treatment with the combination of mepyramine and burimamide, but not with either antagonist alone, increased the rate of non-pregnancy in treated animals and reduced the fetal weight in those pregnancies remaining when compared with saline-treated control animals. Using criteria described by *Edwards* [16] it was estimated that the fetuses from combination-treated animals were developmentally 12–24 h younger than those from controls [4]. Experiments using less frequent, more convenient, dosing regimes did not yield consistent results so an experimental system was sought in which antagonist-induced changes could be detected with greater sensitivity. Since the results described above suggested that treatment could delay implantation (leading to either failure to implant or a 12–24 h delay in implantation) the effect of

Table I. Effect of administration of mepyramine and/or burimamide once every 6 h from early day 2 to late night 5 of pregnancy, on pregnancy assessed on day 18 (mean ± SEM)

Treatment i. p.	NP/N	Uterine weight, g	Fetal weight, g	Fetuses/ pregnant rat
Saline	1/30	22.5 ± 0.8	0.70 ± 0.01	11.6 ± 0.4
Mepyramine 30 μmol kg^{-1}	1/29	23.8 ± 0.8	0.72 ± 0.03	11.5 ± 0.6
Burimamide 300 μmol kg^{-1}	2/29	21.6 ± 0.6	0.68 ± 0.03	10.5 ± 0.5
Mepyramine + burimamide	13/41[1]	18.8[2] ± 1.0	0.59[3] ± 0.02	10.3 ± 0.6

NP = Number of rats non-pregnant at autopsy; N = number of rats in group.
[1] $p < 0.001$, compared to saline-dosed group, chi square test.
[2] $p < 0.01$,
[3] $p < 0.001$, compared to saline-dosed group, Student's t test.

burimamide and mepyramine was assessed at the time of blastocyst attachment [8]. Treatment with mepyramine 12.5 μmol kg^{-1} plus burimamide 300 μmol kg^{-1} on days-to-nights 2–5 of pregnancy (see Materials and Methods), but not with either antagonist alone, was found to (a) reduce the number and intensity of PSB sites observed; (b) increase the number of apparently intact blastocysts which could be flushed from the uterus; (c) increase the number of zona pellucida-encased blastocysts recovered from the uterus and (d) reduce the degree of stromal oedema observed histologically in the vicinity of an attachment site, on night 5 of pregnancy [8]. The greatest numerical degree of change, i. e. a 3-fold increase, was seen in the number of blastocysts recovered from the uterus so this parameter was used to investigate the critical period for treatment. It was found that any dosing schedule which included treatment on day-to-night 5 resulted in an increase in the number of blastocysts recovered later on night 5 [8]. However, treatment with the combination of mepyramine and burimamide did not appear to prevent or to delay ovum implantation so much as to reduce phenomena such as stromal edema, blastocyst adherence and zona lysis. Experiments were conducted using the more potent H$_2$-antagonist metiamide [3] but this compound did not modify ovum implantation when given alone or in combination with mepyramine [7]. The implications of this finding were either that the effects of burimamide were not mediated via H$_2$-receptors or that metiamide did not gain access to the site of action of burimamide. The latter explanation was considered unlikely since the compounds are structurally very similar but nevertheless experiments were conducted to examine the uterine uptake and retention of the 2 drugs on night 5 of pregnancy 5 min to 4 h after i. p. injection. [^{3}H]-burimamide or [^{3}H]-metiamide were given i. p. at 300 μmol (50 μCi) kg^{-1} and the uterine radioactivity assessed 5 min to 4 h later. No differences were found in the uptake or pattern of retention of radioactivity derived from tritiated burimamide or metiamide [Brandon, unpubl. results]. One experiment was conducted in which [^{3}H]-burimamide or [^{3}H]-metiamide were given i. p. at 300 μmol (2.5 mCi) kg^{-1} and PSB sites obtained 5 min to 2 h after treatment were examined using microauto-radiographic techniques [12]. No differences were found in the distribution of radioactivity in sites obtained from animals given burimamide or metiamide [Cross and Brandon, unpubl. results].

To investigate the mode of action of the combination of burimamide and mepyramine an experimental system was sought in which an effect of exogenous histamine could be demonstrated and the receptor involvement investigated. It was found that histamine given i. p. or i. v. augmented the implantation response (that is, the number of PSB sites observed 24 h after an i. v. injection of estradiol) to suboptimal doses of estradiol (< 200 ng kg^{-1}) in the ovariectomized pregnant rat maintained with Depo-Provera [7]. The effect of i. v. histamine was not inhibited by administration of H$_1$- (mepyramine or diphenhydramine) or H$_2$- (burimamide or metiamide) antagonists but was abolished by treatment with a combination of mepyramine and metiamide at doses as low as 5 and 50 μmol kg^{-1}, respectively. The involvement of classical H$_1$- and H$_2$-receptors in this effect of histamine was confirmed by the use of the H$_1$-agonist 2-(2-aminoethyl)-thiazole [15] and the H$_2$-agonist dimaprit [26]. Both agonists increased the implantation response to suboptimal doses of E$_2$. The effect of the H$_1$-agonist was inhibited by mepyramine and the effect of the H$_2$-agonist was inhibited by metiamide but was not modified by the less potent H$_2$-antagonist burimamide [7].

Discussion

The role of histamine in processes involved in ovum implantation has been a source of controversy. Shelesnyak proposed that histamine, released from uterine mast cells by the action of estrogen [30], induced decidualization in the rat. This hypothesis was based on observations that

(a) H₁-antagonists given into the unterine lumen inhibited decidualization [29], (b) systemic histamine or histamine releasing compounds induced decidualization [22], and (c) uterine histamine content fell prior to ovum implantation [23]. The reasons why this hypothesis did not gain general acceptance have been reviewed by *De Feo* [13]. Various lines of evidence have indicated a relationship between histamine and vascular changes in the uterus. Estrogen increases uterine blood flow and vascular permeability and the evidence for an involvement of histamine in these changes has been reviewed by *Spaziani* [31]. Implantation is associated with similar vascular changes in the uterus as well as with decreased uterine histamine content [23] and mast cell degranulation [6, 20, 30]. Reports of the effects of depletion of uterine histamine [13, 17] or stabilization of uterine mast cells [14] on subsequent decidualization or implantation have not yielded consistent results. If histamine is a mediator of vascular, or other, changes in the uterus then treatment with histamine antagonists should modify such changes. H₁-antagonists given intraluminally have been reported to inhibit decidualization [29] and estrogen-induced edema [31] but in both instances their relative potencies in producing these effects did not correspond with their reported relative potencies as H₁-antagonists [e. g. 28]. Systemic administration of a combination of mepyramine and burimamide has been shown to inhibit estrogen-induced edema [5] and to modify ovum implantation [8], but in both instances replacement of burimamide with the more potent H₂-antagonist metiamide resulted in loss of activity [4, 7]. In addition it has been shown that membrane-stabilizing drugs which are *not* H₁-antagonists, such as *d*-propanolol and lignocaine, can replace mepyramine, given in combination with burimamide, in the inhibition of estrogen-induced uterine edema [4]. In summary, the use of histamine antagonists has not clarified the role histamine in implantation.

Exogenous histamine, given intraluminally, has been reported to reverse the effects of mast cell depletion [17] or stabilization [14] in preventing subsequent implantation. Histamine, given systemically, has been reported to increase uterine blood flow by 3,500-fold [19] and to increase the implantation response to suboptimal doses of estradiol [7] but both of these effects were inhibited by the combination of mepyramine and metiamide, a treatment which, however, has no effect in normal pregnancy [7]. Exogenous histamine has been reported to increase estradiol uptake and retention in uterine cell suspensions when given to the intact animal or given *in vitro* [10]. The effect of histamine on estrogen uptake was inhibited by burimamide given *in vivo* but not by the H₁-antagonist diphenhydramine [11]. The effects of metiamide or of a combination of an H_1- and an H_2-antagonist in this system have not been reported. Studies of the effects of exogenous histamine or of modification of the availability of histamine are of great interest but the relevance of observed effects to the normal course of pregnancy must be investigated.

Thus, circumstantial evidence still supports the view that histamine has a role in implantation but even with the aid of new pharmacological tools is has not proved possible to define this role in terms of the receptors involved.

Acknowledgements

The experimental work described in this report was conducted at the Research Institute, Smith Kline & French Research Ltd., Welwyn Garden City, UK. My sincere thanks are due to my colleagues at SK&F for their help and support during the course of this work and to *Jeffrey Jelly, Pravin Raval* and *Robert Wallis* for their assistance with various parts of the work.

References

1   Ash, A. S. F. and Schild, H. O.: Receptors mediating some actions of histamine. Br. J. Pharmacol. *27:* 427–439 (1966).

2   Black, J. W.; Duncan, W. A. M.; Durant, G. J.; Ganellin, C. R., and Parsons, M. E.: Definition and antagonism of histamine H_2-receptors. Nature, Lond. *236:* 385–390 (1972).

3   Black, J. W.; Duncan, W. A. M.; Emmett, J. C.; Ganellin, C. R.; Hesselbo, T.; Parsons, M. E., and Wyllie, J. H.: Metiamide – an orally active histamine H_2-receptor antagonist. Agents Actions *3:* 133–137 (1973).

4   Brandon, J. M.: An investigation of the involvement of histamine in implantation of ova; Ph. D. thesis (Smith Kline & French Labs. Ltd., Welwyn Garden City 1977).

5   Brandon, J. M.: Inhibition of the acute oedematous response to oestradiol in the immature rat uterus by administration of a combination of mepyramine, a histamine H_1-, and burimamide, a histamine H_2-receptor antagonist. J. Endocr. *73:* 42P–43P (1977).

6   Brandon, J. M. and Bibby, M. C.: A study of changes in uterine mast cells during early pregnancy in the rat. Biol. Reprod. *20:* 977–980 (1979).

7   Brandon, J. M. and Raval, P. J.: Interaction of estrogen and histamine during ovum implantation in the rat. Eur. J. Pharmacol. *57:* 171–177 (1979).

8   Brandon, J. M. and Wallis, R. M.: Effect of mepyramine, a histamine H_1-, and burimamide, a histamine H_2-receptor antagonist, on ovum implantation in the rat. J. Reprod. Fertil. *50:* 251–254 (1977).

9 Brimblecombe, R. W.; Duncan, W. A. M.; Durant, G. J.; Emmett, J. C.; Ganellin, C. R., and Parsons, M. E.: Cimetidine – a non-thiourea H₂-receptor antagonist. J. int. med. Res. *3:* 86–91 (1975).

10 Castro-Vázquez, A.; De Carli, D. N.; Martín, J. L.; Denari, J. H., and Rosner, J. M.: The effect of histamine on the earliest steps of oestrogen action. Steroids Lipids Res. *4:* 105–112 (1973).

11 Castro-Vázquez, A.; Gómez, E.; De Carli, D. N., and Rosner, J. M.: Further evidence for histamine facilitating oestrogen action in the uterus. J. Endocr. *68:* 121–126 (1976).

12 Cross, S. A. M.; Alabaster, V. A.; Bakle, Y. S., and Vane, J. R.: Sites of uptake of ³H-5-hydroxytryptamine in rat isolated lung. Histochemistry *39:* 83–91 (1974).

13 De Feo, V. J.: Decidualization; in Wynn, Cellular biology of the uterus, pp. 191–290 (Appleton Century Crofts, New York 1967).

14 Dey, S. K.; Villanueva, C.; Chien, S. M., and Crist, R. D.: The role of histamine in implantation in the rabbit. J. Reprod. Fertil. *53:* 23–26 (1978).

15 Durant, G. J.; Ganellin, C. R., and Parsons, M. E.: Chemical differentiation of histamine H₁- and H₂-receptor agonists. J. med. Chem. *18:* 905–909 (1975).

16 Edwards, J. A.: The external development of the rabbit and rat embryo; in Woollam, Advances in Teratology, vol. 3, pp. 239–263 (Logos Press, London 1968).

17 Ferrando, G. and Nalbandov, A. V.: Relative importance of histamine and estrogen on implantation in rats. Endocrinology *83:* 933–937 (1968).

18 Goth, A.: On the general problem of release of histamine; in Rocha e Silva, Handbook of experimental pharmacology, vol. XVIII/2 Histamine II and antihistamines, pp. 57–74 (Springer, Berlin 1978).

19 Harvey, C. A. and Owen, D. A. A.: Effect of histamine on uterine vasculature in rats. Eur. J. Pharmacol. *56:* 293–296 (1979).

20 Harvey, E. B.: Mast cell distribution in the uterus of cycling and pregnant hamsters. Anat. Rec. *148:* 507–516 (1964).

21 Hesselbo, T.: The pharmacokinetics of burimamide and metiamide; in Wood and Simkins, International symposium on histamine H₂-receptor antagonists, pp. 29–43 (SK&F Labs. Ltd., Welwyn Garden City 1973).

22 Kraicer, P. and Shelesnyak, M. C.: The induction of deciduomata in the pseudopregnant rat by systemic administration of histamine and histamine-releasers. J. Endocr. *17:* 324–328 (1958).

23 Marcus, G. J.; Shelesnyak, M. C., and Kraicer, P.: Studies on the mechanism of nidation. X. The estrogen-surge, histamine-release and decidual induction in the rat. Acta endocr., Copenh. *47:* 255–264 (1964).

24 Orsini, M. W.: Technique for preparation, study and photography of benzylbenzoate cleared material for embryological studies. J. Reprod. Fertil. *3:* 283–287 (1962).

25 Owen, D. A. A.: Histamine receptors in the cardiovascular system. Gen. Pharmac. *8:* 141–156 (1977).

26 Parsons, M. E.; Owen, D. A. A.; Durant, G. J., and Ganellin, C. R.: Dimaprit-[S-[3-(N,N-dimethylamino)]propyl] isothiourea – a highly specific histamine H₂-receptor agonist. Agents Actions *7:* 31–37 (1977).

27 Schayer, R. W.: Histamine and microcirculation. Life Sci. *15:* 391–401 (1974).

28   Schild, H. O.: pA, A new scale for the measurement of drug antagonism. Br. J.
     Pharmacol. *2:* 189–206 (1947).
29   Shelesnyak, M. C.: Aspects of reproduction. Some experimental studies on the
     mechanism of ova-implantation in the rat. Recent Prog. Horm. Res. *13:* 269–
     317 (1957).
30   Shelesnyak, M. C.: Nidation of the fertilized ovum. Endeavour *19:* 81–86 (1960).
31   Spaziani, E.: Accessory reproductive organs in mammals: control of cell and
     tissue transport by sex hormones. Pharmac. Rev. *27:* 207–286 (1975).

J. M. Brandon, PhD, Department of Agricultural Science, Parks Road,
Oxford OX1 3PF (England)

Prog. reprod. Biol., vol. 7, pp. 253–261 (Karger, Basel 1980)

Species Variation in Implantation

C. A. Finn

Department of Veterinary Physiology and Pharmacology,
University of Liverpool, Liverpool

Processes such as the excretion of fluids from the kidney or the pumping of blood around the body, even the transmission of a nervous impulse along a nerve fibre, are fairly similar among mammalian species and it is possible to extrapolate usefully from one species to another. Indeed it is very often possible to carry out work very relevant to mammals on non-vertebrate species, for example neurophysiologists have obtained valuable data using the squid axon. Those of us working on implantation, however, are confronted with the fact that the process only occurs in a very small proportion of animals and even among these there are wide differences in the cellular mechanisms and the control processes.

The attachment of the egg to the wall of the uterus so that an intimate union is attained between the mother and her offspring is a recent evolutionary occurrence. Although there are examples of non-mammalian species in whom the young are nurtured loosely within the mother's body, it is rare outside the mammalian species for an intimate cellular connection to be formed between the mother and offspring.

To some extent the differences in the mechanism of implantation shown among different species appear to represent the evolution of increasing complexity and one can make out a graded series of cellular changes bringing about an increasingly intimate union between the mother and fetus. It is, I think, important to try and establish the main line of evolutionary development so as to get some idea of the relevance, if any, of findings in one species to others. Of course if one is working on mice or rabbits simply as an exercise in curiosity, or to control reproduction in these animals, then the extent to which the results can be extrapolated to other species, for instance the human, does not matter. However, many of us like to think that our results have general relevance and may even lead, we hope, to better control of reproduction in our own species.

As with most developmental processes it is, I think, possible to discern a main line of evolutionary progression, but superimposed on this are many side branches. If we study something at the end of one of these branches, then although the results may be of great interest in themselves, they may be of little general relevance, and if this is not recognised a lot of time and money may be wasted. My own particular personal prejudice here is the research on blastokinin or uteroglobin in rabbits [*Beier,* 1968; *Daiel,* 1968]. A vast amount of work has gone into this interesting protein, but I am unconvinced that the results have any relevance to the vast majority of mammals. Another fallacy regarding implantation is that if species are closely related in the taxonomic table they will necessarily have similar implantation mechanism. This is particularly important with regard to primates. It seems to be assumed that by studying other primates we are coming closer to the situation in the human. One has only to compare implantation in the rhesus monkey, baboon and women to see the fallacy of this [*Ramsay et al.,* 1976].

Reproductive processes have presumably evolved in response to environmental pressures and particular solutions have come about in response to particular environmental problems. It is in fact fairly easy to see how it could be advantageous for a species to be able to control the time of birth of its young so that they are born at the optimum time of the year. This is most commonly achieved by controlling the season of ovulation. A few species, however, are able to vary the duration of pregnancy in response to environmental stimuli by delaying the time of implantation and this is a fairly obvious difference between species [*Sadlier,* 1969]. Delayed implantation has been widely studied in rats and mice and it is fairly clear now that these animals can maintain the blastocysts in a dormant state for a length of time during lactation and achieve control over the time of implantation by using ovarian oestradiol as the triggering mechanism for implantation [*Lataste,* 1891; *Krehbiel,* 1941]. A great deal of effort has been expended into proving whether the oestradiol is secreted as a hump or a surge and the exact time and dose relationships. It appears now that this use of ovarian oestrogen to pre-cipitate implantation is a specialised facet of implantation in the mouse and rat and is probably not applicable to species in which delay of im-plantation does not occur and indeed to many in which delay does occur. The necessity for progesterone in implantation does, however, appear to be generally applicable and the odd species, like the armadillo, which implants without it, presumably represent an evolutionary side branch.

However, if we look at what progesterone actually does to the cells of the endometrium we find some variation between species. Unfortunately the cellular response of the endometrium to ovarian steroids has not been studied in detail in many species. Nevertheless, it is clear that, whereas in the mouse, rat and guinea pig, progesterone causes epithelial cells of the uterine glands to secrete, whilst oestradiol is responsible for cell proliferation [*Finn and Martin*, 1973; *Mehrotra and Finn*, 1974], in the rabbit progesterone in very small doses brings about glandular proliferation [*Lee and Dukelow*, 1972]. This has led to the dogma that progesterone is responsible for glandular proliferation, whereas this is not true in most species and almost certainly not true in the human where we know that gland mitosis is found during the follicular stage of the cycle. Another difference in the action of progesterone is that whereas in rats and mice it prepares the stromal cells for decidualization in response to the blastocyst, in women progesterone alone can cause some decidualization. However, this is more an increase in the extent of response rather than a qualitative difference.

From work, largely carried out on the mouse, we assume that there is a cycle of cell proliferation, differentiation and death in preparation of the uterus for implantation [*Finn and Porter*, 1975]. There is some indication that this is probably true for women [*Noyes et al.*, 1950] but it is important that the process should be studied in a wider variety of species to be sure that it is of general applicability.

Probably the most interesting and widely studied aspects of implantation from the point of view of species variation are the cellular changes in both the endometrium and trophoblast at the time of implantation.

The object of implantation is of course to get the embryo into a sufficiently intimate contact with the maternal tissues so that exchange can take place. At its simplest this involves the passage of substances from the uterine glands to the lumen to be absorbed by the fetal membranes. This would not normally be considered implantation and is very little advance on the chickens egg although there is probably a period of so-called histiotrophic nutrition early in pregnancy in many species. The most simple acceptable type of implantation would be the close contact of the trophoblast with the surface of the uterine luminal epithelial cells. In this type of implantation the blastocyst expands enormously to fill the uterine lumen so that a vast area of contact is attained. Furthermore, the surface of the uterus is covered with numerous microvilli, the surface of which is mirrored by the surface of the trophoblast, so increasing enormously the area of contact [*Bjorkman,* 1972]. This type of implantation is found in many species including most of the farm animals. Although apparently simple, it is obviously very efficient. One has only to look at a new-born foal or 17 or 18 new-born piglets to realise that nutrients are getting through to the embryo very effectively.

This simple picture is of course complicated by various adaptations of the surface of the uterus, for example the caruncles in ruminants which allow specialised areas of attachment. It was at one time thought that the uterine epithelium was eroded at these points so that the trophoblast was in contact with the uterine stroma [*Amoroso,* 1952], later work with the electron microscope however showed that, although the luminal epithelium was very thinned, it remained intact. This thinning does, however, possibly represent a stage in the evolution of the more intimate connection seen later. A further modification in artiodactyls is that the maternal blood vessels may grow up between the epithelial cells thus bringing maternal blood closer to the embryo. Another important stage, found in some species is the extravasation of maternal blood into the space between luminal epithelium and trophoblast. This is taken up by the trophoblast by a process of phagocytosis.

The modification presented so far stop short of removal of any tissue which is the next stage in the evolutionary struggle to get the blastocyst closer to the maternal blood. It might be asked why it was necessary for evolution to proceed further when as mentioned above the simpler system worked adequately, and indeed when closer union brought other problems. With the blastocyst inside the wall of the uterus and a closer contact between maternal and fetal bloodstreams, it is presumably possible to get satisfactory nutrition with a much smaller area of uterus involved. This might appear to be an advantage, especially with polytocous species. However, one of the most polytocous species, the pig, manages perfectly well with a very simple but very large placenta and a correspondingly very large uterus. Furthermore, if size of uterus were important then one would expect it to be more important in herbivorous species where a large proportion of the abdomen must be given over to digestion, whereas most of the latter have fairly simple placentae. There appears to be little relationship between maturity at birth and complexity of attachment. Two animals with the most deeply attached placentae, the guinea pig and

human are at opposite ends of the spectrum as regards maturity of offspring a birth.

So from the nutritional point of view there appears to be little need
to get rid of the part of the barrier between the bloodstreams of the
mother and fetus. I mentioned earlier that reducing the barrier between
the maternal and fetal blood might lead to problems, and I was thinking in
particular of immunological problems. On the other hand, it could be that
this might be an advantage. It is established for several species with
complex placentae that antibodies can pass from the mother to offspring
through the placenta and whilst this can occasionally lead to trouble, for
example for rhesus-negative mothers carrying rhesus-positive babies, its
overall effect could be an advantage by supplying ready-made protection
to enviromental hazards which the offspring is likely to meet after
birth. Animals with simple placentae appear to transfer little antibody
before birth and rely on the passage of antibodies in the colostrum after
birth.

Whatever the advantage, and I assume there must be one, the removal
of the epithelium between the trophoblast and uterine capillaries appears
to occur in most rodents, primates and carnivores, and we can, I think,
consider it to be part of a general evolutionary transformation. The
mechanism whereby the blastocyst comes to lie within the endometrial
stroma, however, differs. Basically, from the information we have, there
appears to be two main mechanisms, with the rabbit possibly providing a
third. In the mouse and rat the uterine epithelial cells surrounding the
blastocyst appear to be removed by a process of programmed cell death
(sometimes called apoptosis). The blastocyst provides the stimulus for the
implantation reaction but once this has been triggered, the blastocyst
remains passive until the surrounding epithelial cells have died, when the
trophoblast can phagocytose them. Thus, if the reaction is stimulated
artificially, for example by a drop of oil, an implantation chamber is
formed on the antimesometrial side of the uterus with breakdown of the
epithelial cells at its base, demonstrating that activity of the trophoblast
is not necessary [*Finn and Hinchliffe*, 1965]. Furthermore, if the drug
actinomycin D is given to mice before implantation then the breakdown
of the epithelium is prevented [*Finn and Bredl*, 1973], again suggesting
that cell death normally involves direction by the nuclear DNA of the
epithelial cells via transcription of RNA.

In the guinea pig, on the other hand, the blastocyst appears to become
extremely active at the time of implantation and to push its way between

the uterine epithelial cells [*Samson and Hill,* 1931]. As far as we can tell from light and electron micrographs the epithelial cells do not die during this passage but are simply parted at their apices by separation of the junctional complexes [*Green,* 1980]. Once the blastocyst has passed between them the junctional complexes are reformed above the blastocyst, which now lies in the stroma. During its passage the trophoblast takes on the appearance of a very active tissue with massive development of rough endoplasmic reticulum and formation of a syncitium. From the few stages we have of human implantation it would appear that the human blastocyst is very invasive; similar to the guinea pig.

Another interesting feature of implantation in the guinea pig is that whereas in the rat and mouse the first stage of implantation consists of close attachment of the trophoblast to the surface of the epithelial cells, in guinea pigs this stage is not pronounced if it exists at all.

The situation in the rabbit, as far as can be ascertained from electron-micrographs, is very odd indeed [*Steer,* 1970, 1971]. The opposing cells of the trophoblast and uterine epithelium not only come into close contact but actually fuse. In preparation for this event both the uterine epithelium and the trophoblast become syncytial and after fusion there is a cellular mass containing both maternal and embryonic nuclei. The former are, however, soon lost so that one is left with tissue presumably under the influence of nuclei from the trophoblast. This is a very odd method of getting rid of maternal epithelium, but then as I have mentioned earlier, the rabbit does have some very odd reproductive habits.

Whether one should look upon the different methods of bringing the fetal vessels into more intimate contact with the maternal as an evolutionary progression must of course be a matter of opinion, but it appears to me that there is a progression from a situation in which the blastocyst is largely passive with the activity in the endometrium (starting with gland secretion, followed by specialised adaptation of the epithelial surface then thinning of the epithelium and finally leading to programmed cell death) to a condition in which the blastocyst actively invades through the epithelial cells with very little help from the uterine epithelium.

It is, I think, reasonable to consider this to be an evolutionary progression because in those species in which the trophoblast is mostly passive during the breaching of the epithelial barrier the trophoblast later becomes more active and invasive once through into the stroma. It is tempting to view the evolution of implantation as a progressive shift in activity from the maternal to fetal side.

In the mouse we see briefly most of these stages ending up with a very invasive blastocyst but not until it is inside the stroma. Interestingly, if in the mouse the death of the uterine epithelium is delayed by the administration of actinomycin D, the blastocyst goes on developing and appears to become invasive, before the breakdown of the epithelium, and starts penetrating between the epithelial cells [*Finn and Bredl,* 1973].

This progressively increasing activity of the blastocyst in implantation is accompanied by an increasing endocrinological control of pregnancy by the blastocyst. In the simplest form, persistence of the production of progesterone is achieved by maintenance of the corpora lutea, controlled by the pituitary gonadotrophins and uterine prostaglandins, but gradually these are rendered redundant, first by the embryonic tissue producing gonadotrophins and ultimately by its producing steroids. In women, for instance, the ovaries can be dispensed with very early in pregnancy and one suspects, in view of the early production of chorionic gonadotrophin, that the pituitary is redundant very early in or even before implantation, although I know of no evidence for or against this.

If we accept this view that the evolution of implantation has involved the acquisition of greater activity on the part of the blastocyst, does this necessitate increased passivity of the uterus? This is probably true for the epithelium and glands but not in respect of the stroma. The increased invasiveness of the trophoblast would presumably present a considerable threat to the integrity of the uterus if it were not for the transformation of stromal cells into decidual cells. In species in which the uterine epithelium remains intact there appears to be little stromal differentiation whereas in those in which the epithelial barrier is breached, stromal differentiation is extensive. In rodents this does not occur unless the blastocyst indicates its presence on the surface of the epithelium. In the human, however, the process appears to have evolved further so that no stimulus to the epithelium is required for decidualization to start.

An interesting feature here ist that decidual transformation starts well before the blastocysts enter the stroma. This would obviously be necessary if the decidual cells are to play any part in controlling the invasiveness of the trophoblast. However, it does make it rather difficult to understand how decidualization can have evolved as a response to invasion!

If this view of the evolution of decidualization is correct then I would expect amount of decidual tissue to be greater in the more actively invading species. I was interested, therefore, to see in a recent study, in which implantation in the rhesus monkey, baboon and women were compared [*Ramsay et a.*, 1976] that such a correlation was clearly established, although I was rather surprised that the authors came to the completely opposite conclusion; because most decidual tissue was found in the species in which trophoblastic invasion was maximum, that is in women, the function of the decidual tissue could not be to control invasion!

References

Amoroso, E. C.: Placentation; in Parkes, Marshall's physiology of reproduction, pp. 127–311 (Longmans, Harlow 1952).

Beier, H. M.: Uteroglobin: a hormone-sensitive endometrial protein involved in blastocyst development. Biochim. biophys. Acta *160:* 289–291 (1968).

Bjorkman, N.: An atlas of placental fine structure (Baillière, Tindall & Cassell, London 1972).

Daniel, J. C.: Comparison of electrophoretic patterns of uterine fluids from rabbits and mammals having delayed implantation. Compar. Biochem. Physiol. *24:* 297–300 (1968).

Finn, C. A. and Bredl, J. C. S.: Studies on the development of the implantation reaction in the mouse uterus: influence of actinomycin D. J. Reprod. Fertil. *34:* 247–253 (1973).

Finn, C. A. and Hinchliffe, J. R.: Histological and histochemical analysis of the formation of implantation chambers in the mouse uterus. J. Reprod. Fertil. *9:* 301–309 (1965).

Finn, C. A. and Martin, L.: Endocrine control of gland proliferation in the mouse uterus. Biol. Reprod. *8:* 585–588 (1973).

Finn, C. A. and Porter, D. G.: The uterus (Elek Science, London 1975).

Green, C. A. L.: An ultrastructural and endocrinological study of early pregnancy in the guinea pig. Thesis, University of Liverpool (1980).

Krehbiel, R. H.: The effects of theelin on delayed implantation in the pregnant lactating rat. Anat. Rec. *81:* 381–392 (1941).

Lataste, F.: Des variations de durée de la gestation chez les mammifères et des circonstances qui déterminent ces variations. Théorie de la gestation retardée. C. r. Séanc. Soc. Biol. *43:* 21–31 (1891).

Lee, A. E. and Dukelow, W. R.: Synthesis of DNA and mitosis in rabbits uteri after oestrogen and progesterone injections, and during pregnancy. J. Reprod. Fertil. *31:* 473–476 (1972).

Mehrotra, S. and Finn, C. A.: Cell proliferation in the uterus of the guinea pig. J. Reprod. Fertil. *37:* 405–409 (1974).

Noyes, R. W.; Hertig, A. T., and Rock, J.: Dating the endometrial biopsy. Fert. Steril. *1:* 3–25 (1950).

Ramsay, E. M.; Houston, M. L., and Harris, J. W. S.: Interactions of the trophoblast and maternal tissues in three closely related primate species. Am. J. Obstet. Gynec. *124:* 647–652 (1976).

Sadlier, R. H. F. S.: The ecology of reproduction in wild and domestic animals (Methuen, London 1969).

Samson, G. S. and Hill, J. P.: Observations on the structure and mode of implantation of blastocysts of cavia. Trans. zool. Soc. *3:* 295–355 (1931).

Steer, H. W.: The trophoblastic knobs of the preimplanted rabbit: a light and electron microscope study. J. Anat. *107:* 315–325 (1970).

Steer, H. W.: Implantation of the rabbit blastocyst: the adhesive phase of implantation. J. Anat. *107:* 315–325 (1971).

Prof. C. A. Finn, Department of Veterinary Physiology and Phosmacology, University of Liverpool, Liverpool 69 3BX (England)

Prog. reprod. Biol., vol. 7, pp. 262–269 (Karger, Basel 1980)

Endocrinology and Timing of Implantation in the Marmoset Monkey, *Callithrix jacchus*[1]

John P. Hearn

Institute of Zoology, The Zoological Society of London, London

Introduction

The vast amount of work on implantation in non-primate species, particularly in laboratory rodents, can be appreciated from many of the papers in this symposium that deal with precise cellular relationships between the blastocyst and the endometrium. In contrast, our knowledge of implantation in primates is still at a superficial stage. There are several good reasons for this state of affairs. Laboratory primates have never been available in the numbers required for such studies, many of them are seasonal breeders and many have long ovarian cycles. The collection of embryos or any precise investigation of the stages of implantation that necessitates hysterectomy or sacrifice of animals is therefore not normally possible.

There is, however, a case for the development of primate models for studies of implantation. The endocrine support of early pregnancy in primates differs from that in non-primates in important aspects. For example the support of the corpus luteum, while implantation and the luteotrophic shift are completed, is thought to depend on chorionic gonadotrophin secreted by the syncytiotrophoblast [5, 24]. The endocrine requirements for implantation are thought to differ in their reliance on progesterone while oestrogens are of less importance [3]. In practical terms, primate models are required for studies of infertility, early embryonic loss, teratology or in understanding the way in which implantation is controlled in the human.

[1] Supported by a programme grant from the Medical Research Council (UK).

In addition any new methods of fertility control using anti-implantation agents require development and testing on primate models.

Problems arise in the choice of primate models for these studies. While the endocrine profiles of the great apes such as the chimpanzee, orangutan and gorilla during early pregnancy resemble closely those of the human, these animals are endangered in the wild, expensive to maintain in captivity and could never be widely available for research. The morphology and endocrinology of implantation in the macaques and baboons differ in important aspects from that in humans [14, 19] and there is still very little knowledge of the physiology of implantation in New World primate species.

In this communication, preliminary data is presented on the endocrinology and timing of implantation in the common marmoset, *Callithrix jacchus*. This small (adult body weight about 400 g) monkey breeds well in captivity, is not a seasonal breeder, shows no lactation suppression of ovulation and usually produces twins [10, 12]. Sexual maturity is reached between 18 and 20 months of age [1] enabling self-sustaining captive colonies to be established in a relatively short time at moderate expense. Consequently we are studying implantation in this species in order to develop a feasible laboratory primate model for early pregnancy in humans.

Material and Methods

The marmoset monkeys used were *C. jacchus,* of proven fertility, caged in stable male-female pairs throughout. Full details of management were published elsewhere [11] as were details of their basic reproductive biology [14] and endocrinology [7]. Pregnancies were timed by monitoring the midcycle rise in luteinising hormone and the luteal rise in progesterone; details of blood sampling, the assays and their validation were published earlier [4]. Groups of 7 females each were immunised, either actively or passively, against hCG-β subunit to examine the effects of circulating antibodies to this hormone on the length of the cycle and early pregnancy [6, 8].

The uteri of marmosets were flushed with 0.1 % saline at 2, 4, 6, 8, 10 and 12 days after the luteal phase rise in progesterone. To date, 2 animals have been examined at each stage. The animals were anaesthetized with 0.5–0.6 ml Saffan anaesthetic (Glaxovet) and the uterus and ovaries were exteriorised through a 1.0 cm midventral laparotomy. The uterus, ovaries and corpora lutea were measured with calipers and 0.3 ml blood samples were taken from each ovarian vein. The point of a 21 G needle on an open syringe was placed in the lumen through the proximal fundus of the uterus and 0.5 ml 1.0 % sterile saline was flushed through the uterus from a 25 G needle on a 1-ml syringe introduced above the cervix. Blastocysts were fixed in glutaraldehyde for electron microscopy. The uterus and ovaries were replaced in the

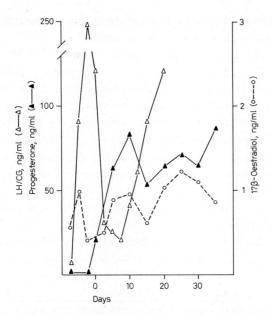

Fig. 1. The peripheral plasma levels of LH/CG, progesterone and 17β-oestra-diol from the midcycle rise in oestradiol to day 35 of pregnancy in a marmoset monkey. Day 0 is taken as the day of the luteal phase rise in progesterone.

abdominal cavity and the incision closed with two interrupted stitches, using 4/0 chromic sutures, through the peritoneum and muscle layers, a continuous subdermal purse stitch around the skin incision and finally two interrupted superficial stitches through the skin. The animals received an injection of 0.2 ml penicillin (Ethacillin, Intervet) and the wound was dusted with penicillin powder. All of them made rapid recoveries and in almost all cases the animals were pregnant again within 2–3 weeks of the operation.

Results

Figure 1 shows the peripheral plasma levels of LH/CG, progesterone and 17β-oestradiol in a marmoset during the midcycle and early pregnancy. Similar profiles were found in 20 conception cycles studied. The preliminary results suggest: (1) In common with the human and in contrast to the rhesus monkey, cynomologous and baboon, there are elevated levels of 17β-oestradiol secreted during the luteal phase over the period when implantation is taking place (see results of uterine flushing, below). (2) The profile of progesterone, which requires further data from more frequent sampling, indicates that the corpus luteum is 'rescued' between days 15–20 of the ovarian cycle. (3) The rise in chorionic gonadotrophin (which in this assay cannot be distinguished from LH) commences by day 10 after the luteal rise in progesterone, suggesting that implantation has been initiated by this time.

Blastocysts were recovered from the uteri of marmosets flushed at 6 and at 8 days after the luteal rise in progesterone. At these stages they were still in their zonae. No blastocysts or embryonic material were recovered in flushings performed at 2, 4, 10 or 12 days, suggesting that the blastocyst has not entered the uterus by day 4 and that implantation has commenced by day 10.

Marmosets immunised actively against the β-subunit of hCG showed apparently normal cycles while antibody titres were high, without any extension of the luteal phase into pregnancy. The animals were cycling, as shown by the levels of progesterone; mating, as shown by sperm in vaginal smears; and becoming pregnant, as shown by the recovery of apparently normal blastocysts at day 6 or 8 after the luteal rise in progesterone. Consequently, antibodies to hCG-β were apparently capable of blocking implantation. Whether this was brought about by a failure of the luteotrophic stimulus or by a more direct effect on the implanting blastocyst has not been determined. In addition, both active and passive immunisation against hCG-β after implantation resulted in rapid termination of pregnancy when performed up to the 6th week, or in resorption of embryos if performed between the 6th and 12th weeks of pregnancy. After the 12th week of pregnancy there was no apparent effect and pregnancy proceeded to term [6, 8].

Discussion

In the human, baboon and rhesus monkey, ovulation normally occurs within 24 and always within 48 h of the LH peak [18]. However, there is some variation in the timing of events before implantation in primates [9] and the day of gestation when embryo attachment occurs and the first elevated levels of chorionic gonadotrophin are reported a summarised in

Table I. The days of gestation when embryo attachment and the first detection of chorionic gonadotrophin are reported for 5 primate species

Species	Days of gestation		
	embryo attachment	CG first detected	References
Human	5$^1/_2$–6	7–8	13, 17
Chimpanzee	9–10	11	20
Baboon	8–9	12	23
Rhesus	8–9	12	2
Marmoset	after 8	10	this paper

table I. An earlier claim that the preimplantation human blastocyst was secreting chorionic gonadotrophin [22] has yet to be confirmed and at present it is probably fair to assume that the first rise in chorionic gonadotrophin signifies attachment and penetration by the syncytiotrophoblast.

There are problems in obtaining precisely timed pregnancies in primates to study implantation. The four events used most are the midcycle rise in 17β-oestradiol or of LH, short-term matings synchronised to these hormonal changes, or the luteal rise in progesterone. Although the mean time relationships between the oestradiol and LH peaks with ovulation are known for the human, baboon and rhesus [18], individual variation and the variation in time between ovulation, conception and the establishment of luteal function probably mean that we are presently limited to a precision of ± 24 h. This would be no great problem were large numbers of such timed pregnancies available, but the recovery of implanted blastocysts *in situ* with a minimum waste of valuable animals does pose some difficulties which we are presently trying to overcome, without using hysterectomy, in the marmoset. It is perhaps not surprising that such studies have yet to be carried out on primates and that the knowledge we have to date in the human and Old World primates rests on very few early implantation sites collected almost by chance.

There is also variation in the morphology of implantation in primates, especially in the degree of invasiveness of trophoblast and the endometrial decidual response. The human and chimpanzee show a close resemblance in their interstitial mode of implantation, whereas all other primates studied so far show only a superficial invasiveness of trophoblast. In the human, where decidualisation is most evident, trophoblastic penetration is greatest, but in the rhesus there is a transient epithelial plaque formed and penetration is comparatively slight; in the baboon there is virtually no decidual reaction and invasiveness is even more limited [19].

There is no point in developing primates as models for implantation in humans if other appropriate animal models are available. The rat and mouse are probably not appropriate since oestradiol is thought to be essential for implantation, but in the hamster and guinea pig (as well as the pig, sheep and ferret) implantation can proceed in ovariectomised animals given exogenous progesterone. However, it is only in the primates that chorionic gonadotrophin is implicated as an essential component of the luteotrophic stimulus and implantation can proceed with progesterone alone [16, 21] although oestrogen may have a facilitatory role. In addition, successful implantation in non-primate species is thought to require precise synchrony of the arrival of the blastocyst in the uterus, whereas *Marsden et al.* [15] have shown in the rhesus monkey that binucleate eggs transferred from the oviduct to the uterus will implant successfully.

If primates are required for studies on implantation and early fetal development, then it makes sense to choose a primate such as the marmoset that will breed rapidly in captivity, has a short cycle, usually produces two blastocysts at ovulation and tolerates surgery well. With its short generation interval it is also possible to establish this species relatively inexpensively in self-sustaining colonies, thereby becoming independent of supplies from the wild. No single animal model, either primate or other species, is likely to prove ideal for studies related to human implantation. However, we feel that our preliminary results on the marmoset suggest it could be as good as any others and better than most.

References

1   Abbott, D. H. and Hearn, J. P.: Physical, hormonal and behavioural aspects of sexual development in the marmoset monkey, *Callithrix jacchus.* J. Reprod. Fertil. *53:* 155–166 (1978).
2   Atkinson, L. E.; Hothckiss, J.; Fritz, G. R.; Surve, A. H.; Neill, H. D., and Knobil, E.: Circulating levels of steroids and chorionic gonadotropin during pregnancy in the rhesus monkey, with special attention to the rescue of the corpus luteum in early pregnancy. Biol. Reprod. *12:* 335–345 (1975).
3   Csapo, A. I.; Pulkinnen, M. O., and Kaihola, H. L.: The regulatory significance of the human corpus luteum. Obstetl gynec. Surv. *31:* 221–225 (1976).
4   Chambers, P. L. and Hearn, J. P.: Peripheral plasma levels of progesterone, oestradiol-17β, oestrone, testosterone, androstenedione and chorionic gonado-

trophin during pregnancy in the marmoset monkey, *Callithrix jacchus*. J. Reprod. Fertil. *56:* 23–32 (1979).

5   Heap, R. P. and Perry, J. S.: The maternal recognition of pregnancy. Br. J. Hosp. Med. *1974:* 8–14.

6   Hearn, J. P.: Immunisation against pregnancy. Proc. R. Soc. *195:* 149–160 (1976).

7   Hearn, J. P.: The endocrinology of reproduction in the common marmoset, *Callithrix jacchus*; in Kleiman, Biology and conservation of marmosets, pp. 163–171 (Smithsonian Institution, Washington 1978).

8   Hearn, J. P.: Immunological interference with the maternal recognition of pregnancy in primates; in Ciba Symp. 64, pp. 353–375 (Elsevier, Amsterdam 1979).

9   Hearn, J. P.: Primate models for early human pregnancy; in Serio, Animal models for human reproduction (Pergamon, New York, in press, 1980).

10 Hearn, J. P. and Lunn, S. F.: The reproductive biology of the marmoset monkey, *Callithrix jacchus*; in Perkins and O'Donoghue, Breeding primates for developmental biology. Lab. Anim. Handbk *6:* 191–202 (1975).

11 Hearn, J. P.; Lunn, S. F.; Burden, F. J., and Pilcher, M. M.: Management of marmosets for biomedical research. Lab. Anim. *9:* 125–134 (1975).

12 Hearn, J. P.; Abbott, D. H.; Chambers, P. L.; Hodges, J. K., and Lunn, S. F.: Use of the common marmoset, *Callithrix jacchus,* in reproductive research. Prim. Med., vol. 10, pp. 40–49 (Karger, Basel 1978).

13 Landesman, R. and Saxena, B. B.: Results of the first radioreceptor assays for the determination of human chorionic gonadotrophin. Fert. Steril. *27:* 357–368 (1976).

14 Luckett, P. W.: Comparative development and evolution of the placenta in primates. Contr. Primatol., vol. 3, pp. 142–233 (Karger, Basel 1974).

15 Marsden, J. H.; Penn, R., and Sivelle, P. C.: Successful autotransfer of tubal eggs in the rhesus monkey, *Macaca mulatta*. J. Reprod. Fertil. *49:* 175–176 (1977).

16 Meyer, R. K.; Wolfe, R., and Arslan, M.: Implantation and maintenance of pregnancy in progesterone treated ovariectomised monkeys *(Macaca mulatta)*. Proc. 2nd Int. Cong. Primatology, vol. 2, pp. 30–35 (1968).

17 Ortiz, M. E. and Croxatto, H. B.: Observations on the transport, ageing and development of ova in the human genital tract; in Talwar, Recent advances in reproduction and fertility, pp. 307–318 (Elsevier, Amsterdam 1979).

18 Pauerstein, C. J.; Eddy, C. A.; Croxatto, H. D.; Hess, R.; Siler-Khodr, T. M., and Croxatto, H. B.: Temporal relationships of estrogen, progesterone and luteinising hormone levels of ovulation in women and infrahuman primates. Am. J. Obstet. Gynec. *130:* 876–886 (1978).

19 Ramsey, E. M.; Houston, M. L., and Harris, J. W. S.: Interactions of the trophoblast and maternal tissues in three closely related primate species. Am. J. Obstet. Gynec. *124:* 647–652 (1976).

20 Reyes, F. I.; Winter, J. S. D.; Faiman, C., and Hobson, W. C.: Serial serum levels of gonadotrophins, prolactin and sex steroids in the non-pregnant and pregnant chimpanzee. Endocrinology *96:* 1447–1455 (1975).

21 Ross, G. T.: hCG and the hormonal milieu in early human pregnancy; in Maternal recognition of pregnancy. Ciba Fdn Symp. 64, pp. 191–208 (Elsevier, Amsterdam 1978).

22 Saxena, B. B.; Hasen, S. H.; Haour, F., and Schmidt-Gollwitzer, M.: Radio-receptor assay of hCG: detection of early pregnancy. Science, N. Y. *184:* 793–795 (1974).
23 Shaikh, A. A.: Animal models for research in human reproduction. NIH Invited Report (1978).
24 Short, R. V.: Implantation and the maternal recognition of pregnancy; in Foetal anatomy. Ciba Fdn. Symp. 23, pp. 2–26 (Churchill, London 1969).

J. P. Hearn, PhD, Director, Institute of Zoology, The Zoological Society of London, Regent's Park, London NW1 4RY (England)

Prog. reprod. Biol., vol. 7, pp. 270–283 (Karger, Basel 1980)

Morphological Basis of Implantation in the Rhesus Monkey

A. C. Enders and A. G. Hendrickx

Department of Human Anatomy, School of Medicine,
University of California, Davis, Calif.

Introduction

All of us create visual images of events, even those which we are not privileged to see in any aspect. In the case of implantation of the blastocyst, its sequestration within the uterus at the time of implantation prevents continuous observation. The first concerted effort to study implantation in the rhesus monkey took advantage of a number of skills that had been developed at the Carnegie Institution of Embryology nearly four decades ago. The material was collected there using the method of *Hartman* [1933] to determine when ovulation had occurred by palpation of the ovary and the excellent histological techniques of *Heuser* in preparing serially sectioned material for light microscopy [*Heuser and Streeter,* 1941]. The excellent material prepared at that time is still available for study, and is currently housed at the University of California at Davis. It provides the greatest single store of information available on the intrauterine events of implantation with regard to interaction of the trophoblast of the blastocyst and the endometrium in the earliest stages of implantation in the human [*Hertig et al.,* 1956] and rhesus monkey [*Heuser and Streeter,* 1941].

Before initiating the series of studies summarized here we asked a number of questions. What aspects of implantation can be obtained by morphological methods? Is the zona pellucida lost before implantation? Where is the blastocyst situated at implantation? What is the nature of the uterine lumen at this time? Is it distended or is it apposed to the blastocyst? What is the condition of the trophoblast at the time of first adhesion? Is it entirely cellular or is there syncytial trophoblast as well? How is the first incursion of trophoblast through the uterine epithelium achieved? Does the trophoblast enter where a cell has sloughed? Is cell death involved? Is there evidence of lysis of the uterine cells? Is there evidence of fusion of uterine cells to the trophoblast, or of intrusion of trophoblast cells or syncytium between uterine epithelial cells? Is there any evidence of either surface proteolytic or glycolytic activity or extracellular proteolytic or glycolytic activity? What are the first local modifications of the endometrium that can be observed during implantation? Is there any evidence of inflammatory response to the blastocyst? Are there any specialized reactions of the endometrium that occur at implantation, but not in relationship to the cycle or to infection? Can partial information concerning mechanisms of implantation be obtained on how adhesion might be brought about? What is the nature of the glycocalyx coats? How does syncytium formation occur? What is the incidence and importance of cell death in the process of blastocyst formation and of implantation? Unfortunately, some of these types of information cannot be obtained with routine methods. Methods most appropriate for one approach may render the specimens useless for others. For this reason it is particularly appropriate that there is a background of serially sectioned materials from sequential stages. The report presented here constitutes a preliminary survey of the morphological information that our group is currently accumulating on the basis of implantation in the rhesus monkey [*Hendrickx and Enders,* 1980; *Enders and Hendrickx,* 1980], information that will be presented in a series of detailed studies elsewhere.

Materials and Methods

The materials for this study were collected at the California Primate Research Center in Davis. The different types of materials were obtained by three procedures. During the preimplantation period, the uterus and oviducts were flushed in phosphate-buffered saline at laparotomy and the flushing fluid collected and examined. Cleavage and blastocyst stages were then prepared for electron microscopy by fixation in glutaraldehyde, dehydration in alcohol, and embedding in araldite resin, thin sectioning and subsequent staining with lead citrate and uranyl acetate. The embryos were also sectioned at approximately one μm and stained with azure B for light microscopic examination. Daily serum samples had been collected from animals that were selected for the flushing procedure. If an embryo was found, the blood was analyzed by radioimmunoassay to determine the estrogen peak. The day following the estrogen peak, which occurs 24–48 h before ovulation, was designated day 0 of pregnancy. It should be noted that with this system the material should be comparable to that of *Carl Hartman* who palpated the ovary to determine the collapse

of follicle indicating that ovulation had occurred. The early postimplantation animals were selected on the basis of elevation above cycle levels of both estrogen and
progesterone in the peripheral blood. This method could be used successfully after
day 13. For collection of implantation stages on days 9–11, peripheral blood levels
of progesterone were used to determine ovulation but whether or not the females
were pregnant was a matter of chance since there was no significant elevation of
ovarian steroids above cycle levels during that time period. Both at the time of implantation and in the later stages of implantation, fixative was injected directly into
the abdominal aorta after a brief saline lavage while the animals were anesthesized.
This procedure effectively hardened the uterus stopping all contraction prior to removal of this organ from the animal. Tissue was then examined to determine the
presence of an implantation site.

For the stages of implantation from day 13–16, cross sections of the uterus
were carefully hand cut under a dissecting microscope until the implantation site
was encountered. This procedure permitted direct observation of the extent of the
uterine lumen without distortion while enabling us to achieve appropriate conditions
for rapidly continuing the fixation process.

Results

Preimplantation Embryos

With the uterine flush method, it is not entirely certain that only unattached blastocysts were removed from the uterus. The flushing process
introduced fluid into the lumen while simultaneously draining through a
second catheter. In a few instances elongated processes from the trophoblast cells were observed suggesting that adhesion might have begun, but
in no instance did we find adherent uterine epithelial cells, nor were torn
regions found on any of the flushed blastocysts. Important observations
are possible using this electron microscopic approach which are not
available from other methods. Not only is the endoderm present prior to
implantation as has been reported both by light microscopy [*Luckett*,
1978] and electron microscopy [*Hurst et al.*, 1978], but in addition there
is a well-formed basal lamina under all of the trophoblast and hence, interposed between the trophoblast and the endoderm. It is clear, therefore,
that the endoderm develops as an extension of the hypoblast and indeed it
is continuous with it, although it does not form a complete sac within the
trophoblast prior to implantation. After the loss of the zona pellucida,
there tends to be a collapse of the blastocyst. This results in rounding up
of all of the trophoblast cells, not just those overlying the embryonic cell
mass as was originally reported by *Heuser and Streeter* [1941]. There is
considerable variation in the cells of the trophoblast which do not appear

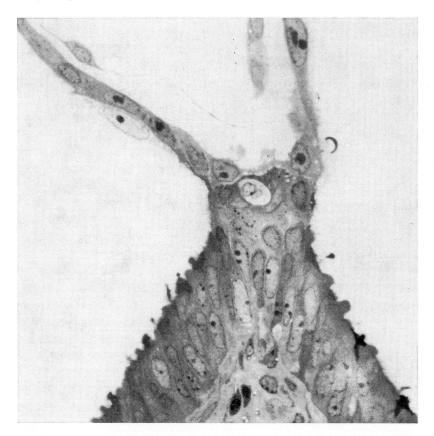

Fig. 1. Light micrograph of the first invasion of the uterus by the trophoblast. The syncytial trophoblast surrounds some uterine cells and intrudes between others, but does not reach the basal surface of the uterine epithelium. Day 9.5.

as mature as in many other species in that there are fewer cytological changes from the cleavage stages until after the loss of the zona pellucida. In the preliminary survey of our specimens there is no evidence of syncytium formation in the blastocyst after the loss of the zona prior to implantation, but the earliest implantation specimen has appreciable syncytium.

Implantation

In the earliest of the three specimens, day 9.5, there is only a single site of association of the trophoblast and the uterus (fig. 1, 2). At this site

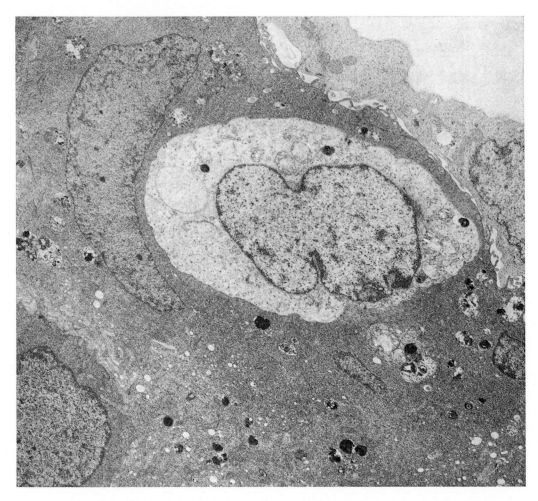

Fig. 2. Electron micrograph from the previous implantation site. Note the basal lamina on the cytotrophoblast and the portions of three nuclei in the syncytial trophoblast. A uterine epithelial cell is in the lower left.

the trophoblast has partially penetrated the uterine epithelium which still has an intact basal lamina. However, some of the epithelial cells have been displaced and the trophoblast is intruded between epithelial cells. Interestingly, not only does trophoblast intrude between cells, but is sends processes into the uterine cells, indenting them (fig. 3). Whether these aid in

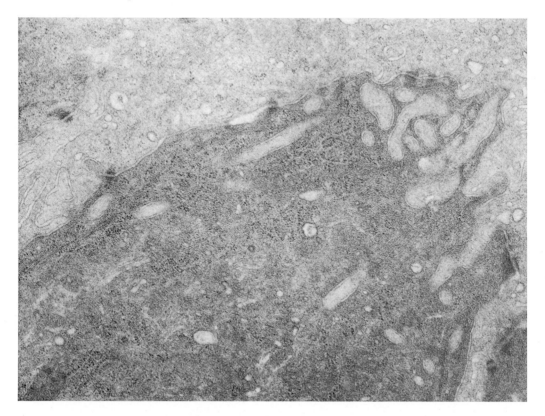

Fig. 3. Syncytial trophoblast (above) intrudes on a uterine epithelial cell (below, center) with which it shares desmosomes at the margin of the same implantation site shown in the previous figures.

adhesion or in encouraging autolysis of the cells is not known. These processes are rather similar to the processes of the trophoblast cells that invade the endometrium of the horse to produce the endometrial cup tissue [*Allen et al.,* 1973]. The invasive trophoblast is syncytial. Cytotrophoblast is abundantly associated with it, but it restricted to a position bordering the cavity of the blastocyst (fig. 2, 4). The endoderm extends well beyond the ICM, but is incomplete abembryonically. At this stage there is not yet any epithelial hypertrophy in the uterus adjacent to the blastocyst (epithelial plaque reaction). The uterine stroma contains numerous leukocytes, particularly neutrophilic granulocytes and macrophages,

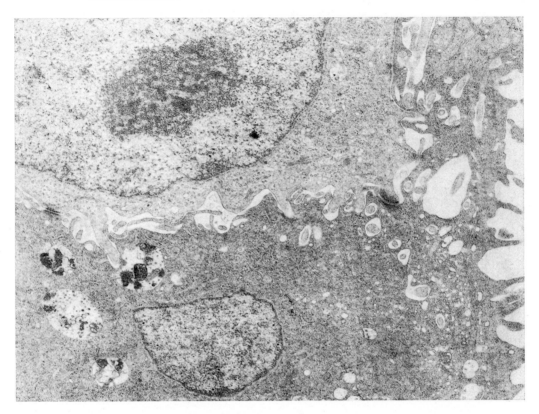

Fig. 4. The cytotrophoblast (above) also shares desmosomes with the syncytium (below). However, cytotrophoblast tends to have processes intruding into syncytium rather than vice versa. Note the large multivesicular bodies characteristic of syncytium at this stage. Day 9.5.

and occasional mitotic figures are seen in the stroma. One mitotic figure was present in a vessel near, but not directly underlying, the implantation site.

With further development of the implantation site, the epithelial plaque reaction of the endometrium is initiated. The first cells to respond are apparently the basal cells within the epithelium (fig. 5). Presumably, these are less differentiated cells than the surface cells. The plaque reaction produces a pad of epithelioid cells rich in glycogen under and surrounding the area of invasion but only incompletely interposed between

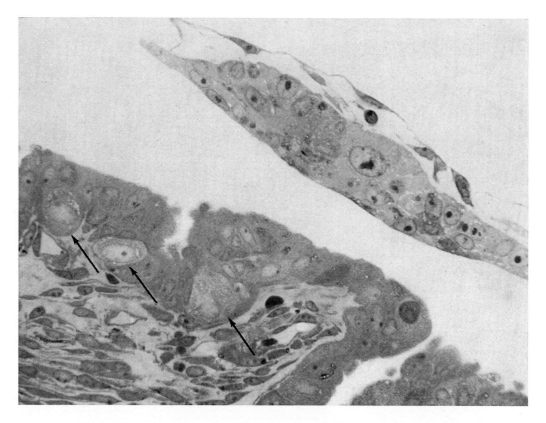

Fig. 5. Light micrograph of trophoblast and endometrium marginal to the secondary implantation site on day 13. Note the transforming basal epithelial cells (arrows) and the trophoblastic knob underlain by endoderm.

the trophoblast and the maternal vessels (fig. 6, 7). While the epithelial plaque reaction is still developing, a pronounced edema develops peripheral to the invasion site and the trophoblast enters the enlarging vessels of the underlying endometrium. These vessels enlarge not only by distension, but also by cell division. Unusual hypertrophy and differentiation of the endothelial cells proceed throughout the area of implantation. Thus, reorganization of the endothelial cells includes an aggregation and 're-versal' of the Golgi apparatus (fig. 8). At an early stage (prior to the completion of the cellular hypertrophy of the endometrial capillaries), trophoblast enters these vessels without surrounding them.

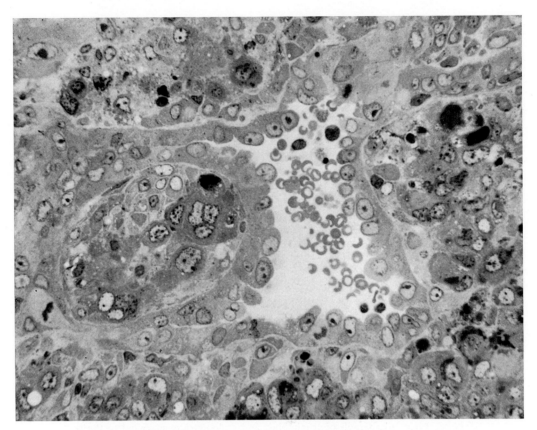

Fig. 6. Light micrograph of a hypertrophied endometrial vessel surrounded by epithelial plaque cells. The endothelium shows apical displacement of the nuclei in addition to occasional mitotic figures.

Subsequent development of the implantation site results in a thickening of the epithelial pad or plaque reaction, and the elevation of the trophoblast in part by blood lacunae. The lacunae do not develop in pre-existing spaces in the trophoblast, but seem to be the result of pooled blood from vessels tapped by the trophoblast occupying space surrounded by trophoblast in a manner similar to hematoma formation. Subsequent development of villi by the trophoblast and fetal mesoderm results in the lining of the maternal blood spaces with syncytial trophoblast, the establishment of cell columns at the maternal-fetal interface and the exploitation of the blood-filled spaces as the intervillous space.

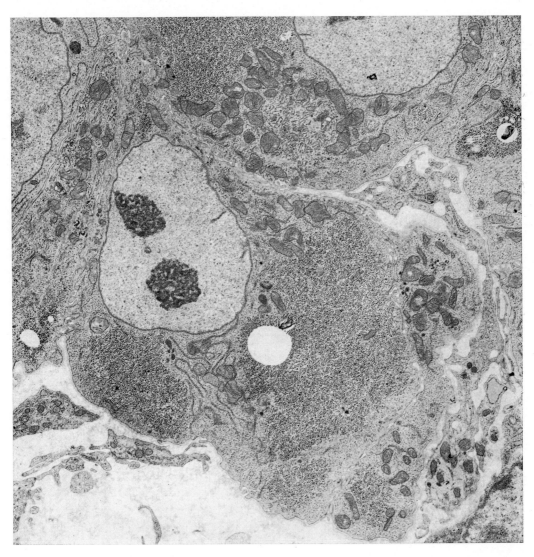

Fig. 7. Electron micrograph of epithelial plaque cells. Note the prominent nucleoli and that much of the cytoplasm of these cells is occupied by glycogen deposits.

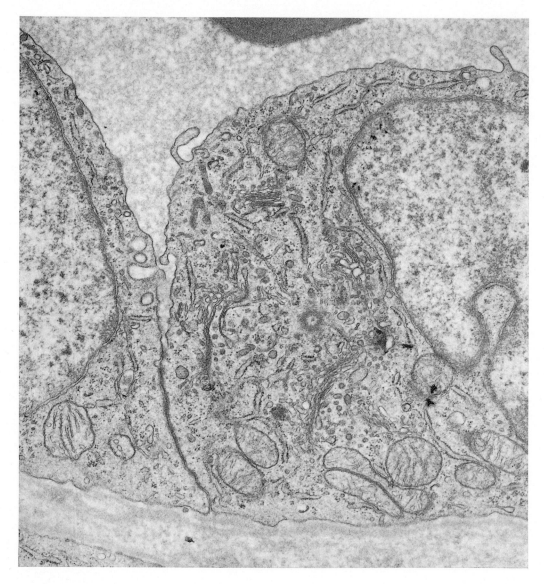

Fig. 8. Electron micrograph of endothelial cells from a hypertrophied endometrial capillary. Note the large Golgi zone in the cell to the right. In some endothelial cells from this stage (day 14) the nuclei are apical and the Golgi zone basal.

Discussion

If we put together our current results using electron microscopy with those previously available [*Hurst et al., 1978*], it is evident that the rhesus monkey embryo loses its zona pellucida prior to the invasion of the epithelium and quite probably prior to adhesion to the uterine surface. It appears common for the blastocyst to undergo partial collapse after losing the zona pellucida. The formation of the endoderm in the blastocyst is conventional as suggested recently by *Luckett* [1978], and is similar to that of other species, being essentially a delamination and extension of the hypoblast. At the time of implantation the uterine lumen is a closed slot, but there is not any indication of interdigitation of microvilli such as in the uterine epithelial 'attachment reaction' in the rat [*Ljungkvist, 1972*]. The trophoblast intrudes processes between cells identing these cells and surrounds them. There is no evidence of fusion of trophoblast and uterine epithelial cell such as occurs in the rabbit [*Enders and Schlafke, 1971; Larsen, 1961*] or nuclear migration from the epithelium to the trophoblast as suggested by *Heuser and Streeter* [1941], but not by *Wislocki and Streeter* [1938] who examined some of the same material. From a comparative viewpoint epithelial penetration in this primate appears to be more similar to that in the ferret than the mechanism in some other species [*Schlafke and Enders, 1975*]. Our hypothesis at the moment is that the intrusion of trophoblast processes into the apical ends of the epithelial cells may cause these cells to dissociate their apical junctional complexes, and allow for the intrusion of the trophoblast. Even in the early stages of epithelial penetration it is syncytial trophoblast that is the intrusive tissue as *Wislocki and Streeter* [1938] originally suggested.

The adjacent endometrium responds in three different ways to the trophoblast: by cell proliferation in both the basal cells of the epithelium and in the endothelial cells of the vessel; secondly, by a generalized edema which includes a leukocytic response; and by vascular hypertrophy. The vascular hypertrophy eventually includes pronounced cytological changes in these cells including Golgi reversal and must therefore be considered a specialized phenomenon, not simply part of an inflammatory type response. The original invasion of one area by the trophoblast is followed by invasion in several other areas, even in the formation of the primary placenta. Interestingly enough, not only are regions of syncytium formed in relationship to the opposed side of the uterus where the secondary placenta forms, but some occur paraplacentally. Presumably these 'knobs' do not attach because they do not form a proper association with the epithelium. It is also important that the second invasion of the epithelium not only occurs in an endometrium that is older and has a different endocrine environment, but also occurs after formation of the epithelial plaque rather than prior to formation of the epithelial plaque as is the case on the primary placenta.

Ramsey et al. [1976] have recently pointed out some of the similarities and differences in the later stages of implantation and placentation in the rhesus monkey, human and baboon. Since very little information is available concerning early stages of human implantation [e. g. *Knoth and Larsen,* 1972] and there is little likelihood that the situation will change markedly in the near future, it is useful to speculate on what can be learned from other species [*Enders,* 1976]. From data collected to date it appears that the rhesus monkey may well be similar to the human in zona loss, syncytium formation, mechanism of epithelial penetration and elicitation of a mild inflammatory response in the endometrium. The rhesus monkey is clearly peculiar in formation of the plaque reaction, in not forming trophoblastic lacunae around maternal vessels and possibly in the eliciting of an endothelial differentiation response from endometrial vessels.

References

Allen, W. R.; Hamilton, D. W., and Moor, R. M.: The origin of equine endometrial cups. II. Invasion of the endometrium by trophoblast. Anat. Rec. *177:* 485–502 (1973).

Enders, A. C.: Cytology of human early implantation. Res. Reprod. *8:* 1–3 (1976).

Enders, A. C. and Hendrickx, A. G.: Implantation in non-human primates. I. Morphology. Relevance of research on non-human primates to the understanding of human reproduction (Karger, Basel 1980).

Enders, A. C. and Schlafke, S.: Penetration of the uterine epithelium during implantation in the rabbit. Am. J. Anat. *132:* 219–240 (1971).

Hartman, C. G.: Pelvic (rectal) palpation of the female monkey, with special reference of the ascertainment of ovulation time. Am. J. Obstet. Gynec. *26:* 600–608 (1933).

Hendrickx, A. G. and Enders, A. C.: Implantation in non-human primates. II. Endocrinology. Relevance of research on non-human primates to the understanding of human reproduction (Karger, Basel 1980).

Hertig, A. T.; Rock, J., and Adams, E. C.: A description of 34 human ova within the first 17 days of development. Am. J. Anat. *98:* 435 (1956).

Heuser, C. H. and Streeter, G. L.: Development of the macaque embryo. Contrib. Embryol. *29:* 15–55 (1941).

Hurst, P. R.; Jeffries, K.; Eckstein, P., and Wheeler, A. G.: An ultrastructural study of preimplantation uterine embryos of the rhesus monkey. J. Anat. *126:* 209–220 (1978).

Knoth, M. and Larsen, J. F.: Ultrastructure of a human implantation site. Acta obstet. gynec. scand. *51:* 385–398 (1972).

Larsen, J. F.: Electron microscopy of the implantation site in the rabbit. Am. J. Anat. *109:* 319–334 (1961).

Ljungkvist, I.: Attachment reaction of rat uterine luminal epithelium. IV. The cellular changes in the attachment reaction and its hormonal regulation. Fert. Steril. *23:* 847–865 (1972).

Luckett, W. P.: Origin and differentiation of the yolk sac and extraembryonic mesoderm in presomite human and rhesus monkey embryos. Am. J. Anat. *152:* 59–98 (1978).

Ramsey, E. M.; Houston, M. L., and Harris, J. W. S.: Interactions of the trophoblast and maternal tissues in three closely related primate species. Am. J. Obstet. Gynec. *124:* 647–652 (1976).

Schlafke, S. and Enders, A. C.: Cellular basis of interaction between trophoblast and uterus at implantation. Biol. Reprod. *12:* 41–65 (1975).

Wislocki, G. B. and Streeter, G. L.: On the placentation of the macaque *(Macaca mulatta)* from the time of implantation until the formation of the definitive placenta. Contrib. Embryol. *27:* 1–66 (1938).

A. C. Enders, Ph. D., Department of Human Anatomy, School of Medicine, University of California, Davis, CA 95616 (USA)

Prog. reprod. Biol., vol. 7, pp. 284–296 (Karger, Basel 1980)

Human Implantation and Clinical Aspects[1]

J. Falck Larsen

Herlev University Hospital, Department of Obstetrics and Gynaecology,
University of Copenhagen, Copenhagen

Our information about the mechanism of the human implantation is still incomplete. The fertilized ovum arrives at the uterine cavity 4 days after fertilization as a morula or early blastocyst. *Hertig et al.* [9] described a morula consisting of 58 cells, 5 embryonic and 53 trophoblastic cells.

They also described a blastocyst aged 4½ days with 107 cells, 8 embryonic cells and 99 trophoblastic cells. At this stage an embryonic pole and an embryonic disc was formed.

Then, there is a gap in our information up to the last part of the 7th day, from which we have a specimen of an ovum in the process of implantation [9]. The blastocyst is in broad contact with the uterine epithelium, which is penetrated in many areas.

The following stages are well described with the light microscope by *Hertig et al.* [9] and other investigators. On the 9th day, the blastocyst has penetrated the uterine epithelium completely and the place of invasion has been closed by a coagulum. The surrounding trophoblast has been transformed into a syncytium but is has not yet invaded the maternal vessels.

2 days later, on the 11th day after the fertilization, the trophoblast has penetrated the maternal vessels and the trophoblastic lacunae have been filled with maternal blood. The human implantation site will appear as a red ring formed by the lacunae (fig. 1). This specimen was recovered from an uterus removed on the 25th day of a menstrual cycle. It was fixed for electron microscopy. Figure 2 shows a montage made

[1] Supported by a grant from *P. Carl Pedersens* Foundation and the Danish Medical Research Council.

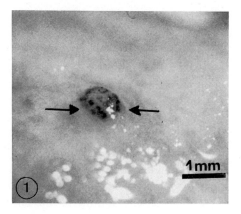

Fig. 1. Human implantation site on the 11th day after fertilization. Note the dark lacunae with maternal blood.

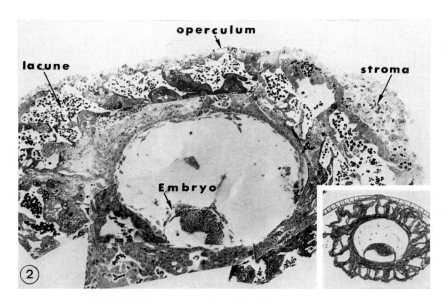

Fig. 2. Light micrographic survey-montage of the implanted ovum.

from micrographs of 1-μm sections stained with toluidine blue. The large trophoblastic lacunae are well demonstrated. It should be noted that the uterine stroma shows very little decidual transformation at this stage. The decidual reaction does not seem to have any importance in the human implantation process. However, the decidual reaction develops in the human uterine mucosa at a later stage, and the decidua may be of importance in the limitation of the trophoblastic invasion.

Which information do we have about the mechanism of the human implantation? The first condition of implantation is the transport of the fertilized ovum to an appropriate location for nidation. The site of attachment seems to be rather critical for each particular species. Most of the human implantation sites were found near the midline on the anterior or posterior wall of the uterus [8]. Due to the fusion of the Müllerian ducts in the human, the implantation may be regarded as antimesometrial.

When the blastocyst is transported to the right position, it rotated until the embryonic pole is in contact with the endometrium. This rotation and polar orientation is also species specific. In the rabbit the implantation occurs at the embryonic pole, in the guinea pig at the anti-embryonic pole. A defective rotation and incorrect orientation may lead to an unsuccessful implantation.

The next step in implantation is fixation of the blastocyst at the surface of the uterine epithelium. The blastocyst is covered with the zona pellucida and a protein layer, the 'gloiolemma' [1]. The zona pellucida may be penetrated by the processes of the trophoblast making contact with the uterine epithelium. From the available human material, we cannot tell if the zona pellucida is dissolved or penetrated. The zona pellucida is present at the surface of the morula at about 96 h developmental age (Carnegie No. 8794). However, it seems to be incomplete, and no zona pellucida is found around the free blastocyst of 4$^{1}/_{2}$ days developmental age (Carnegie No. 8663). This indicates that in the human the zona pellucida dissolves before the contact between the cell membranes of the epithelium and the trophoblast is established. This is not the case in all species. In the guinea pig the trophoblastic processes penetrate the zona and make contact with the epithelium [3, 14].

We know from the animal experiments that a very intimate contact is established between the cell membranes of the trophoblast and the uterine epithelium as soon as the zona pellucida has disappeared. This very first stage of contact has been called the stage of apposition by *Enders and Schlafke* [3]. The contact is so close that the blastocyst is kept in position

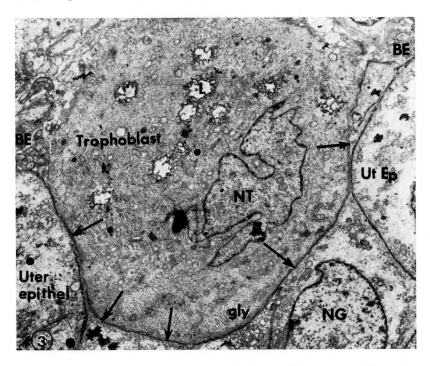

Fig. 3. Electron micrograph of trophoblastic giant cell in contact with uterine epithelium in a human implantation site. It has invaded the stroma and destroyed the basal membrane of the epithelium (BE). The cell membrane of the giant cell and those of the uterine epithelial cells (Uter epithel and Ut Ep) are closely apposed (arrows). NT = Nucleus of the trophoblast; gly = glycogen; NG = nucleus of the uterine gland cell.

when the uterus is opened and during fixation procedures. The microvilli of the two sites interdigitate in a manner similar to cogs in a set of cogwheels. This stage has been described in many animals, the armadillo [2], the mouse [12], the rat [13], and the hamster [15]. Later, the number and length of the microvilli are reduced and the surfaces become smooth. Soon the junctional zone takes form of a serpentine of two parallel membranes interrupted by intervals of few microvilli. This stage was called the attachment stage by *Potts* [12].

We do not have human material from these early stages. However, the intimate contact between the cell membranes of the trophoblast and uterine epithelium could be demonstrated in a later stage of human implantation (fig. 3).

The attachment between the blastocyst and the uterine epithelium becomes increasingly firm by establishment of submicroscopic intercellular contacts. *Potts* [12] described these contacts as 'septate desmosomes', *Tachi et al.* [13] called the connections in the rat implantation site 'tight junctions' and *Enders and Schlafke* [4] found 'punctate desmosomes' and 'primitive junctional complexes' in the implantation site of the rabbit, the ferret and the bat.

The nature and significance of intercellular contacts are still matters for discussion. Therefore, at present, it is not possible to explain their significance in the implantation process. Are the contacts simply mechanical, or do they have other purposes (pathways for exchange)?

This stage has not been observed in the human implantation process either. It is very difficult to preserve the contact without using a perfusion technique, which is impossible in the human. However, subcellular contacts between human trophoblasts and other cells have been observed by *Falck Larsen et al.* [6]. They used human choriocarcinoma transplanted into hamster liver as model of the human implantation process. The invading trophoblast established submicroscopic contacts with the liver cells (fig. 4). Furthermore, *Knoth and Falck Larsen* [11] showed that trophoblastic giant cells established desmosomes with the uterine epithelium during later stages of trophoblastic invasion (fig. 5).

It has been shown that in some species the next stage is rupture of the membranes of the trophoblast and uterine epithelium followed by a fusion of cytoplasms at the borderline [3, 5, 12].

The earliest stage of implantation in primates studies with the light microscope corresponds to this stage of fusion. It has been described in the macaque monkey by *Heuser and Streeter* [10] and in the human by *Hertig* [8] (Carnegie No. 8020). In both specimes the uterine epithelium is in part degenerating with some of the cells showing loss of their boundaries and indistinct nuclear details. In some areas it is difficult to distinguish clearly between the affected epithelium and the related trophoblast. Maternal nuclei seem to have been incorporated in the syncytial trophoblast. At this stage there is no decidual change in the endometrium.

The fusion of trophoblastic and epithelial cytoplasms during human implantation has not been confirmed with the electron microscope. The earliest human specimens studied with the electron microscope are from day 11 [11] and day 22 [7] after ovulation. In the latter specimen fusion was observed in a few places (fig. 6), but only when the epithelium was

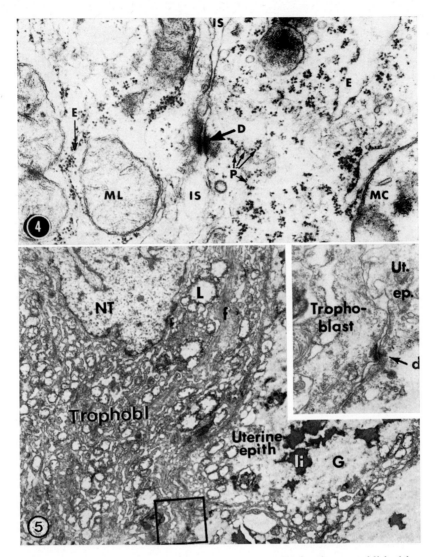

Fig. 4. A submicroscopic connection, a desmosome (D) has been established between a human trophoblastic cell and a hamster liver cell. ML = Mitochondrian of
the liver cell; IS = intercellular space; MC = mitochondrion of a choriocarcinoma
cell; E = endoplasmic reticulum; P = polyribosomes. Electron micrograph.

Fig. 5. A connection between trophoblast and uterine epithelium during human
implantation. The trophoblast (Trophobl) and uterine epithelium (Uterine epith) is
in very close contact. *Insert:* The area indicated in the large figure. A desmosome
(d) connects the trophoblast and the epithelial cell (Ut. ep.). li = Lipid of the uterine epithelium; L = lipid of the trophoblast; NT = nucleus of the trophoblast; gly =
glycogen of the epithelium. G = Glycogen; f = fibrils. Electron micrograph.

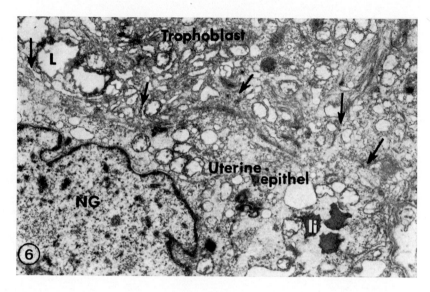

Fig. 6. Electron micrograph from a human implantation site. The cell membranes have disappeared between the trophoblast and the uterine epithelium in many areas (arrows). NG = Nucleus of an uterine gland cell; L = lipid in the trophoblast; li = lipid in tle uterine gland cell.

degenerating. Fusion was also observed between trophoblasts and liver cells when choriocarcinoma was transplanted into hamster liver [6].

The basement membrane of the uterine epithelium seems to resist the invasion by the trophoblast for some time [5]. However, as soon as the penetration of the basement membrane occurs, it is broken at many places by club-shaped processes. Then the trophoblast makes its way into the intercellular substance among the fibroblast, which at this stage are transformed into decidua cells. The trophoblast does not seem to 'attack' the maternal cells (fig. 7, 8). The destruction may be caused by blocking the oxygen supply of the invaded tissue.

The syncytial trophoblast penetrates the maternal vessels at the 10th day and in the specimen from the 11th day we have found maternal vessels limited in some areas with maternal endothelium, in other areas with syncytial trophoblast (fig. 9). It is very interesting that the endothelium and uterine trophoblast is in a very close contact at this stage.

The trophoblastic invasion continues up to the end of the 3rd week of

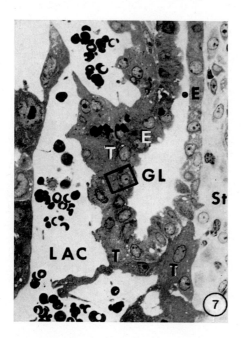

Fig. 7. Light micrograph from a human implantation site on the 11th day. An uterine gland (GL) has been surrounded partly by the trophoblast (T) containing lacunae (LAC). The stroma (St) has no decidual change. The uterine epithelial cells (E) is in close contact with the trophoblastic cells.

gestation, or longer [7]. Then, the demarcation takes place by formation of a non-cellular area, after which the trophoblastic invasion is arrested.

How can we relate this information to the clinical problems of infertility? We have learned that the implantation process in all species, including man, is very complicated. Therefore, it is a very vulnerable step in reproduction. From a theoretical point of view one could concentrate on some special stages of implantation.

It is reasonable to suggest that the early contact requires a perfect synchronization between the blastocyst and endometrium. Therefore, a too slow or too fast transport of the fertilized ovum through the Fallopian tube may result in a lack of implantation. *Hertig* [8] has described a morula with five blastomeres in the uterine cavity, an example of an ovum, which arrived too early. The transport of the ovum through the Fallopian tube is regulated by oestrogens and progesterone. It is accelerated by oestrogens and delayed by progesterone. Therefore, the ratio between oestrogen and

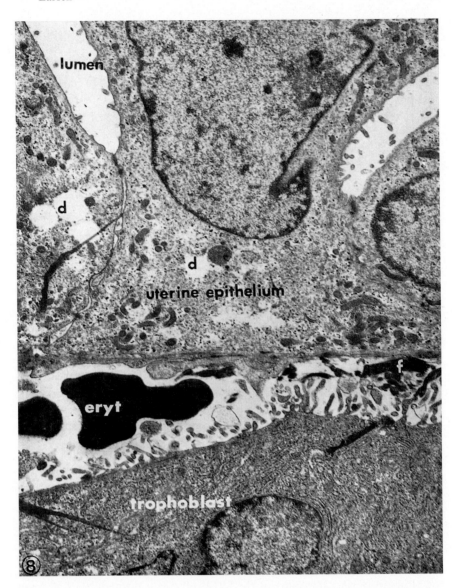

Fig. 8. Electron micrograph of an area similar to that indicated in figure 7. The trophoblast is in close contact with the uterine epithelium in which a degeneration (d) is initiated. Erythocytes (eryt) and fibrin (f) are found in the intercellular space. Lumen = lumen of the uterine gland.

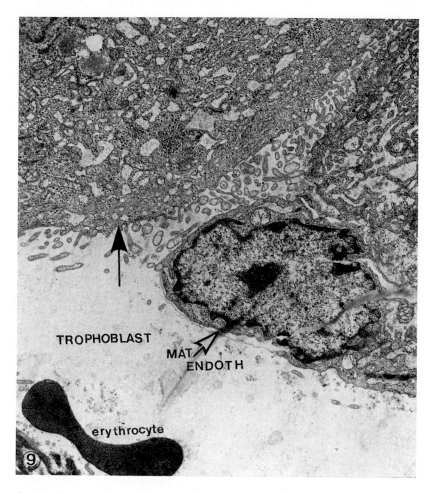

Fig. 9. Electron micrograph from a human implantation site. The maternal vessel is lined in part by maternal endothelial cells (mat endoth) and in parts by trophoblast.

progesterone may be important for the transport and synchronization. Perhaps we should look more on the oestrogen/progesterone ratio in cases of infertility of unknown aetiology.

Further, a delay in the development of the uterine epithelium caused by an insufficient corpus luteum, will result in a non-receptive stage of the endometrium. The new knowledge about the importance of prolactin in luteal insufficiency has thrown some light on this problem. Increased values of prolactin lead to luteal insufficiency and may cause an inhibition of the implantation.

The orientation of the blastocyst has been proved to be important in animal experiments. It is reasonable to believe that this factor is of importance in the human nidation. In some of the pathological human implantation sites described by *Hertig* [8] the embryonic disc was malpositioned (Carnegie No. 8290 and 8299) indicating that the ovum was incorrectly oriented at the time of attachment.

One of the possible actions of the intrauterine contraceptive devices (IUDs) may be a disturber of the rotation and orientation of the blastocyst before its initial attactment. Fibromyomas in the submucosa may have the same effect.

If the blastocyst does not penetrate the uterine mucosa sufficiently, the trophoblast will not have enough contact with the maternal tissue to establish the necessary exchange of substances. Some of the pathological ova described by *Hertig* [8] were found on the surface of the uterine epithelium.

Unsuccessful implantation may occur more often than generally presumed. The Danish pathologist *Vesterdal Jørgensen* has described a large number of 'frustrated ova'. He screened microscopical specimens from curettages for trophoblastic tissue. Methodically, he filtered all the material before embedding and isolated large particles. He called the particles 'the fluffs'. When he embedded and examined the 'fluffs' he often found trophoblastic tissue or parts of a degenerated blastocyst. I have seen his collection of 'frustrated ova', it was quite impressive.

The clinical diagnosis of infertility caused by repeatedly unsuccessful implantation is difficult to establish. In these cases, the conceptus is lost at such an early stage that an uterine bleeding occurs at the expected time of menstruation. Therefore, the pregnancy may not be recognized. However, the diagnosis of pregnancy is possible at the time of implantation by the sensitive HCG-receptor assay. The possibility of defective implantation should be considered in cases of inexplicable infertility.

Which treatment do we have to offer in these cases? It is hard to say. If we are able to find an increased plasma prolactin or an unbalanced oestrogen/progesterone ratio a treatment with bromocriptine or steroid hormones may correct the hormonal status. On the other hand, if the failure of implantation is caused by a defect in the blastocyst, a positive treatment of the condition is not very promising nor desirable.

The human implantation process is still an interesting object for theoretical and clinical research.

References

1 Böving, B. G.: Blastocyst-uterine relationship. Cold Spring Harb. Symp. quant. Biol. *19:* 9 (1954).
2 Enders, A. E.: Electron microscopic observations on the placenta of the armadillo. J. Anat. *94:* 205 (1960).
3 Enders, A. A. and Schlafke, S.: Cytological aspects of trophoblast-uterine interaction in early implantation. Am. J. Anat. *125:* 1 (1969).
4 Enders, A. E. and Schlafke, S.: A morphological analysis of the early implantation in the rat. J. Anat. *120:* 185 (1967).
5 Falck Larsen, J.: Electron microscopy of the implantation site in the rabbit. Am. J. Anat. *109:* 319 (1961).
6 Falck Larsen, J.; Ehrmann, R. L., and Bierring, F.: Electron microscopy of human choriocarcinoma transplanted into hamster liver. Am. J. Obstet. Gynec. *99:* 1109 (1967).
7 Falck Larsen, J. and Knoth, M.: Ultrastructure of the anchoring villi and trophoblastic shell in the second week of placentation. Acta obstet. gynec. scand. *50:* 117 (1971).
8 Hertig, T. T.: Human trophoblast (Thomas, Springfield 1968).
9 Hertig, A. T.; Rock, J.; Adams, E. C., and Mulligan, W. J.: On the preimplantation stages of the human ovum. Contrib. Embryol. *35:* 199 (1954).
10 Heuser, C. H. and Streeter, G. L.: Development of the macaque embryo. Contr. Embryol. *29:* 15 (1941).
11 Knoth, M. and Falck Larsen, J.: Ultrastructure of a human implantation site. Acta obstet. gynec. scand. *51:* 385 (1972).
12 Potts, M.: The ultrastructure of egg implantation. Am. J. Obstet. Gynec. *96:* 1122 (1966).
13 Tachi, S.; Tachi, C., and Lindner, H. R.: Ultrastructural features of blastocyst attachment in the rat. J. Reprod. Fertil. *21:* 37 (1967).
14 Von Spee, F. G.: Die Implantation des Meerschweineneis in die Uteruswand. Arch. Anat. Physiol. *7:* 44 (1883).
15 Young, M. P.; Whicher, J. T., and Potts, M.: The ultrastructure of implantation in the golden hamster. J. Embryol. exp. Morph. *19:* 341 (1968).

J. Falck Larsen, MD, Herlev University Hospital, Department of Obstetrics and Gynaecology, University of Copenhagen, Copenhagen (Denmark)

Prog. reprod. Biol., vol. 7, pp. 296–301 (Karger, Basel 1980)

Medical Aspects of Human Ovo-Implantation

P. O. Hubinont

Université Libre de Bruxelles, Clinique de Gynécologie, Bruxelles

Introduction

Any step of the human reproductive cycle may be at stake by anomalies of the biological processes leading to infertility. Correcting inter-
ventions aim at reestablishing the normal sequence of several events
necessary for implantation and normal development of the blastocyst. On
the other hand, a normal reproductive cycle may be interrupted or de-
flected by mechanical or therapeutic interferences at all its successive
levels: gametogenesis, gamete release and transport, blastocyst transport,
implantation and intrauterine development. The better known example is
given by ovulation: when absent for reasons pertaining to neuroendo-
crinological mechanisms, provided the ovary is anatomically and physiologically normal, ovulation can be induced by suitable and well controlled hormonal therapies. On the contrary, when present, it can be
suppressed by pharmacologically active steroids.

In the case of implantation, however, things are less clear. There are
many points where knowledge is relatively poor, both in its anomalies
leading to infertility and in the methods to be used for its prevention.

The facts at hand are still scarce and sometimes hardly related to
the subject. It is quite clear for instance that successful intrauterine
pregnancies obtained through preovulatory ovum collection, *in vitro*
fertilization and embryonic transfer in the uterine lumen, represent a
major advance in bypassing the tubal causes of infertility. But would not
lead to the elucidation of some facts listed below.

Conceptus Wastage

Estimates of the probability of a conception leading to viable pregnancy in a given ovulatory cycle where intercourse with a fertile partner happens at the appropriate time [15] show that it would lie between 23 and 28 % in married women who had never had premarital sex and never used any form of contraception. The figures in women abandoning oral contraception in order to become pregnant are even lower [17]. Such a low yield can be explained only in part by loss due to anomalies of the embryo taking place before or after implantation. From the studies of *Hertig* [6] and *French and Bierman* [2] this type of event would account for only 53 % of pregnancy failures. Thus, the 20 % difference remains unexplained. Some data [1, 7] suggest that implantation may occur, as shown by the short-lived appearance of β-HCG immunoreactive material during the second half of ovulatory cycles where menses occurred around the expected date. As it is postulated that rising HCG values coincide with the implantation of the blastocyst and the vascularization of the implantation site [14], one has to understand why, under certain conditions, the implanted embryo so rapidly fails to develop that its presence would have remained unnoticed if clinical observation only had been available.

The rate of success (i. e. pregnancies proceeding to a fair stage) of embryo transfer is indeed very poor since in the series of 32 attempts by *Steptoe* [16], 28 blastocysts did not implant at all, or failed to develop beyond the observation cycle. One may regret that the hormonal profiles were not studied more systematically.

It is common experience for all gynecologists that an embryo can implant on structures hardly resembling an endometrium, as tube (fimbria, ampulla or isthmus), ovary or even any location in the peritoneal cavity. Strangely enough, in classical books one reads statements as: 'The etiology is not entirely understood and diverse mechanisms are invoked . . . ' [4]. Actually, if there are some epidemiological relations between ectopic implantation and for instance tubal anomalies (either congenital or acquired), it is very difficult to understand how it can happen – in view of the most probably very selective conditions inducing endometrial receptivity to attachment of a developing fertilized ovum.

From data obtained in rodents [9] it appears that endometrial implantation is possible only during a short period where refractoriness is lifted, under the combined action of progesterone and estrogens. The mechanisms involved are likely to be: (1) either the epithelium secretes a substance that activates the blastocyst to implant; (2) or the epithelium ceases to synthetize a substance that prevents nidation; (3) or the blastocyst in some way provokes receptivity in the endometrium.

If these ideas can be applied to the human species, which is apparently the only one where ectopic implantation can take place, either the properties attributed to the endometrial epithelium are shared by the ectopic surfaces whatever their site, or the two first mechanisms are not applicable and we are left with the likelihood that the blastocyst itself induces receptivity in the nonendometrial surface of attachment.

Hormonal Profile

We have nowadays a relatively clear view of the plasmatic concentration patterns and the feedback mechanisms of pituitary gonadotropes and ovarian sex steroids. It seems relatively clear that implantation occurs approximately on day 21 of a normal 28-day ovulation cycle, at the peak of corpus luteum activity corresponding with the simultaneous maximal plasma concentrations of progesterone and 17β-estradiol. At the same time, large zones of ignorance exist, mainly due to ethical difficulties encountered in the collection of human data. The gaps in our knowledge have been reviewed recently by *Gautray et al.* [3] and from the literature at hand it can be stated that we have the most imperfect information (not to say no information at all) on secretion patterns of steroids by the ovary, clearance rate of the plasmatic pool, receptivity of ovarian steroid-secreting cells to gonadotropes, on some aspects of steroids receptors biology in the human endometrium, on metabolic capacities of endometrial cells under the influence of, or in relation to, steroids, on the determinants of 'refractoriness' lifting.

In defining the luteal phase defect, *Jones* [8] refers to serum progesterone, urinary pregnanediol, endometrial biopsy, basal body temperature and vaginal cytology. She does not seem to take into consideration the significance of 17β-estradiol. By using these criteria in diagnosing luteal insufficiency, she treated 15 sterile patients with progesterone alone, 12 became pregnant – out of 34 patients with repeated pregnancy wastage, 31 delivered live babies. However, in a recent French study by *Gautray et al.* [3] where the criteria for establishing luteal deficiency were

BBT, endometrial biopsy, serum progesterone and 17β-estradiol assays, they have found regularly low values of the preovulatory 17β-estradiol peak, and significantly low values of progesterone and 17β-estradiol during the luteal phase. Measurements of specific receptors to 17β-estradiol and to progesterone have been performed on the group of *Beaulieu*. Recent reports [11, 12] definitely show that the highest concentration of these receptors is found at the end of the proliferative period and coincides with the preovulatory plasma estradiol peak. The receptor of progesterone in the endometrium increases with the increment of 17β-estradiol in the plasma and this association exhibits a highly significant correlation coefficient (r = 0.77). In luteal phase defects the concentration of receptors for progesterone was found relatively low and this is ascribed to low 17β-estradiol values throughout the preovulatory period. The authors would recommend carefully monitored estrogen-supplementation at the time of the preovulation 17β-estradiol surge in some cases of sterility where the concentration of progesterone receptors would appear to be low in postovulatory endometrium samples. On the contrary they suggest that progestogens administered at an ill chosen moment of the cycle could prevent implantation through premature decay of progesterone receptors. These human data correlate with the extensive information obtained in animal experiments and are in line with the long established observation that continuous administration of low dose progestins exert a contraceptive effect without influencing the mechanism of ovulation – and with doubtful effect on sperm migration [11].

Control of Implantation

In her studies on the mode of action of medicated IUDs, *Hagenfeldt* [5] had found that if there was an apparently normal development of the endometrium, both in the proliferative and in the secretory phase, the endometrium of copper-T bearers, in comparison with inert-T bearers, showed some significant differences in the steroid content with a higher concentration of estradiol in the secretion phase and a lower progesterone content in the proliferative phase. If this is related with the synthesis of receptors it would of course suggest that the copper IUD might interfer with the hormonal determinants of endometrium receptivity. It is indeed known that Ca^{2+} inhibits the binding activities of progesterone and its myometrial receptor [9]. The problems related with the mechanical and

Hubinont 300

pharmacological control of implantation will be reviewed extensively by *Hagenfeldt* and by *Haspels* [this volume].

Conclusions

1. Natural history of reproductive performances in the human shows that there is a significant wastage of conceptuses which either do not implant or do not develop.

2. Knowledge on the conditions of successful embryo transfer in the human is still limited and incomplete.

3. Human blastocysts have a unique capacity to implant on non-endometrial surfaces and to develop in extrauterine organs.

4. First hand information is needed on molecular aspects of endocrine endometrium stimulation in the human.

References

1 Chartier, M. et Roger, M.: Les avortements menstruels spontanés; in du Mesnil du Buisson, Psychoyos and Thomas, L'implantation de l'œuf, pp. 147–156 (Masson, Paris 1978).

2 French, F. F. and Bierman, J. M.: Probabilities of fetal mortality. Publ. Hlth Rep. *77:* 835–847 (1962).

3 Gautray, J. P., et al.: Physiologie et physiopathologie du cycle menstruel humain à l'époque de l'implantation. Aspects hormonaux et endométriaux; in du Mesnil du Buisson, Psychoyos and Thomas, L'implantation de l'œuf, pp. 35–53 (Masson, Paris 1979).

4 Gompel, C. and Silverberg, S.: Pathology in gynecology and obstetrics, p. 387 (PAE, 1969).

5 Hagenfeldt, K.: The modes of action of medicated intrauterine devices. J. Reprod. Fertil. *25:* suppl., pp. 117–132 (1976).

6 Hertig, A. T.: Implantation of the human ovum: the histogenesis of some aspects of spontaneous abortion; in Behrman and Kistner, Progress in infertility, 2nd ed., pp. 411–438 (Little, Brown, Boston 1975).

7 Hodgen, G. D., et al.: Transitory HCG-like activity in the urine of some IUD users. J. clin. Endocr. Metab. *46:* 698–701 (1977).

8 Jones, G. S.: Luteal phase defects; in Behrman and Kistner, Progress in infertility, 2nd ed., pp. 299–324 (Little, Brown, Boston 1975).

9 Kontula, K., et al.: Progesterone-binding protein in human myometrium. Inference of metal ions on binding. J. clin. Endocr. Metab. *38:* 500–503 (1974).

10 Porter, A. and Finn, C.: The biology of the uterus; in Greep and Koblinsky, Frontiers in reproduction and fertility control, pp. 146–156 (MIT Press, Cambridge 1977).

11   Psychoyos, A.: Hormonal control of uterine receptivity for nidation; in J. Reprod. Fertil. (Suppl. 25), Implantation and the mechanism of action of IUDs, pp. 17–23 (1976).

12   Robel, P., et al.: Estradiol and progesterone receptors in normal and abnormal human endometrium. 8th Meet. Int. Study Group for Steroid Hormones (Academic Press, New York 1978).

13   Robel, P. et Levy, C.: Les récepteurs des hormones sexuelles – leur intérêt dans les études physiopathologiques de l'endomètre humain; in du Mesnil du Buisson, Psychoyos and Thomas, L'implantation de l'œuf, pp. 71–80 (Masson, Paris 1979).

14   Ross, G. T.: Human chorionic gonadotropin and maternel recognition of pregnancy; in Whelan, Maternal recognition of pregnancy, pp. 191–201 (Excerpta Medica, 1979).

15   Short, R. V.: When a conception fails to become a pregnancy; in Whelan, Maternal recognition of pregnancy, pp. 377–387 (Excerpta Medica, 1979).

16   Steptoe, P.: Establishing pregnancies by replacing human embryos grown in culture; in Leroy, Finn, Psychoyos and Hubinont, Blastocyst-endometrium Relationships (Karger, Basel 1980).

17   Vessey, M., et al.: A long-term follow-up study of women using different methods of contraception – an interim report. J. biosci. Sci. *8:* 373–427 (1976).

Prof. P. O. Hubinont, Université Libre de Bruxelles, Clinique de Gynécologie, 322, rue Haute, B-1000 Bruxelles (Belgium)

Prog. reprod. Biol., vol. 7, pp. 302–309 (Karger, Basel 1980)

Mechanism of Action of Medicated Intrauterine Devices

K. Hagenfeldt

Department of Obstetrics and Gynecology, Karolinska sjukhuset, Stockholm

Introduction

During the last 20 years, when the intrauterine device (IUD) has been increasingly used as a contraceptive method, both animal and clinical research have been performed to elucidate the mechanism of action of the method.

The so called inert or nonmedicated IUDs, usually manufactured in plastic or polyethylene, exert their contraceptive effect mainly by producing a foreign body reaction with an outpouring of acute and chronic inflammatory cells into the uterine fluid. This effect is sterile the first month after insertion. The inflammatory cells undergo cytolysis and the cellular products are toxic for the preimplantation blastocyst and partially to sperm. The foreign body reaction is positively correlated to the surface area of the device. Hence, larger devices give a higher protection against pregnancy but unfortunately often a higher frequency of side effects, especially bleeding [18].

The advantage of the medicated device over its inert counterpart is the potential it offers for controlling both effectiveness and side effects through the addition of a proper dosage of drugs or a combination of drugs.

The medicated IUDs may have anyone or a combination of at least five mechanism of action. The drug or the metal ion diffusing from the IUD may (1) disturb the pituitary-ovarian function inhibiting ovulation; (2) have a direct action on sperm, inhibiting fertilization; (3) have an effect on the preimplantation blastocyst; (4) act directly on the endometrium inhibiting implantation, and (5) disrupt the implanted ovum acting as an abortifacient.

What Types of Medicated IUDs Are Available Today?

Theoretically, any drug which has an antifertility effect and can be released into the uterine fluid can be used in an intrauterine device. During the last decade several different drugs have been tested but only a few have been more extensively used. The copper IUD first designed by *Zipper et al.* [33] and now available in different models with different surface area of the copper thread is by far the most widely used. The progesterone-releasing IUD (USP system or Progestasert) designed by *Pharriss et al.* [24] after experimental work by *Scommegna et al.* [28] is now available in some countries. Recently, the Population Council and the World Health Organization have started clinical studies on an IUD releasing Norgestrel [19, 31]. Our own experience concerning the mechanism of action of medicated IUDs is restricted to the copper IUD and the progesterone-releasing device. The five different possibilities mentioned above will be discussed with some comments on work performed by other groups.

Disturbance of Pituitary-Ovarian Function

More than one hundred women have been observed during one years use of the copper-T device. Repeated endometrial biopsies have shown a normal secretory endometrium in the second half of the cycle and serial serum progesterone determination have indicated a normal corpus luteum function [8]. Recently, 16 women were studied during six consecutive cycles and only three out of the 96 cycles were anovulatory as judged by serum progesterone on cycle days 22–28. Our conclusions are therefore that the copper released into the uterine cavity does not effect pituitary ovarian function.

28 women were observed during 1 years use of the USP system releasing 65 μg of progesterone per day during one year. Repeated endometrial biopsies and serial serum progesterone determinations in the second half of the cycle indicated a normal pituitary-ovarian function [11]. It should be noted that of the two new experimental devices releasing

Norgestrel, the release rate of the Population Council device is high enough to give an inhibition of ovulation in some women contributing to the mechanism of action for that device [19]. The levo-Norgestrel device developed by the WHO releases a low amount of the steroid and does not inhibit ovulation [31].

Effect on Spermatozoa Inhibiting Fertilization

Experimental work on the effect of copper on sperm motility *in vitro* was reviewed by *Jecht and Bernstein* [14]. Most studies indicate an inhibition of sperm motility in the presence of a copper concentration of around 10^{-3} M and several hours of incubation are needed. In our own studies on copper concentration in uterine fluid and cervical mucus in women using the copper-T device we found the highest concentration in the late proliferative mucus during the first six months after the insertion of the IUD; the concentration did not exceed 7×10^{-5} M. The copper concentration in the uterine fluid was calculated to approximately the same levels [8]. We therefore conclude that the contraceptive effect of the copper releasing IUD in women does not involve an immediate inhibition of sperm motility as a primary mechanism of action. It has, however, been suggested by *Oster* [22] that the rheological properties of the uterine fluid and the cervical mucus could be changed by copper released from the IUD and that this could affect the metabolism and progress of the spermatozooa. In vivo studies performed by other investigators [5, 16] do not give support to this theory. *Rosado et al.* [26] studied the effect of intrauterine release of progesterone on sperm metabolism and found a dose-dependent inhibitory action.

Destruction of the Blastocyst before Implantation

Experimental data in animals have given diverging support to the theory of an effect of copper on the preimplantation blastocyst. No human data are available out of obvious reasons. The effect of copper could be mediated through an intense increase in inflammatory cells which then, when undergoing autolysis, exert a toxic effect on the blastocyst [23]. Secondary, through an uptake of copper directly into lysosomes in the blastocyst followed by autolysis and cell death as shown by *Abraham et al.*

[1]. Thirdly it was shown by *Brinster and Cross* [2] that coppersalt solutions in a concentration of 2.5×10^{-5} M destroyed mouse blastocysts. Increasing amounts of bovine serum albumin had a protective effect on the blastocyst in this experimental model.

One or all of these theories suggested by animal work could be effective in the human giving the explanation to the mechanism of action of the copper-releasing IUD.

No experimental or clinical data are available on the effect of intrauterine release of progesterone or Norgestrel on the blastocyst survival.

Changes in the Endometrium Resulting in an Inhibition of Implantation

At light microscopy, the endometrium influenced by a copper IUD looks grossly normal [9]. Scanning and transmission electron microscopy reveal changes attributable to a disrupted glycogen metabolism and secretion [6, 20, 21].

In contrast to the situation with the copper IUD, the morphology of the endometrium under influence of intrauterine release of 65 μg progesterone per day is grossly abnormal; showing a depressed proliferation, inhibition of glandular development and a decidual reaction in the stroma [11, 17].

The copper influenced endometrium at the time of implantation displays severe metabolic disturbances as indicated by studies on trace elements, enzymes, DNA, RNA, proteins by several investigators [4, 8, 13, 25, 32].

The grossly abnormal endometrium under influence of progesterone is also metabolically changed, e. g., enzymes, proteins, DNA, RNA [10, 11, 15].

The role of prostaglandins in the process of implantation has been discussed earlier during this meeting.

In 1973, *Chaudhury* [3] suggested a role for prostaglandins in the mechanism of action of IUDs. When discussing species in which the IUD acts by inducing luteolysis (e. g., the sheep) it is obvious that the luteolysis can be accelerated by an increased production of prostaglandin $F_{2\alpha}$ in the endometrium [30].

The situation in the human seems to be more complex. Several investigators, among them our own group, have determined prostaglandin levels in the endometrium during the normal cycle and during the use of nonmedicated and medicated devices. Although there are differences between methodologies, most research workers seem to agree upon the fact that the human endometrium at the time of implantation is capable of producing both prostaglandin $F_{2\alpha}$ and prostaglandin E_2 in higher amounts than during the proliferative phase and that the production is controlled by ovarian hormones [for review see 12].

The insertion of a large nonmedicated device or a copper-T 200 did not induce any significant variation in the production of $PGF_{2\alpha}$ or its main metabolite in plasma 15-keto-13,14-dihydro-$PGF_{2\alpha}$ in the endometrium at the time of implantation [7].

The prostaglandin production in the uterus of women using a progesterone-releasing device has been studied by *Scommegna et al.* [29]. No difference was found in the progesterone-influenced tissue with specimens obtained during the normal cycle or in women using a nonmedicated device.

The involvement of prostaglandins and/or related compounds in the process of implantation and their possible role in the mechanism of action of medicated or nonmedicated IUDs are still incompletely understood and more research in this area is needed.

Early Abortifacient Effect on the Implanted Ovum

During recent years there have been several reports on the occurrence of pregnancy specific proteins such as HCG in urine and serum of IUD using women indicating a varying degree of early implantations not leading to a clinical pregnancy. Both medicated and nonmedicated devices have been studied. We choose to investigate 16 married women with regular menstrual cycles of 26–32 days and earlier proven fertility and having regular sexual intercourse. The women were studied during a period of six menstrual cycles using the copper-T 200. 3–4 blood samples were obtained between days 20 and 30. Plasma progesterone indicated ovulation in 93 of the 96 cycles. β-HCG could not be detected in any of the luteal phase samples obtained in each ovulatory cycle indicating that disruption of an early implanted blastocyst is not the main mode of action of copperbearing IUDs. In a recent review *Saxena and Landesman* [27] conclude that the disagreement between different researchers seem to be due to the use of nonspecific assays at the limit of their sensitivity.

There are not data available with any of the progesterone or progestin releasing IUDs in this area.

Summary and Conclusions

Intrauterine devices releasing copper into the uterine cavity most probably exert their contraceptive effect through several mechanisms of which the most important are the changes in the metabolic processes in the endometrium at the time of implantation and the effect on the uterine secretion which could have secondary effects on the preimplantation blastocyst.

The progesterone-releasing device and the low dose 1-Norgestrel device (WHO) exert their major effect through a depression of the development of the endometrium rendering it unsuitable for implantation while the d-Norgestrel releasing IUD (Population Council) having a higher release rate has a combined effect on the endometrium and on the pituitary-ovarian axis.

References

1   Abraham, R.; Mankes, R.; Fulfs, J.; Goldberg, L., and Coulston, F.: Effects of intrauterine copper wire on blastocyst and uterine lysosomes in the rabbit. A cytochemical and ultrastructural study. J. Reprod. Fertil. 36: 59–67 (1974).

2   Brinster, R. L. and Cross, P. C.: Effect of copper on the preimplantation mouse embryo. Nature, Lond. 238: 388–389 (1972).

3   Chaudhury, R.: Release of prostaglandins by the IUCD. Prostaglandins 3: 773–784 (1973).

4   De la Osa, E.; Hagenfeldt, K., and Diczfalusy, E.: Effect of the Cu-T device on the glycogen content of the human endometrium. Contraception 6: 449–457 (1972).

5   Elstein, M. and Daunter, B.: Trace elements in cervical mucus; in Elstein, Moghissi and Borth, WHO Symp. Cervical Mucus in Human Reproduction, pp. 122–127 (Scriptor, Copenhagen 1973).

6   Gonzales-Angulo, A. and Aznar-Ramos, R.: Ultrastructural studies on endometrium of women wearing TCu-200 intrauterine devices by means of transmission and scanning electron microscopy and X-ray dispersive analysis. Am. J. Obstet. Gynec. 125: 170–179 (1976).

7   Green, K. and Hagenfeldt, K.: Prostaglandins in the human endometrium. Gas chromatographic-mass spectrometric quantitation before and after IUD insertion. Am. J. Obstet. Gynec. 122: 611–614 (1975).

8 Hagenfeldt, K.: Studies on the mode of action of the copper T device. Acta endocr., Copenh., suppl. 69 (1972).

9 Hagenfeldt, K.; Johannison, E., and Brenner, P.: Intrauterine contraception with the copper T device. III. Effect upon endometrial morphology. Contraception 6: 207–217 (1972).

10 Hagenfeldt, K. and Landgren, B. M.: Contraception by intrauterine release of progesterone. Effects on endometrial trace elements, enzymes and steroids. J. Steroid Biochem. 6: 895–898 (1975).

11 Hagenfeldt, K.; Landgren, B. M.; Edström, K., and Johannison, E.: Biochemical and morphological changes in the human endometrium induced by the Progestasert device. Contraception 16: 183–197 (1977).

12 Hagenfeldt, K.: Prostaglandins and related compounds and their metabolism in normal and steroid exposed endometrium. WHO Symp. Steroid Contraception and Mechanism of Endometrial Bleeding, Geneva 1979.

13 Hernandez, O.; Ballesteros, L.; Mendez, J. D., and Rosado, A.: Copper as a dissociating agent of liver and endometrial polysomes. Fert. Steril. 25: 108–112 (1974).

14 Jecht, E. W. and Bernstein, G. S.: The influence of copper on the motility of human spermatozooa. Contraception 7: 381–401 (1973).

15 Johannison, E.; Landgren, B. M., and Hagenfeldt, K.: The effect of intrauterine progesterone on DNA content in isolated human endometrial cells. Acta cytol. 21: 441–446 (1977).

16 Liedholm, P. and Sjöberg, N. O.: Migration of spermatozooa in cervical mucus from women using copper intrauterine devices. Acta obstet. gynec. scand. 53: 375–376 (1974).

17 Martinez-Manautou, J.; Maqueo, M.; Aznar, R.; Pharriss, B., and Zaffaroni, A.: Endometrial morphology in women exposed to uterine systems releasing progesterone. Am. J. Obstet. Gynec. 121: 175–179 (1975).

18 Moyer, D. L. and Shaw, S. T., jr.: Intrauterine devices. Biological action; in Hafez and Evans, Human reproduction conception and contraception; 1st ed., pp. 309–334 (Harper & Row, New York 1973).

19 Nilsson, C. G.: A d-Norgestrel releasing IUD; thesis Helsinki (1977).

20 Nilsson, O. and Hagenfeldt, K.: Scanning electron microscopy of human uterine epithelium influenced by the Cu-T intrauterine contraceptive device. Am. J. Obstet. Gynec. 117: 469–472 (1973).

21 Nilsson, O.; Hagenfeldt, K., and Johannison, E.: Ultrastructural signs of an interference in the carbohydrate metabolism of the human endometrium produced by the intrauterine copper-T device. Acta obstet. gynec. scand. 53: 139–149 (1974).

22 Oster, G. K.: Chemical reactions of the copper intrauterine device. Fert. Steril. 23: 18–23 (1972).

23 Parr, E. L. and Shirley, R. L.: Embryotoxicity of leukocyte extracts and its relationship to intrauterine contraception in humans. Fert. Steril. 27: 1967–1977 (1976).

24 Pharriss, B. B.; Erickson, M. A.; Bashaw, J.; Hoff, S.; Place, V. A., and Zaffaroni, A.: Progestasert: a uterine therapeutic system for long-term contraception. Fert. Steril. 25: 911–921 (1974).

25 Robles, F.; De La Osa, E.; Lerner, U.; Johannison, E.; Brenner, P.; Hagenfeldt, K., and Diczfalusy, E.: α-Amylase, glycogen-synthetase and phosphorylase in the human endometrium; influence of the cycle and of the Cu-T device. Contraception 6: 373–384 (1972).

26 Rosado, A.; Hicks, J. J.; Aznar, R., and Mercado, E.: Intrauterine contraception with the progesterone-T device. Contraception 9: 39–51 (1974).

27 Saxena, B. B. and Landesman, R.: Does implantation occur in the presence of an IDU?: Res. Reprod. 10: 1–2 (1978).

28 Scommegna, A.; Pandaya, G. N.; Christ, M.; Lee, A. W., and Cohen, M. R.: Intrauterine administration of progesterone by a slow releasing device. Fert. Steril. 21: 201–210 (1970).

29 Scommegna, A.; Ilekis, J.; Ramaa, R.; Dmowski, W. P.; Rezai, P., and Auletta, J. J.: Endometrial prostaglandin F content in women wearing non-medicated or progesterone-releasing intrauterine devices. Fert. Steril. 29: 500–504 (1978).

30 Spilman, C. H. and Duby, R. T.: Prostaglandin mediated luteolytic effect of an intrauterine device in sheep. Prostaglandins 2: 159–165 (1972).

31 WHO Special Programme of Research, Development and Research Training in Human Reproduction. Annual Report, pp. 71–72 (1979).

32 Wilson, E. W.: The effect of metallic copper on human endometrial alkaline phosphatase activity. Br. J. Obstet. Gynec. 80: 648–650 (1973).

33 Zipper, J.; Tatum, H. J.; Pastene, L.; Medel, M., and Rivera, M.: Metallic copper as an intrauterine contraceptive adjunct to the T-device. Am. J. Obstet. Gynec. 105: 1274–1278 (1969).

K. Hagenfeldt, MD, Department of Obstetrics and Gynecology,
Karolinska Sjukhuset, 10401 Stockholm (Sweden)

Prog. reprod. Biol., vol. 7, pp. 310–323 (Karger, Basel 1980)

Interfering with Implantation by Postcoital Estrogen Administration

I. Endometrium Histology[1]

M. R. Van Santen and A. A. Haspels[2]

University Clinic for Obstetrics and Gynecology, Utrecht

Introduction

On the subject of interfering with implantation by postcoital estrogen administration, one must state that despite the effectiveness of estrogens as postcoital antifertility agent, the mechanism of action is still not understood [*Haspels,* 1976]. In the human species and the rhesus monkey, estrogens neither interfere with fertilization nor do they interrupt an established implantation [*Heckel,* 1939; *Van Wagenen and Morse,* 1944; *Board,* 1970; *Bacic et al.,* 1970]. In the rabbit [*Greenwald,* 1957, 1963] estrogens do interfere with early pregnancy before implantation and in the mouse and rat [*Emmens and Finn,* 1962] estrogens inhibit implantation with a single local (in the cornual part of the tube) or parenteral injection. In mice a number of synthetic antiestrogens have been shown to interrupt early pregnancy [*Martin et al.,* 1963]. A clear relation to a mechanism related to dimethylstilbestrol (DMS) and 17β-estradiol-3 could not be established.

During the period of tubal transport, however, it seems easy to influence the period of time commonly used for transferring the egg from the ampulla to the uterus. Estrone terminates pregnancy by this mechanism in rats [*Edgren,* 1961] and in the rabbit estrogens and other antifertility drugs such as antiestrogens are demonstrated to be effective [*Emmens,* 1962].

---

[1] The microscopic slides were prepared by the Academic Institute for Pathology, we wish to thank the Head of institute, Prof. Dr. *J. A. M. Van Unnik,* for his clear judgements of the endometrium pictures presented.

[2] Head of department, University Clinic for Obstetrics and Gynecology.

The transport of eggs through the tube in rabbits becomes very irregular by administration of estrogens [*Noyes et al.*, 1959]. Ova may be ejected and lost at the ampullar site, or transported so fast that they are lost into the uterus instead of retained in the tube for the proper time. Asynchronic arrival of fertilized ova in the uterus, e. g. on an endometrium with an earlier stage or later stage of the menstrual cycle leads the ova to perish [*Chang*, 1950]. It is demonstrated that if blastocysts arrive over 2 days too early or later than 1 day this is beyond the critical time to develop to young. Outside of this range it is apparently unsuitable for ova to arrive into the uterine cavity in an attempt to nidate into the endometrium.

On the duration of egg transport in the human tube [*Croxatto*, 1972] the normal time is found to be no less than 3 and no more than 4 days, although unfertilized ova are recovered from human uteri by the 2nd and 3rd postovulatory day. Extensive studies have been performed in the rabbit [*Greenwald*, 1961]. Exogenous estrogens caused retaining of the eggs for as long as 5–6 days at the ampullary-isthmic junction (AIJ). This is the same level where in the physiological cycle the delay occurs after arrival within 2 h after ovulation, lasting for 2 days. In women, pregnancy incidences are significantly higher following an ampullary reconstructive salpingostomy rather than following an isthmic one, probably due to the preservation of the functional mechanism of the AIJ [*Aref*, 1973].

This suggests that the same mechanism is involved in the estrogen-influenced cycle. Denervation of endogenous sympathic neurotransmitters by surgical procedure, 6-hydroxydopamine injection or systemic administration of reserpine did not interrupt normal ovum transport [*Pauerstein et al.*, 1974]. However, estrogen-induced 'tube locking' was partially antagonized by depletion of neurotransmitter from the intrinsic adrenergic nerves of the oviduct, suggesting an effect of estrogen on ovum transport mediated through adrenergic processes. Major antagonization of the estradiol tube locking is achieved by progesterone or phenoxybenzamine [*Pauerstein*, 1973]. During transport of eggs from the tubes to the uterus, the uterus undergoes certain developmental changes which are essential for survival and implantation of embryos. The progestational reaction of the endometrium should be in line with the stage of the fertilized ovum to ensure proper implantation. A late ovum or an early endometrium makes conditions unfavourable for the egg. Close phase relationship is required, early arrival or later by more than 1 day, does result in nidation failure [*Noyes*, 1960]. As early as 1950 [*Chang*, 1950] it has been suggested asynchronization of blastocyst and endometrium to be a cause of ova to perish and making implantation of an intact ovum impossible.

The endometrium is controlled by estrogen in the proliferative stage and after the LH surge, by estrogen and progesterone as well. As described by the basic work of *Noyes* [1950] the endometrium changes every day with the menstrual cycle, which is reflected in the histological epithelial pattern. High dosages of estrogens interfere with this daily sequence of alterations [*Egger*, 1974]. The proliferative stage is extended by the use of estrogens. Biopsies show a fixing in a stage of early secretion, when taking ethinylestradiol 5 mg/day on the postovulatory days [*Haspels*, 1969a, b, 1973, 1977]. The retronuclear vacuoles, containing glycogen, ment for apical secretion on the 6th postovulatory day, stay at the basal line. The physiological change of moving onto the apex of the cells, while the nucleus returns and becomes more round in shape [*Dallenbach-Hellweg*, 1975] does not take place. Others have found these glycogen vacuoles to persist up to the 8th postovulatory day or even up to the menstruation [*Morris*, 1966; *Egger*, 1974].

Methods

In our clinic, informed consent was obtained from 6 patients, willing to volunteer in a study to investigate possible changes in the endometrium after taking ethinylestradiol, they are coded as Mrs. Smart, according to Van *Santen Morning after*

Table I.

	Postovulatory day in	
	control cycle	experimental cycle
Mrs. V. R. Smart I	+ 6	+ 13
Mrs. W. Smart II	+ 6	+ 7
Mrs. M. Smart III	+ 6	+ 9
Mrs. A. Smart IV	+ 9	+ 13
Mrs. B. Smart V	+ 11	anovulatory
Mrs. D. Smart VI	+ 6	+ 12

Research Trial, as the project is named. BBT recordings were made in the first menstrual cycle and in the following month, while from the 2nd through the 6th day after the expected ovulation (e. g. day + 2 up to + 6 inclusive) 5 mg EE2 was taken at 9 a. m. All persons served as their own control, on the previous cycle.

Mrs. B. Smart V appeared to have had an anovulatory cycle in her experimental month, and is excluded from further comparative study. Larger multiple-purpose endometrial samples could be collected in the experimental cycle since surgical procedures for nonendocrine disorders had to be carried out (uterine prolapse, cervical stenosis following exconization, etc.).

On dating, the recommendation of *Dallenbach-Hellweg* [1975] was taken into account. The histological pattern, hormonally induced, is known not to be equally spread, a uniform appearance is apparently not achieved, a greater differentiation in certain parts is common in comparison to adjacent cell groups. Such differences are said to be related to many variable factors, such as differences in blood supply, in amounts of hormones reaching the target cells and cellular nutrition and metabolism. For dating the endometrium one should be guided by those regions showing the most advanced changes or by the majority of the cells [*Dallenbach-Hellweg*, 1975].

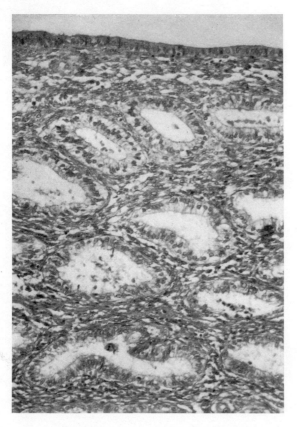

Fig. 1. V. R. Smart I. Low magnification, haematoxylin and eosin, day +13. Photography by da Graça.

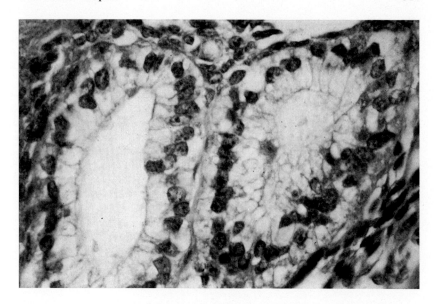

Fig. 2. V. R. Smart I. High magnification, haematoxylin and eosin, day +13. Photography by da Graça.

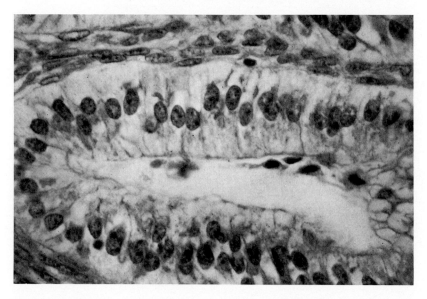

Fig. 3. W. Smart II. High magnification, haematoxylin and eosin, day +7. Photography by da Graça.

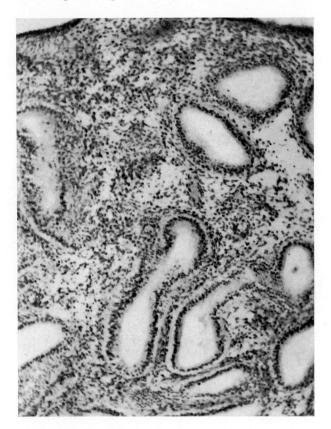

Fig. 4. M. Smart III. Low magnification, haematoxylin and eosin, day +9. Photography by Van Santen.

Results

Mrs. V. R. Smart I: In the experimental cycle, the sample is taken on day + 13 (fig. 1). Glands should have been collapsed. Predecidual changes starting as mantles around the vessels from the 10th day onwards, appear not to have occurred. The earliest signs of corpus luteum regression as in the physiological cycle can be seen as shrinkage of the endometrial epithelium, lacks completely in these pictures. In an attempt for dating, it is dated as day + 6 or + 7. Stroma is compact, no edema is found. The nuclei are rounded, returning to their basal positions. With higher magnification vacuoles are found at the apical rim of the cell (fig. 2). A retardation of ca. 6 days is established.

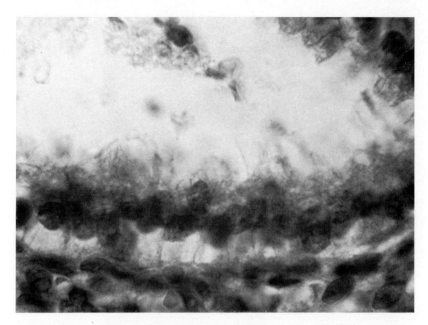

Fig. 5. M. Smart III. High magnification, haematoxylin and eosin, day + 9. Photography by Van Santen.

Mrs. W. Smart II: The same result has been seen here. BBT indicates this to be day + 7. On this day the greatest stromal edema is to be expected. On dating the epithelial glandular cells one sees elongated nuclei in the cylinder-shaped cells, less rounded up as in the Smart I pictures (fig. 3). Basal vacuoles are not yet fading away, resembling the 3rd day of the ideal menstrual cycle. So a delay of 5 days is documented. There are no signs of stromal edema.

Mrs. M. Smart III: These biopsies have been taken at day + 9 in the experimental cycle (fig. 4). Administration of estrogens is known to induce stromal edema [*Diczfalusy and Lauritzen, 1961*]. Stromal edema is physiologically found around the 8th postovulatory day. In this patient on day + 9 one could expect gross stromal edema and nearly empty epithelial cells. The glycogen should have been disposed of into the lumen. However, we do observe some edema in the stroma, and rounded apical nuclei, all in a single row with extensive basal vacuoles, hampered in their way to the apex of the cell (fig. 5). Therefore, this has been classified as day 3 postovulatory, a retardation of 5 days. Spiral arteries with beginning

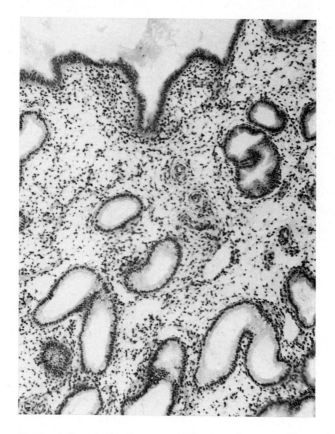

Fig. 6. A. Smart IV. Low magnification, haematoxylin and eosin, day +9.
Photography by Van Santen.

decidualization is to be seen which usually characterizes day + 7 approximately and onwards.

Mrs. A. Smart IV: In the control cycle, pictures of the 9th postovulatory day are shown (fig. 6, 7). Although we expect a more expressed
tortuousity of the endometrial glands and shrinkage due to the lessening
stroma, we see a bit of edema in the stroma characteristic to the 8th day.
The epithelial nuclei all are lined up at the basal side of the secretory cells,
and no glycogen is left in the cell. Few traces in the glandular lumen show
presence of the glycogen produced. The biopsy is in phase, with day 8–9,
according to the BBT. In the experimental cycle, on day + 13, this biopsy
is taken, after EE2 medication on days 8, 9, 10, 11 and 12 postovulatory.

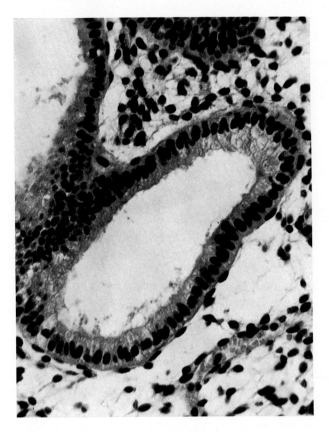

Fig. 7. A. Smart IV. Magnification 400×, haematoxylin and eosin, day +9. Photography by Van Santen.

Some edema can be seen in the stroma, with glandular distention, resembling day + 8 (fig. 8). The nuclei are fading away from the apex, leaving space to the glycogen vacuoles above them (fig. 9). The lumen contains a mucoid-glycogen substance. The nuclei are markedly round. This resembles day + 6. To conclude, in this controlled pair a delay in the stage of the endometrium of over 5 days is seen.

Our last volunteer, Mrs. D. Smart VI, had her control endometrial biopsy taken on day + 6. The biopsy shows elongated nuclei in the epithelial cells, with distinct basal vacuoles in all glandular cells. The stroma shows no edema resembling the regular 3rd day, suggesting a

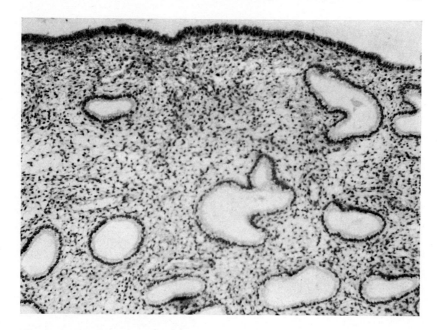

Fig. 8. A. Smart IV. Low magnification, haematoxylin and eosin, day +13. Photography by Van Santen.

discrepancy of 3 days or more. In the experimental cycle she appeared to have finished her EE2 intake on day +10, so the biopsy is according to her BBT day +12. We see no marked stromal edema, some into the upper compacta with distended glandular lumina, resembling partially glandular biopsy taken on day +6. The biopsy shows elongated nuclei in the cystic hyperplasia (fig. 10). The epithelial cells, however, show vacuoles at the basal line, and nearly all round nuclei fixed to the apex of the cells (fig. 11). This is usually the picture at day +14 of the progestational stage, just before the nuclei are returning to their basal places.

Discussion

Postcoital interception possibly is so effective, since it interferes with the synchronization of ovum and endometrium. Through the altered tubal transport in either way, backward or forward [*Aref,* 1973] and by retardation of the endometrium, the ovum arrives, in the uterine cavity

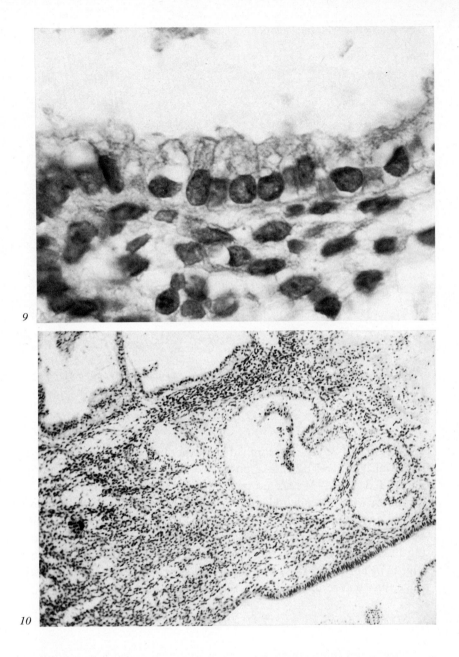

9

10

Fig. 9. A. Smart IV. High magnification, haematoxylin and eosin, day +13.
Photography by Van Santen.

Fig. 10. D. Smart VI. Low magnification, haematoxylin and eosin, day +6.
Photography by Van Santen.

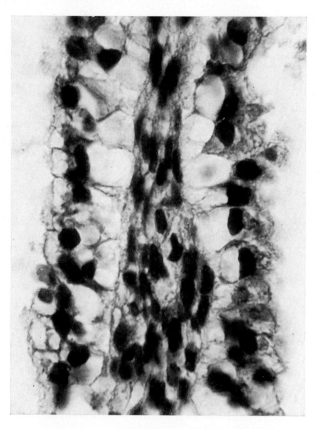

Fig. 11. D. Smart VI. High magnification, haematoxylin and eosin, day +6. A lot of glycogen-containing vacuoles are to be seen in the retronuclear position. An asynchronization seemed to be achieved of approximately 6 days. Photography by Van Santen.

in a more advanced stage than the endometrium is, as shown in this study. A human fertilized ovum reaches the uterus between day + 3 and + 4 [*Croxatto,* 1972]. As suggested by our study, postovulatory estrogens administered from day + 2 up to + 6 inclusive, keep the endometrium fixed in an earlier secretory stage. It is unable to follow the hormonal stimulation as usually by the corpus luteum. Consequently the ovum might perish, since studies in animal experiments showed, ova to arrive over 1 day older in stage and not to survive an attempt to nidate [*Noyes,* 1960].

Other studies in the human with large doses of diethylstilbestrol found depression of progesterone secretion of the corpus luteum [*Johansson*, 1973]. This supports the phenomenon of endometrium retardation as shown in the present study. The mechanism of the AIJ, whose anatomical presence in women is still a hypothesis [*Aref*, 1973], could possibly give an incentive for future development of postcoital drugs. The postcoital use of prostaglandins, even by vaginal route, influences the pressure in the human tube as demonstrated by Rubin tests [*Eliasson*, 1960].

Summary

Ethinylestradiol, given in the postovulatory period of the menstrual cycle in a dosage of 5 mg/day for 5 consecutive days, withholds the endometrium to proceed in its normal consequential changes. Endometrial retardation of 5 days is demonstrated in all cases. In all likelihood this is a mechanism through which postcoital interception by estrogens is so effective [*Haspels*, 1976].

References

Aref, I. and Hafez, E. S. E.: Utero-oviductal motility with emphasis on ova transport. Obstet. gynec. Surv. 28: 679–703 (1973).

Bačič, M.; Wesselius de Casparis, A., and Diczfalusy, E.: Failure of large dosis of ethinylestradiol to interfere with early embryonic development in the human species. Am. J. Obstet. Gynec. 170: 531–534 (1970).

Board, J. A.; Bhatnagar, A. S., and Bush, C. W.: Effect of oral diethylstilboestrol on plasma progesterone. Fert. Steril. 24: 95–97 (1973).

Chang, M. C.: Development and fate of transferred rabbit ova or blastocyt in relation to the ovulation time of recipients. J. exp. Zool. 114: 197–216 (1950).

Croxatto, H. B., et al.: Studies on the duration of egg transport in the human oviduct. I. The time interval between ovulation and egg recovery from the uterus in normal women. Fert. Steril. 23: 447–458 (1972).

Dallenbach-Hellweg, G.: Histopathology of the endometrium (Springer, Berlin 1975).

Diczfalusy, E. und Lauritzen, C.: Oestrogene beim Menschen, pp. 140–148 (Springer, Berlin 1961).

Edgren, R. A. and Shipley, G. C.: A quantitative study of the termination of pregnancy in rats with oestrone. Fert. Steril. 12: 178–181 (1961).

Eliasson, R. and Posse, N.: The effect of prostaglandin on the nonpregnant human uterus. Acta obstet. gynec. scand. 39: 112–126 (1960).

Emmens, C. W.: Action of oestrogens and anti-oestrogens in early pregnancy in the rabbit. J. Reprod. Fertil. 3: 246–249 (1962).

Emmens, C. W. and Finn, C. A.: Local and parenteral action of estrogens and anti-oestrogens on early pregnancy in the rat and mouse. J. Reprod. Fertil 3: 239–245 (1962).

Egger, H. and Kindermann, G.: Effect of estrogens at high dosage on the human endometrium. Arch. Gynaek. *216:* 399–408 (1974).

Greenwald, G. S.: Interruption of pregnancy in the rabbit by administration of estrogen. J. Exp. Zool. 1957 (135), 461–478.

Greenwald, G. S.: Interruption of early pregnancy in the rabbit by a single injection of oestradiol cyclopentylpropionate. J. Endocr. *26:* 133–138 (1963).

Greenwald, G. S.: A study of the transport of ova through the rabbit oviduct. Fert. Steril. *12:* 80–95 (1961.

Haspels, A. A.: The morning-after pill, a preliminary report. IPPF med. Bull. *3:* 6 (1969a).

Haspels, A. A.: The morning-after pill. 9th Abstr., 6th Wld Congr. Obst. & Gyn. New York. Int. J. Obstet. Gynec. *8:* 113 (1969b).

Haspels, A. A. and Andriesse, R.: The effect of large dose of estrogens post coitum in 2000 women. Eur. J. Obstet. Gynec. reprod. Biol. *3:* 113–117 (1973).

Haspels, A. A.: Interception. Post-coital estrogens in 3016 women. Contraception *14:* 375–381 (1976).

Haspels, A. A.; Linthorst, G. A., and Kicovic, P. M.: Effect of postovulatory administration of 'morning-after' injection on corpus luteum function and endometrium. Contraception *15:* 105–112 (1977).

Heckel, G. P. and Allen, W. M.: Maintenance of the corpus luteum and inhibition of parturition in the rabbit by injection of estrogenic hormone. Endocrinology *24:* 137–148 (1939).

Johansson, E. D. B.: Inhibition of the corpus luteum function in women taking large doses of diethylstilboestrol. Contraception *8:* 27–35 (1973).

Martin, L.; Cox, R. J., and Emmens, C. W.: Further studies on the effects of early estrogens and anti-estrogens on the pregnancy in mice. J. Reprod. Fertil. *5:* 239–247 (1963).

Morris, J.; McLean, G., and Van Wagenen, G.: Compounds interfering with ovum implantation and development. III. Role of estrogens. Am. J. Obstet. Gynec. *96:* 804–813 (1966).

Noyes, R. W.; Hertig, A. T., and Rock, J.: Dating the endometrial biopsy. Fert. Steril. *1:* 3–25 (1950).

Noyes, R. W.; Adams, C. E., and Walton, A.: The transport of ova in relation to the dosage of estrogen in ovariectomized rabbits. J. Endocr. *18:* 108–117 (1959).

Noyes, R. W. and Dickmann, Z.: Relationship of ovular age to endometrial development. J. Reprod. Fertil *1:* 186–196 (1960).

Pauerstein, C. J., et al.: The promise of pharmacologic modification of ovum transport in contraceptive development. Am. J. Obstet. Gynec. *116:* 161–166 (1973).

Pauerstein, C. J.: Effect of sympathetic denervation of the rabbit oviduct on normal ovum transport and on transport modified by estrogen and progesterone. Gynecol. Invest. *5:* 121–132 (1974).

Van Wagenen, G. and Morse, A. H.: Estrogen tolerance in pregnancy. Yale J. Biol. Med. *17:* 301–309 (1944).

M. R. Van Santen, MD, University Clinic for Obstetrics and Gynecology, Utrecht (The Netherlands)

Prog. reprod. Biol., vol. 7, pp. 324–336 (Karger, Basel 1980)

Establishing Pregnancies by Replacing Human Embryos Grown in Culture

Patrick Steptoe

Oldham Group Hospitals, Manchester

Since 1972 a number of patients with sterility due either to intractable tubal disease or to tubal absence following surgery, have been subjected to laparoscopy for the purpose of recovering oocytes. The majority of these patients had normal ovulatory cycles, and superovulation was induced by the injection of gonadotrophic hormones. These were given in the form of human menopausal gonadotrophin (HMG) in three intramuscular injections on day 6, 8 and 10 of the cycle, consisting of five ampoules of Pergonal (Serono Pharm. Co. Ltd.) followed by human chorionic gonadotrophin (HCG) in the form of Pregnyl (Schering) 5,000 IU. The purpose of recovery of oocytes in these patients was for fertilization *in vitro* as developed by *Edwards et al.* [1970] and the subsequent replacement of the resulting embryo into the uterus in the hope that implantation would occur with development of a normal pregnancy. The justification for proceeding with this experimental clinical treatment was based on the previous animal experimentation reported by *Jones and Bodmer* [1974] which had failed to show any increased risk of embryonic abnormalities, and on the clinical pressure exerted by the situation of these unfortunate patients. 4–6 preovulatory oocytes were collected from many of these patients. The collection was timed very carefully because the ova had to be obtained just before ovulation could occur, and yet in a mature state of advanced meiosis. This would enable the natural barriers to polyspermic fertilization to be set up. The timing was based on the oestrogenic response to the HMG, as measured in 24-hour urine samples and on the basis of experience of laparoscopic collection at timed intervals following the HCG injections. 77 patients were subjected to these treatments with only three pregnancies occurring (table I). One of these resulted in an ectopic pregnancy previously

Table I. Replacement of embryos into the uterus of patients primed with HMG and HCG

Stimulation	Luteal phase	Number of patients	Number becoming pregnant
HMG/HCG	none	13	0
HMG/clomiphene/HCG	none	2	0
Natural cycle/HCG	HCG and progesterone	7	0
HMG/HCG	HCG repeatedly	9	0
HMG/HCG	bromocryptine	12	0
HMG/HCG	clomiphene	3	0
HMG/HCG	HCG and progesterone	5	0
HMG/HCG	Primolut and HCG	16	2
HMG/HCG	HCG, progesterone and Primolut repeatedly	10	1

reported [*Steptoe and Edwards,* 1976] and the other two ended very early in spite of supportive progesterone therapy. The luteal phases in most of the other patients were invariably shortened. Analyses of the follicular aspirates of these cases were compared with those of patients subjected only to injections of HCG in anticipation of the natural surge of luteinizing hormone (LH). Nine steroids were identified and compared by dendographic methods of analysis according to the techniques described by *Fowler et al.* [1978]. These showed an intolerable situation in the luteal phase of patients subjected to superovulation (fig. 1). Asynchronous development of follicles occurred, with some producing high levels of oestrogens even after laparoscopy so that the luteal phase of the cycle was grossly disturbed. The development of a secretory endometrium was invariably delayed, and implantation did not occur frequently enough. In 1977 it was decided to abandon superovulatory methods in those patients with an intact hypothalamic, pituitary ovarian axis, and to establish a technique for recovery of a mature oocyte in the natural cycle. A high rate of fertilization and cleavage had been achieved in the group of superovulated patients, and the technique of laparoscopic recovery had been perfected.

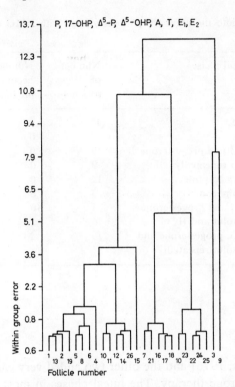

P, 17-OHP, Δ^5-P, Δ^5-OHP, A, T, E_1, E_2

Within group error

Follicle number

Fig. 1. Dendograms of steroids in follicular fluid after HMG and HCG.

Selection and Preparation of Patients

These techniques, however, can only be applied in couples in whom certain clinical conditions are satisfactory. The female patient must have at least one, preferably two, ovaries *readily accessible* at laparoscopy. She must also be ovulating regularly, and possess a normal uterus. If one or both tubes are still present although diseased, then uterotubal occlusion is essential to reduce the risk of ectopic implantation. Her genital tract should be free of pathogenic organisms, including T-strain mycoplasmas, and it is preferable for her to be resistant to rubella and toxoplasma viruses. Steps have to be taken to procure these conditions and one of these will invariably include an exploratory laparoscopy. The male consort should be able to produce readily by masturbation a clean specimen of semen, free of pathogens, and potentially fertile.

Laparoscopy is not easy in these patients. Usually they have had several operations to establish tubal function by different techniques, all of which have failed. These efforts may have left extensive adhesions in the lower abdomen and pelvis barring the approach to the ovaries, as well as multiple abdominal scars. The exploratory laparoscopy may be accompanied by laparoscopic adhesolysis of moderate adhesions using two additional punctures for instrumentation, and also by operative occlusion of the tubo-uterine junction to reduce the risk of subsequent tubal ectopic pregnancy following embryo replacement. Severe adhesions may necessitate an immediate laparotomy to clear the ovaries and to remove hopelessly diseased tubes. Whether lysis of adhesions is performed by laparoscopy or laparotomy, suitable steps should be taken to reduce the possibility of repetitive adhesions formation by the use of corticosteroids.

Patients

77 couples were assessed clinically for suitability for attempted oocyte recovery, including laparoscopy of the wife. 10 patients were subjected to immediate laparotomies to clear effective access to the pelvis, and 6 were treated by laparoscopic thermal coagulation and division of adhesions. All 77 patients were then accepted into the programme for oocyte recovery, 2 of them on two occasions separated by an interval of 2 months, and 1 month, respectively.

Techniques

Assessment of the onset of the LH surge was attempted on 79 cycles using the special testing kits supplied by the Mochida Pharmaceutical Company under the trade name of Hi-Gonavis. The method is designed to measure the amount of LH in a 24-hour specimen of urine and is based on the agglutination of erythrocytes sensitized to LH and HCG [*Kumasaka et al.,* 1973]. The method was modified by *Edwards* [unpublished] to measure the amounts of LH/HCG in 3-hourly specimens of urine, after establishment of the follicular phase basic level. Patients were brought into the special unit on the 9th, 10th or 11th day of their menstrual cycles according to their menstrual cycles, and previous performances as assessed clinically by basal temperature charts, cervical mucous and/or previous 'ovulatory' laparoscopies, and assessment of luteal phase serum progesterone. Comparisons were made between LH levels in urine by the Hi-Gonavis

Table II. Aspiration of preovulatory oocytes at various intervals after the beginning of the LH surge during the menstrual cycle

Interval LH surge-laparoscopy	Number of patients	Number with ovulatory eggs	Egg not aspirated
15–18	9	7	2
18–21	12	9	3
21–24	14	12	2
24–27	19	14	5
27–30	5	2	3
30–35	6	1	5
	65	45	20

method, and those of radioimmunoassay of serum. The radioimmunoassays in many cycles closely resembled the urinary assays, but the blood results were not available in time to assess the onset of the LH surge.

Aspiration of the Preovulatory Oocyte during the Natural Cycle

The detection of the LH surge within a few hours gave ample time to prepare the patients for laparoscopy for oocyte recovery. Previous experience had shown that the optimum time to aspirate follicles was approximately 32–34 h, after injection of HCG. The optimum interval for aspirating oocytes after the LH surge began during the natural cycle was assumed to be slightly less, allowing for the time needed for the surge of plasma LH to accumulate in urine before it could be detected by assay [*Saxena et al.,* 1977]. Accordingly, attempts to aspirate preovulatory oocytes were spread over a period of time from 15 to 35 h after the beginning of the LH surge was detected in urine. There was no difficulty in identifying the single preovulatory follicle in most patients. It was sometimes obscured by adhesion in the ovary, and the problems arising during this will be considered in a subsequent paper [*Steptoe and Edwards,* 1980]. Details of the number of patients, and the number with ovulatory eggs at various intervals after the LH surge are shown in table II. The optimal time for aspirating oocytes was between 15 and 24 h after the surge of LH began. Results were still excellent up to 27 h, but became progressively worse after this time. Aspirations carried out at 30 h were often accompanied by the aspiration of large amounts of blood or haemorrhagic tissue from the ovary,

Table III. Causes of the failure to collect preovulatory oocytes at laparoscopy

Cause	Number of patients
Ovaries partly covered by adhesions	3
Ovaries wholly covered by adhesions	4
Errors in diagnosing LH surge	4
Laparoscopy too late after the LH surge	1
Failure of aspiration or collection	11

presumably as the preovulatory changes in follicles were reaching an advanced state. This made the laparoscopy difficult, and also caused problems in identifying the oocyte amongst the amounts of tissue and fluid aspirated from the follicle.

The onset of the surge was missed in 11 cycles so that only 68 patients were submitted to laparoscopy. Many factors influence the time of onset on the LH surge, including any physical or psychological stress, air and land travel and coitus. Some of these factors may also inhibit or delay the onset of the surge. In 44 of the 68 patients preovulatory ova were recovered. Of the 23 patients in which an egg was not aspirated, the inspection of the endocrine data some time after laparoscopy indicated that errors in the timing of the LH surge were probably responsible for the failure in 5 patients. It was due to adhesions or endometriosis in 7 others. Failure to aspirate the follicle successfully, or to find the preovulatory oocyte in the collecting chamber was the most likely cause of failure in 11 patients (table III). The average time of recovery was 100 sec from the insertion of the laparoscope, usually not more than 10 min after the start of the anaesthetic. This minimal disturbance may well be an important factor in the subsequent success of implantation after replacement of the embryo.

Fertilization and Cleavage of the Egg Aspirated at Mid-Cycle

Fertilization and cleavage *in vitro* were carried out in exactly the same way as described in our earlier publication [*Edwards and Steptoe, 1975*]. The preovulatory oocyte was transferred through 3 or 4 washes of culture medium, usually under microdrops of paraffin, before being placed in a suspension of spermatozoa. The highest concentration of spermatozoa was 10^6/ml, and this was reduced depending on the proportion of motile spermatozoa in the semen sample and on the motility of individual spermatozoa. In many cultures, the numbers of spermatozoa in the fertilization

Table IV. Estimates of time (in hours) from insemination to attain specified stages of development

Cell stage	Estimate of mid-point of cleavage stages ($\pm$ SE)	Upper 95 % point of distribution[1]
2-cell	34.9 $\pm$ 1.9	46
4-cell	51.2 $\pm$ 1.9	63
8-cell	67.9 $\pm$ 2.5	86
16-cell	84.6 $\pm$ 3.4	112
Morula[2]	100.2 $\pm$ 3.0	120
Early blastocyst	112.7 $\pm$ 3.8	132

[1] These figures give the estimated time from insemination by which 95 % of embryos will have attained the specified stage.
[2] Estimates for the 2-, 4-, 8-, 16-cell stages were obtained from the exponential growth curves; those for morula and early blastocyst by direct computation.

droplets were reduced when the majority of corona radiata cells had been dispersed from around the oocyte. Approximately 18 h after insemination, the oocytes were placed in a growth medium, most often Ham's F10, containing human serum. They were left in this medium for a period of growth, and were examined briefly at occasional intervals during their cleavage. Preovulatory oocytes were collected from 44 patients. Of these, 34 cleaved through their 2-cell stage and beyond. Three of the embryos did not cleave well, two of them coming from the same patient on two different occasions. The timing of cell division was similar to that described in our earlier work, and the blastomeres in the embryos were of even size and regular (table IV). The 32 embryos which developed normally were used for re-implantation into their mother. A small number of embryos had also been cultured beyond the early blastocyst stage, and hatching of the embryo was observed in one 9-day blastocyst. This embryo compared well morphologically and in electron-microscopic features with Ender's 9-day embryo found after natural conception. Of the 44 preovulatory oocytes recovered, 32 were fertilized and developed normally to the 8-cell stage and beyond. Of the 32 patients receiving embryos, replacements were started at the 8-cell stage (23 patients) and then proceeded to the 16-cell stage (9 patients) (table V). At first, timing of replacement was based solely on embryo growth, but detailed analysis showed that successful implantation followed in 4 patients where transfer was carried out at night (table VI).

Table V. Pregnancies after the replacement of embryos in various stages of growth into women during their menstrual cycle

Embryonic stage	Number of patients	Number becoming pregnant
6- or 7-cell	6	0
8-cell	16 + 1?	2
16-cell	7 + 1?	2
Morula	1	0

In 3 patients who became pregnant after treatment with HMG and HCG, one embryo was replaced in the 8-cell stage, and two as blastocysts.

Table VI. Pregnancies after the replacement of embryos at various times of day

Time of day when embryo replaced	Number of patients	Number becoming pregnant
Morning	7	0
Early afternoon	4	0
Late evening	21	4

No implantation occurred when transfer was carried out in the daytime. The series is too small for conclusions to be drawn, but it does encourage further investigation of diurnal rhythms and optimum timing of transfer.

The Pregnancies

The first pregnancy occurred in a lady of 31 years who had been infertile for 12 years. 8 years before she was first seen in Oldham, she had been operated on for tubal occlusion. The diseased ampullae were removed on each side and an attempt was made to refashion an abdominal ostium. Tubal occlusion reoccurred soon after this operation. Preliminary laparoscopy showed marked adhesions with the omentum partly screening off the pelvis from the abdominal cavity, together with distorted hydrosalpinges heavily adherent to their respective ovaries and the uterus. In August 1977, laparotomy was carried out and the adhesions were removed together with both diseased tubes. The inflammatory origin of the disease was confirmed by histological examination. The patient recovered rapidly from this procedure and ovulation and menstruation were re-established satisfactorily. In November 1977, she was admitted for assessment of the onset of the LH surge for oocyte recovery. The husband's semen was both potentially fertile and free of bacterial and T-strain mycoplasma, count 72 millions/ml, 60 % motility and 5 % abnormals. In November 1977, she was admitted to the special unit at Kershaw's Hospital and the LH surge was recognized by examination of 3-hourly specimens of urine for 10th day of her cycle. At 8.00 a. m. on November 10th her husband produced a clean specimen of semen and at 9.00 a. m. (26 h after the estimated onset of the LH surge) anaesthesia was commenced for laparoscopic oocyte recovery. At 9.10 a. m. the ovaries were visualized and it was seen that the left ovary was surrounded by moderate adhesions. The right ovary was found to be accessible but of poor mobility with some adhesions to the pelvic wall. A large follicle some 3.5 cm in diameter was found in the left ovary and aspirated with some difficulty. The fluid was amber coloured, of moderate viscosity and of 4 ml volume. It was found to contain an excellent preovulatory oocyte within a 'nest' of cumulus cells. The oocyte was quickly washed through two collections of culture medium and transferred to a previously prepared droplet of culture medium containing some 100,000 spermatozoa from the husband. It was placed in the incubator in a dessicator gassed with nitrogen 90 %, oxygen 5 % and carbon dioxide 5 % at atmospheric pressure at 37 °C. Fertilization was observed to have occurred by 8.00 p. m. on the 10th November and the zygote was transferred to Ham's F10 culture medium. Cleavage occurred in a normal regular pattern and at 11.30 p. m. on November 12th the embryo was replaced into the uterus via the cervix at the 8-cell stage. Implantation followed and pregnancy hormones appeared (fig. 2). The gestation sac was found to be normal by ultrasonic scan at 12 weeks, and blood α-fetoprotein was also within normal limits. Amniocentesis at 15 weeks showed normal levels of α-fetoprotein and a normal karyotype 46 XX was found after culture of amniotic cells. The pregnancy was complicated by a pre-eclamptic toxaemia in the last 3 weeks so that delivery was accomplished in the 39th week by Caesarean section. A normal female weighing 2,600 g was safely delivered. Extensive examination by paediatricians revealed no abnormalities [Hilson and Bruce, 1978].

The second pregnancy was unfortunately associated with a chromosome imbalance. This patient, who was 37 years old at the time of oocyte recovery and extracorporeal fertilization, had cornual and ampullary tubal occlusions. The embryo was replaced in the uterus at the 16-cell stage. Spontaneous abortion occurred at 7 weeks gestation of a poorly developed embryo which proved to have a karyotype of 69 XXX – a triploid.

The third pregnancy occurred in a woman of 31 years of age with 9 years infertility. At the time of laparoscopy she was found to have no oviducts and a left ovary only, freely accessible. The husband's semen fulfilled the criteria required, and on three occasions this patient had laparoscopy for oocyte recovery at 2-monthly intervals. On the first occasion, the oocyte was not found, on the second normal fertilization and culture occurred but replacement was not followed by implantation. On the third occasion, embryo transfer at the 8-cell stage was followed by implantation. The pregnancy proceeded normally, and successful amniocentesis and cell culture was achieved. This showed a karyotype 46 XY with pleomorphism at 15D and a large Y chromosome. A similar karyotype was found in the father. Unfortunately the patient had a spontaneous abortion at 21 weeks in spite of treatment to try to conserve the pregnancy. A living male weighing 600 g was delivered in a poor state, and he died 2 h later. Postmortem examination showed him to be normal anatomically in every system.

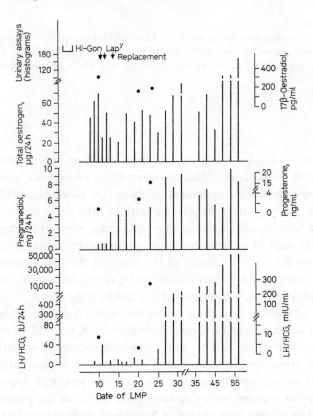

Fig. 2. Hormone profile of first pregnancy after replacing embryo during natural cycle. ● = Plasma assay.

The fourth pregnancy occurred in a lady of 31 years with a long history of infertility associated with endometriosis and tubal occlusions. Surgical attempts to restore tubal patency had failed. Distorted occluded hydrosalpinges were removed and adhesolysis was performed. The isthmus of each tube was found to be blocked by endometriosis. Oocyte recovery was performed in 1978, followed immediately by fertilization, cleavage and oocyte replacement at the 16-cell stage. Normal pregnancy was established, amniocentesis revealed a normal 46 XY karyotype, and pregnancy proceeded to 36 weeks. Premature rupture of the membranes then occurred followed by establishment of premature labour. A low forceps delivery of a normal male weighing 2,600 g occurred on January 14th, 1979. No abnormalities were found in the child.

Both of these children have grown and flourished normally ever since birth. Of the 28 patients, who also had embryo replacement, the hormone profiles showed no evidence of implantation or pregnancy whatsoever, and normal cyclical menstruations occurred in all of them.

Contraindications

Contraindications to this procedure are few. They include absence of normal function of ovaries and uterus, inaccessibility of the ovaries, medical unfitness for pregnancy. There are reservations about patients over the age of 36 because of increased risk of abnormalities.

Discussion

Although several questions about the factors controlling implantation remain unanswered, it would appear that recovery of the natural ovulatory oocyte offers the best chance of successful implantation following fertilization *in vitro*. This recovery must be effected with as little disturbance to the ovum donor as possible so that the hormonal balance remains normal. Alternative methods such as radioreceptor assays would be valuable adjuncts to Hi-Gonavis for early detection of the *onset* of the LH surge so that laparoscopic recovery of a mature ovum can be timed accurately. Improvements in fertilization rates of recovered ova can be expected.

The technique of embryo replacement into the uterus demands a high degree of skill and development of special instruments. A method of detecting the exact position of the embryo at replacement should be an early target for further improvement in technique. The process of *in vitro* fertilization and embryo replacement in humans should offer some real opportunity for pregnancy in women otherwise hopelessly sterile at the present time. It is indicated in women with absent or hopelessly diseased oviducts. It may also be applied in cases where the male consort has some degree of oligospermia, in cases of prolonged idiopathic infertility, and possibly in some couples with immunological factors in their fertility.

The growth of human embryos *in vitro* offers the opportunity to examine them before they are replaced in the mother. It is possible that the birth of children with some inherited disorders could be averted by replacing a normal embryo, and discarding those carrying chromosomal and genetic malformations. This approach depends on identifying traits in the pre-implantation embryo. One approach would be to sex the embryo, in an attempt to avert sex-linked diseases and disorders. Currently, the sexing of blastocysts has been achieved in rabbits and cattle, but the procedures appear too drastic to permit the techniques to be applied to the human embryo. There are various other methods which are of potential value in typing early embryos, but most of these are still in the earliest stages of development.

The scientific analysis of early human embryology could lead to considerable advances in clinical medicine. There is virtually no information on the factors underlying cytodifferentation and morphogenesis. Such information is urgently needed because it is likely to lead to considerable insights into the nature of abnormal growth, including tumours. Analysis on cells composing the primary germ layers and undergoing the early stages of organogenesis could assist in the introduction of new insights and treatments designed to alleviate various human anomalies.

Conclusions

Fertilization *in vitro* and reimplantation of embryos in the mother is a proven safe method for the alleviation of human infertility. It can now be offered to patients who are at present not being treated at all. The method may supplant techniques currently being used to repair or reconstruct the oviduct, because it could prove to be more simple, acceptable and successful. The success rate must be raised, but improvements will almost certainly arise as more studies are carried out. The procedure can be applied repeatedly, with minimum harm to the patient. Amniocentesis and other tests should be used to examine the fetuses, at least initially, until the risks of chromosomal imbalance are assessed. The major causes of anomalies are likely to be those inherited or developmental disorders which carry risks for all pregnancies whether arising through fertilization *in vitro* or *in vivo*.

Some women cannot conceive because of adhesions or other causes preventing ovulation or the aspiration of preovulatory oocytes. These women could receive an oocyte from a donor, to be fertilized by their husband's spermatozoa either *in vivo* or *in vitro*. This situation is analogous to artificial insemination by donor (AID), currently being practised in many countries. It is, in a sense, more acceptable than AID because the recipient must carry the pregnancy, and she is thus involved in the gestation and birth of the child. The adoption of this procedure would best be left to the conscience of the individual patient and practitioner.

The use of surrogate mothers to carry embryos for another couple should not be practised. This procedure could involve considerable legal and social difficulties over the paternity of the child. There could be conflicting claims on the child at birth, and there have been indications of the type of difficulties in some recent cases following the use of AID. In effect, the medical situation is replaced by a more complex legal situation.

Attempts to type embryos during their earlier stages of growth should be encouraged. The implantation of a normal embryo would be preferable to the current practice involving the identification of fetuses by amniocentesis during mid-gestation and aborting those with severe anomalies.

References

Edwards, R. G.; Steptoe, P. C., and Purdy, J. M.: Fertilization and cleavage *in vitro* of preovulatory human oocytes. Nature, Lond. *227:* 1307–1309 (1970).

Edwards, R. G. and Steptoe, P. C.: Induction of follicular growth, ovulation and luteinization in the human ovary. J. Reprod. Fertil. *22:* suppl., pp. 121–156 (1975).

Fowler, R. E.; Edwards, R. G.; Walters, D. E.; Chan, S. T. H., and Steptoe, P. C.: Steroidogenesis in preovulatory follicles of patients given human menopausal and chorionic gonadotrophins as judged by the radioimmunoassay of steroids in follicular fluid. J. Endocr. *77:* 161–169 (1978).

Hilson, D.; Bruce, R. L., and Sims, D. G.: Successful pregnancy following *in vitro* fertilization. Lancet *ii:* 473 (1978).

Jones, A. and Bodmer, W. F.: Our future inheritance: choice or chance, pp. 35–40 (Oxford University Press, London 1974).

Kumasaka, T.; Kotoh, H.; Yaoi, Y.; Koyama, T.; Nishi, N.; Okhora, T., and Saito, M.: A rapid sensitive assay for urinary HCG/LH by haemagglutination reaction. J. Jap. Soc. Obstet. Gynecol. *25:* 1237 (1973).

Saxena B. B.; Saito, T.; Said, N., and Landeyman, R.: Radioreceptor assay of luteinizing hormone-human chorionic gonadotrophin in urine: detection of luteinizing hormone surge and pregnancy. Fert. Steril. *28:* 163–167 (1977).

Steptoe, P. C. and Edwards, R. G.: Reimplantation of a human embryo with subsequent tubal pregnancy. Lancet *i:* 880–882 (1976).

Steptoe, P. C. and Edwards, R. G.: Clinical aspects of pregnancies established by the Oldham procedure with embryos cultured *in vitro* (in press, 1980).

P. Steptoe, MD, 38 Caxton End, Bourne, Cambridgeshire (England)